# Seismic Monitoring in Mines

# Seismic Monitoring in Mines

Edited by

Dr A.J. Mendecki

*Managing Director and Head of Research at ISS International, Welkom, South Africa*

CHAPMAN & HALL

London · Weinheim · New York · Tokyo · Melbourne · Madras

**Published by Chapman & Hall, 2–6 Boundary Row, London SE1 8HN, UK**

Chapman & Hall, 2–6 Boundary Row, London SE1 8HN, UK

Chapman & Hall GmbH, Pappelallee 3, 69469 Weinheim, Germany

Chapman & Hall USA, 115 Fifth Avenue, New York, NY 10003, USA

Chapman & Hall Japan, ITP-Japan, Kyowa Building, 3F, 2-2-1 Hirakawacho, Chiyoda-ku, Tokyo 102, Japan

Chapman & Hall Australia, 102 Dodds Street, South Melbourne, Victoria 3205, Australia

Chapman & Hall India, R. Seshadri, 32 Second Main Road, CIT East, Madras 600 035, India

First edition 1997



Production control and typesetting by Anne Hordley

Printed in Great Britain at the University Press, Cambridge

ISBN 0 412 75300 6



A catalogue record for this book is available from the British Library

Library of Congess Catalog Card Number: 96–86555

∞ Printed on acid-free text paper, manufactured in accordance with ANSI/NISO Z39.48-1992 (Permanence of paper)

# Contents

# Contributors

Dr A.H. Dzhafarov
ISS International, South Africa

Mr C. Funk
ISS International, South Africa

Dr S.C. Maxwell
Department of Geophysics, Keele University, UK

Dr A.J. Mendecki
ISS International, South Africa

Dr P. Mountfort
ISS International, South Africa

Dr J. Niewiadomski
ISS International, South Africa

Dr S. Radu
ISS International, South Africa

Mr M. Sciocatti
ISS International, South Africa

Dr G. van Aswegen
ISS International, South Africa

Prof R.P. Young
Department of Geophysics, Keele University, UK

# Preface

Routine seismic monitoring in mines was introduced over 30 years ago with two main objectives in mind:

- immediate location of larger seismic events to guide rescue operations;
- prediction of large rockmass instabilities.

The first objective was achieved fairly quickly, but with the subsequent development of mine communication systems, its strategic importance has diminished. The very limited success with prediction can, at least partially, be attributed to three factors:

- seismic monitoring systems based on analogue technology that provided noisy and, frequently, poorly calibrated data of limited dynamic range;
- the non-quantitative description of a seismic event by at best its local magnitude; and
- the resultant non-quantitative analysis of seismicity, frequently through parameters of some statistical distributions, with a somewhat loose but imaginative physical interpretation.

The introduction of modern digital seismic systems to mines and progress in the theory and methods of quantitative seismology have enabled the implementation of realtime seismic monitoring as a management tool, quantifying rockmass response to mining and achieving the first tangible results with prediction.

A seismic event, being a sudden inelastic deformation within the rockmass, can now routinely be quantified in terms of seismic moment, its tensor, and radiated seismic energy, so that the overall size of, and stress released at, the seismic source can be estimated.

Thus seismicity, being the intermittent momentum flux due to the sudden motion of discrete lumps of rock, and its associated stress and strain changes in the rock, can be quantified. This brings seismology into the realms of rock mechanics and rheology, where changes in stress, strain rate, flow viscosity and diffusion are fundamental in determining the stability of the deforming structures.

However, from seismological observation one can only measure that portion of stress, strain or rheology of the process which is associated with recorded seismic waves. The wider the frequency and amplitude range, and the higher the throughput, of the seismic monitoring system, the more reliable and more relevant the measured values of these parameters become.

The objectives of seismic monitoring in mines then become:

- to verify the parameters and assumptions of mine design while mining;
- to predict larger instabilities;
- to backanalyse, so that we learn from history.

Seismic monitoring in mines consists of sensors, data acquisition, signal processing, seismological analysis, quantification of the seismic response of the rockmass to mining, and finally, interpretation in terms of the potential for instability. Since each of these stages must be conducted with great care for meaningful results, this book has been structured accordingly.

The principal objective of the book is to suggest, but not prescribe, possible solutions to the problems encountered in achieving the above objectives and to show examples of successful applications. No claim is made for completeness. The emphasis has clearly been placed on parameters describing seismic sources as opposed to changes in wave velocity or attenuation to infer the state of the rockmass. Nevertheless, we hope all users of mine seismic systems will be inspired to gain more insight from their data and increase the value of their systems as management tools.

Aleksander J. Mendecki, Editor.

# Acknowledgements

The majority of this book describes the results of work performed in two projects: GAP 017, 'Seismology for rockburst prevention, control and prediction', and GAP 211, 'Nonlinear seismology'. These were the two major seismological projects awarded by the Department of Mineral and Energy Affairs, on behalf of the South African mining industry, on the recommendation of the Safety in Mines Research Advisory Committee (SIMRAC).

The authors and editor wish to express their thanks for the following contributions to various chapters (in chapter order):

Chapter 1 is based on a far more detailed report by Dr R.W.E. Green, retired Professor at the Bernard Price Institute, University of the Witwatersrand, South Africa, and work performed by him and Mr A.v.Z. Brink, ISS Pacific. Dr A. McGarr, USGS, Menlo Park, California, reviewed an early form of the manuscript and offered useful suggestions concerning ground motions.

Chapter 2 owes much to a far more detailed report by Dr R.W.E. Green, and many fruitful discussions with him.

Professor A. Hanyga, University of Bergen, Norway, provided guidance and useful discussions for Chapter 3.

Professor K. Aki, Observatorie du Piton de la Fournaise, La Réunion, and University of Southern California, offered constructive comments on Chapters 7 and 8.

# 1 Seismic Transducers

The transducer is the key element of any seismic monitoring system. Once the ground motion is transformed into an electrical signal by the transducer, the rest of the system is simply a problem of calibration and data acquisition.

## 1.1 Requirements imposed by ground motion

The type of transducer to be used is determined by the ranges of amplitude and frequency to be measured which, in turn, depend on the magnitude range and distance from the sensor of seismic events occurring in the volume of interest.

The largest events experienced in rockburst prone mines range between moment magnitude $m_M$=3 and $m_M$=5. The smallest events that are useful in determining the state of the rock in seismically quiet periods have magnitudes between $m_M$=−4 and $m_M$=−3 depending on noise levels and other environental factors. The minimum range of frequencies that must be recorded for meaningful seismological processing is determined from the expected corner frequencies of events occurring in the volume to be monitored. The corner frequency is the predominant frequency on the spectrum of instrument and attenuation corrected ground velocity; see Fig. 8.8 in Chapter 8 of this book. To correctly measure the seismic moment we need frequencies down to at least an octave or five spectral points, whichever is lower, below the corner frequency of the largest event to be analysed. To correctly measure the radiated seismic energy we need frequencies at least five times above the corner frequency of the smallest event to be analysed (Mendecki, 1993).

To estimate the range of amplitude and frequency of ground motion which sensors will experience due to seismic events in the above magnitude ranges, we start with the relation, for a circular fault, between the source radius $r$, stress drop $\Delta\sigma$ and seismic moment $M$ (Keilis-Borok, 1959):

$$r^3 = \frac{7M}{16\Delta\sigma} \tag{1.1}$$

By assuming the Brune model for the source (Brune, 1970), we get a corner frequency from the source radius:

$$f_0 = \frac{K V_S}{2\pi r} \tag{1.2}$$

where $K$ is a constant 2.34 for the Brune model, $V_S$ is the S wave propagation velocity and $f_0$ is the corner frequency. $V_S$ depends on the rock type, as given in Table 1.1.

**Table 1.1** Properties of extreme rock types encountered in mining

| | Density $\rho$ [kg/m$^3$] | Shear modulus $\mu$ [GPa] | S wave velocity $V_s = \sqrt{\mu/\rho}$ [m/s] |
|---|---|---|---|
| Hard rock | 2700 | 37 | 3700 |
| Soft rock | 1800 | 7.2 | 2000 |

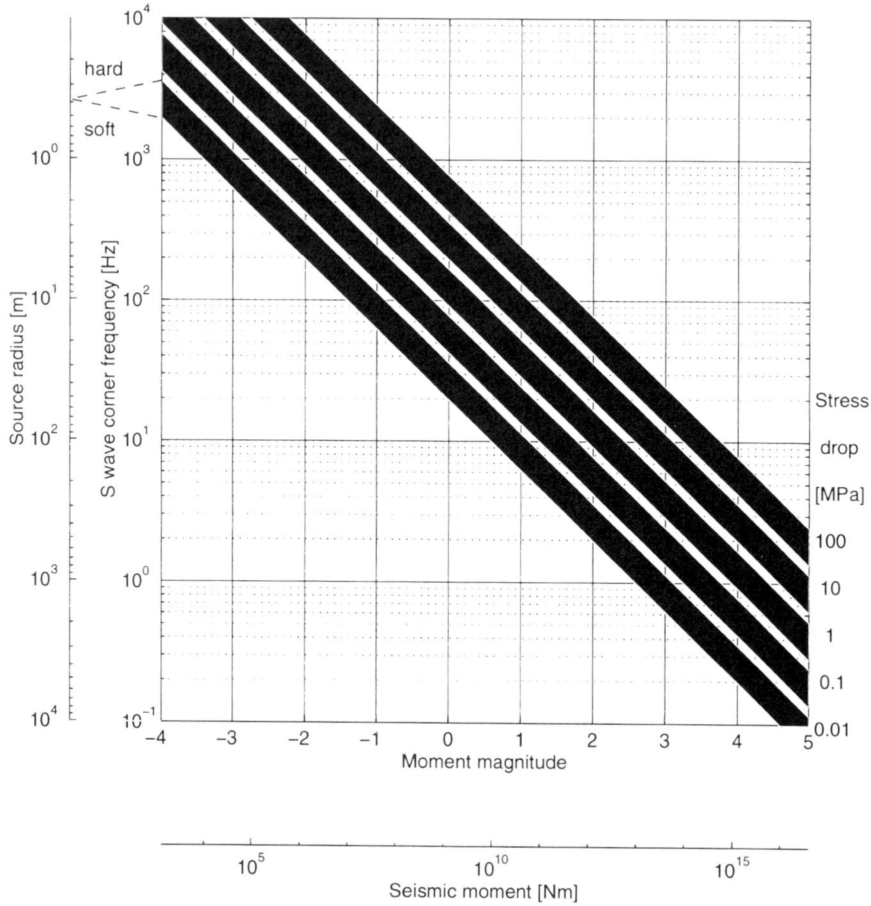


**Fig. 1.1** Expected source radius $r$ and S wave corner frequency $f_0$ as a function of seismic moment for a range of stress drops. The general relations are given by equations (1.1) and (1.3). Moment magnitude $m_M$, equation (1.4), is also shown. The width of each band shows the variation with rock properties as shown in Table 1.1. For a given event, frequencies up to $5f_0$ must be recorded for accurate energy determination and frequencies down to $f_0/2$ for seismic moment.

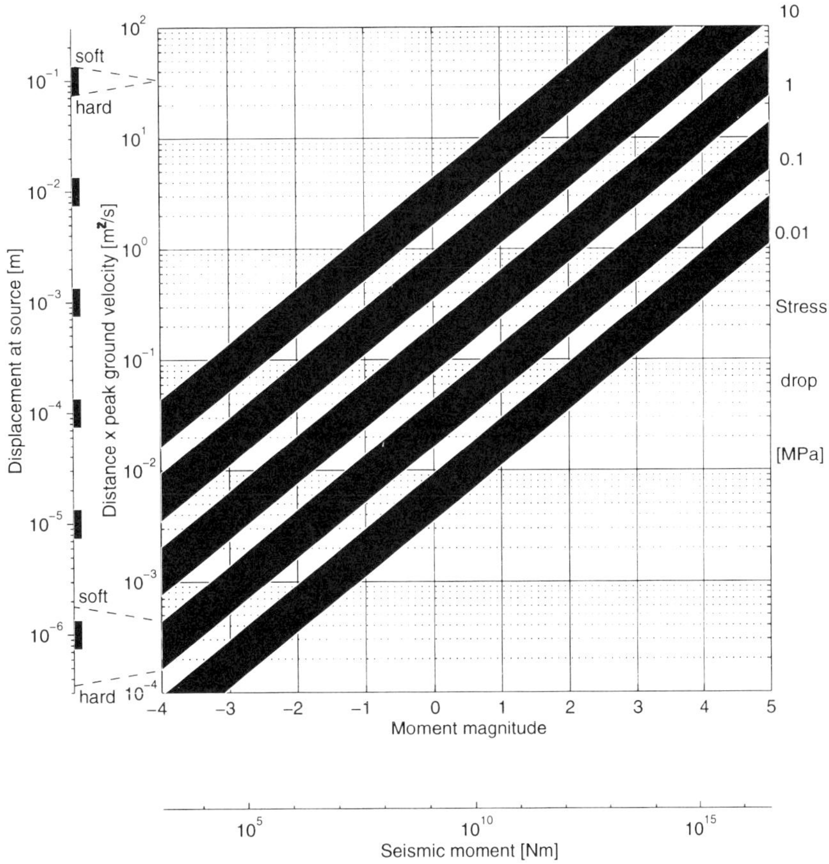


**Fig. 1.2** The product of distance from the source and the far field peak ground velocity, $Rv_{max}$, and displacement at the source, $D$, plotted as a function of moment for a range of stress drops. The general relations are given in equations (1.6) and (1.7). Moment magnitude $m_M$, equation (1.4), is also shown. The width of the lines represents the variation due to the range of rock types listed in Table 1.1.

Substituting equation (1.1) into equation (1.2) we obtain:

$$f_0 = \frac{K V_S}{2\pi} \sqrt[3]{\frac{16\Delta\sigma}{7M}} \qquad (1.3)$$

describing corner frequency as a function of moment, stress drop and S wave velocity. This relation is illustrated in Fig. 1.1 for a wide range of stress drops. While stress drop and moment are independent parameters, in general small events do not occur at large stress drops, which lowers the high frequency requirements somewhat. The moment magnitude, $m_M$, is also shown (Hanks and Kanamori, 1978) where:

$$m_M = \frac{2}{3}\log M - 6.1 \qquad (1.4)$$

To estimate the amplitude of the ground motion we follow the model of McGarr (1991):

$$Rv_{max} = \frac{0.57}{4\pi \rho V_s^3} \cdot 2\pi f_0^2 \cdot M \tag{1.5}$$

where $R$ is the distance from the source, $v_{max}$ is the far field peak ground velocity, ρ is the rock density and 0.57 is the median value of the S wave radiation pattern. Substituting for $f_0$ from equation (1.3) and retaining $K$=2.34, yields:

$$Rv_{max} = \frac{0.0686}{\rho V_s} \cdot \sqrt[3]{\Delta\sigma^2 M} \tag{1.6}$$

McGarr (1991) uses the definition $M = \mu\pi r^2 D$, where $D$ is the amplitude of the displacement at the source and μ is the rigidity or shear modulus, in equation (1.5) to derive the expression:

$$D = \frac{8.1\, Rv_{max}}{V_s} \tag{1.7}$$

and also shows that the slip velocity

$$\dot{D} = \frac{KV_s D}{\pi r} \tag{1.8}$$

and

$$Ra_{max} = 0.4\, \Delta\sigma/\rho \tag{1.9}$$

where $a_{max}$ is the amplitude of the far field ground acceleration associated with the corner frequency. Measured peak acceleration tends to increase with increasing sensor bandwidth because the acceleration spectrum of an event is flat for some interval above the corner frequency.

It must be noted that McGarr used an inhomogeneous model of the source in deriving these relations, with a number of asperities within a larger source region. In this case, the frequency–amplitude relationships still hold, but the moment and source radius are those of a single asperity which is responsible for $v_{max}$, and are less than those of the whole source. Where the source is homogeneous, or, more probably, the internal structure is not resolved, then the moment and radius are those of the entire source as implied above.

The relations (1.6) and (1.7) are plotted in Fig. 1.2, while Fig. 1.3 illustrates the relationship between ground acceleration and distance from the source for several stress drops from equation (1.9). The line thickness again denotes the range of rock properties as shown in Table 1.1.

It must be stressed that these models are not very accurate in the mining environment in that attenuation, near field, propagation and site effects have been ignored. However, we are now in the position, given an event characterized by moment and stress drop, to estimate the dominant frequency, peak ground velocity and peak ground acceleration at a given distance from the source in the far field, as well as the displacement and slip velocity in the source region. This is sufficient to gauge sensor specifications for suitability.

The quantitative description of seismic sources requires that the direction of ground motion be resolved and all sites should have three orthogonal directional sensors (transverse sensitivity < 5%).

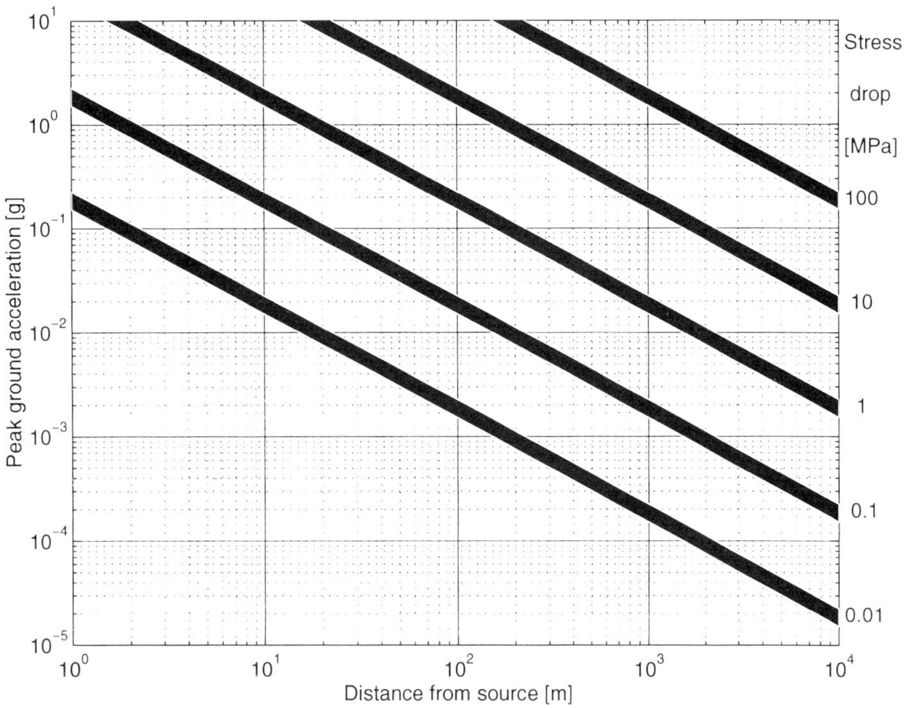


**Fig. 1.3** The far field peak acceleration at the source corner frequency, $a_{max}$, plotted as a function of distance from the source, $R$, for a range of stress drops. The general relation is given in equation (1.8). The width of the lines represents variations in rock density, ρ, as listed in Table 1.1. Attenuation due to transmission through the rock is ignored. Conversely, higher values may be registered by wideband accelerometers, as the source spectrum is flat for a range of frequencies above the corner frequency.

## 1.2 Theory of inertial sensor operation

Most seismic transducers operate on the principle of measuring the ground motion relative to that of an inertial mass. The mass is, for practical reasons, suspended by a spring, so we end up with a variation of the classical mass/spring/damper problem, shown schematically in Fig. 1.4.

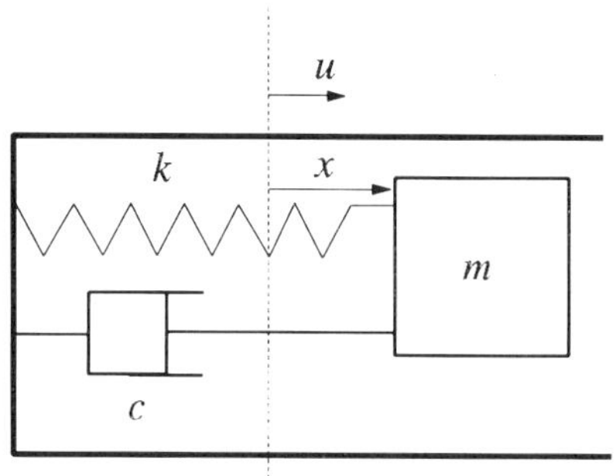


**Fig. 1.4** A schematic diagram of an inertial sensor, showing the mass *m* free to move unidirectionally within the case under the influence of a spring and damper.

Applying Newton's second law to the mass

$$m(\ddot{u} + \ddot{x}) = -kx - c\dot{x} \tag{1.10}$$

where:

| | |
|---|---|
| $u$ | ground displacement to be measured |
| $x$ | displacement of the mass relative to the ground and case |
| $m$ | inertial mass |
| $k$ | spring constant |
| $c$ | damping coefficient |

By applying the Laplace transform to equation (1.10) (see for example Etkin, 1972), we arrive at the transfer function:

$$\frac{\bar{x}(s)}{\bar{u}(s)} = \frac{-s^2}{s^2 + 2b\omega_n s + \omega_n^2} \tag{1.11}$$

where:

| | |
|---|---|
| $\bar{x}(s)$ | Laplace transform of $x(t)$ |
| $\bar{u}(s)$ | Laplace transform of $u(t)$ |
| $s$ | complex frequency variable |
| $\omega_n$ | $= 2\pi f_n$, the natural frequency, $\omega_n^2 = k/m$ |
| $b$ | relative damping factor, $2b\omega_n = c/m$. |

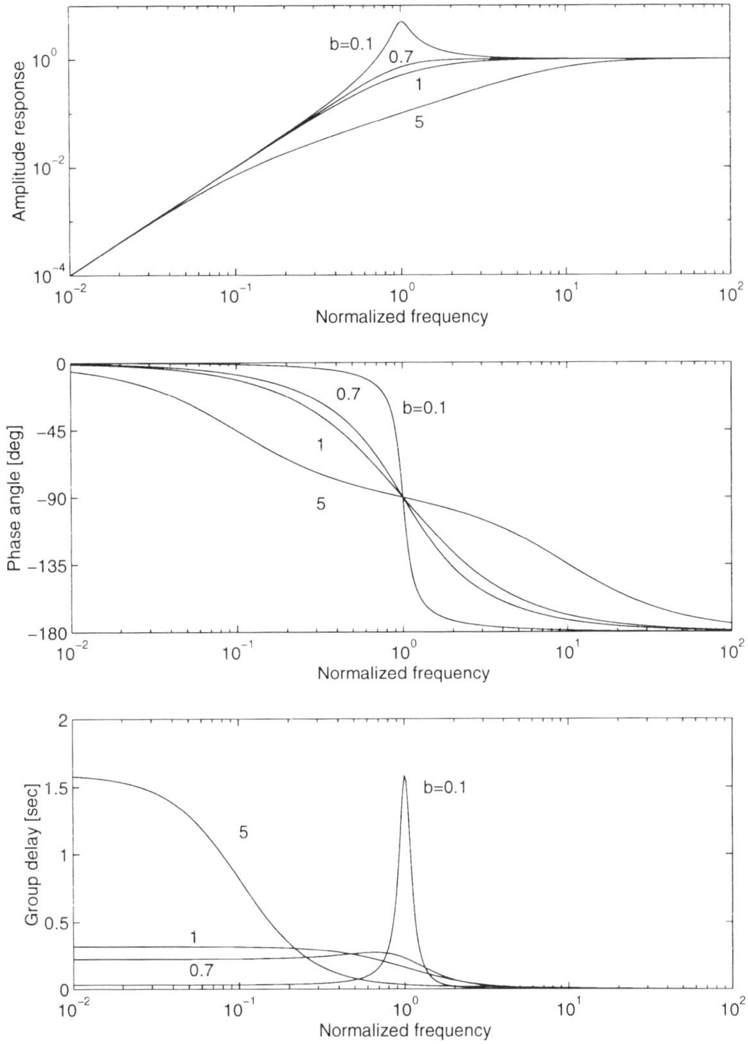


**Fig. 1.5** Inertial sensor transfer function, equation (1.11), for several values of the relative damping coefficient $b$ as a function of normalized frequency $f/f_n$.
The complex function is illustrated by separate graphs for amplitude and phase response, and group delay. On the amplitude response graph, the upper curve shows the strong amplification at $f/f_n$.
The next lower curve shows the slightly extended low frequency response over critical damping $b$=1. At $b$=5 two separate corner frequencies become apparent.

Since $x=0$ for $t<0$, we may obtain the Fourier domain version of the transfer function by restricting the independent variable to the imaginary axis, setting $s = i\omega$.

The first point to note is the existence of a natural frequency and a damping factor. The effect of damping is best illustrated graphically and Fig. 1.5 shows the normalized transfer functions for a range of values in $b$. Lightly damped sensors ($b << 1$), with their exaggerated response at $f_n$, can be problematic. A value of $b$ = 0,7 gives the flattest frequency response.

By examining the extremes of frequency, we can see the basis for two types of sensor. When $s >> \omega_n$ the transfer function is approximately $-1$, i.e. the mass does not move with respect to an inertial frame of reference and the relative motion of the case and the mass perfectly mirrors the ground motion. This is the classical seismometer mode of operation. When $s << \omega_n$ the relative motion of mass and case is very small and proportional to $s^2$. This represents a double differentiation in the time domain, so the displacement of the mass relative to the case is proportional to ground acceleration. In physical terms the motion of the mass and case are almost identical and the extension of the spring is a measure of the force necessary to accelerate the mass. The accelerometer operates on this principle.

The sensor impulse response may be obtained by transforming equation (1.11) back to the time domain. The numerator may be set to 1, so that the result corresponds to displacement of the mass relative to the case for an impulse in ground acceleration. This is useful because the calibration inputs on real sensors usually provide the capacity of displacing the equilibrium position of the mass within the case. To apply the standard transform tables, the zeros of the quadratic denominator of equation (1.11) need to be found:

$$-b\omega_n \pm \omega_n\sqrt{b^2 - 1}$$

Let $n = -b\omega_n$ and distinguish between

$$\omega = \omega_n\sqrt{1 - b^2} \quad \text{for } b<1 \tag{1.12}$$

and

$$\omega' = \omega_n\sqrt{b^2 - 1} \quad \text{for } b>1$$

Then, the right–hand side of equation (1.11) may be rewritten as

$$\frac{1}{(s-n)^2 + \omega^2} \quad \text{for } b<1$$

$$\frac{1}{(s-n)^2} \quad \text{for } b=0$$

$$\frac{1}{(s-n)^2 - \omega'^2} \quad \text{for } b>1$$

which transform to

$$\frac{1}{\omega} e^{nt} \sin \omega t; \quad te^{nt}; \quad \frac{1}{\omega'} e^{nt} \sinh \omega' t$$

respectively. The impulse responses for a natural frequency of 1 Hz and a range of damping factors are plotted in Fig. 1.6.

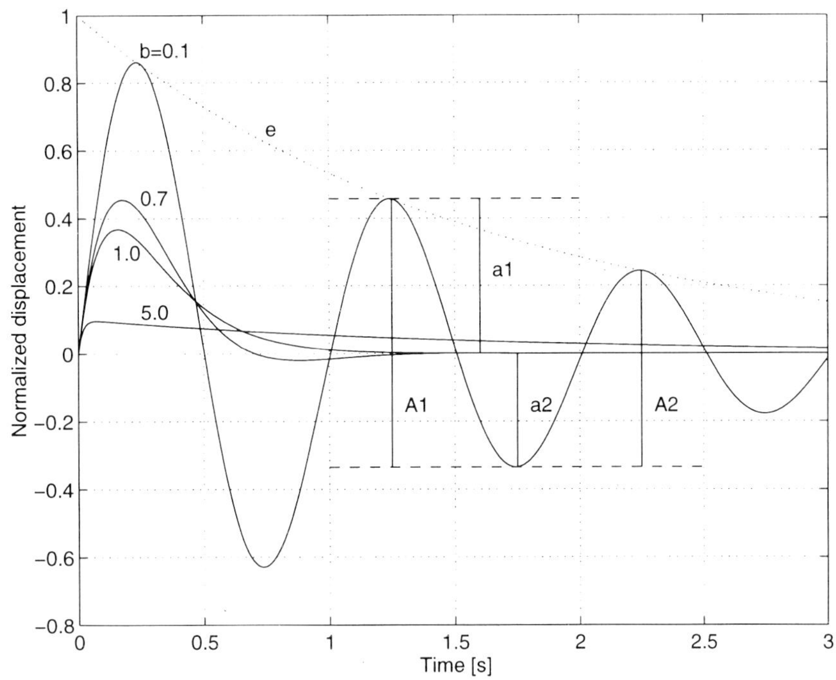


**Fig. 1.6** Response of an inertial sensor to an impulse in acceleration. The same damping factors are illustrated as used for the frequency response in the previous figure. The curve for $b=0.1$ has been enhanced with its exponential decay envelope marked e, and amplitudes of successive swings for the calculation of the logarithmic decrement. The difference between $b$=0.7, which crosses the time axis once and critical damping $b$=1.0, which does not, can clearly be seen.

When the damping is less than critical, $b<1$, and therefore the response is oscillatory, it is useful to consider the ratio of the amplitudes of two successive peaks in opposite directions, denoted by $a_1$ and $a_2$ in Fig. 1.6. The so-called logarithmic decrement $d$ is defined as

$$d = \ln \frac{a_1}{a_2} = \ln \frac{e^{nt}}{e^{n(t+\pi/\omega)}} = \frac{-\pi n}{\omega} = \frac{\pi b}{\sqrt{1-b^2}}$$

Inverting for $b$

$$b = \frac{d}{\sqrt{\pi^2 + d^2}}$$

providing a fairly direct method of measuring $b$ independently of $\omega_n$. In Fig. 1.6, it may also be shown that

$$\frac{A_1}{A_2} = \frac{a_1}{a_2}$$

where $A_1$ and $A_2$ are independent of the signal zero level.

## 1.3 Realizable sensor characteristics

The range of ground velocities and frequencies that can be measured with commonly available sensors is illustrated in Fig. 1.7. The noise level is illustrated not as noise density as a function of frequency, but as it would be measured using a long term average (LTA) in the time domain. The frequency dependence is only due to the sensor frequency response. Dynamic range is the ratio of the amplitudes of the maximum measurable signal, $A_{max}$, and the noise level, $A_{noise}$, expressed in decibels:

$$DR = 20\log(A_{max}/A_{noise}) \quad [\text{dB}] \tag{1.13}$$

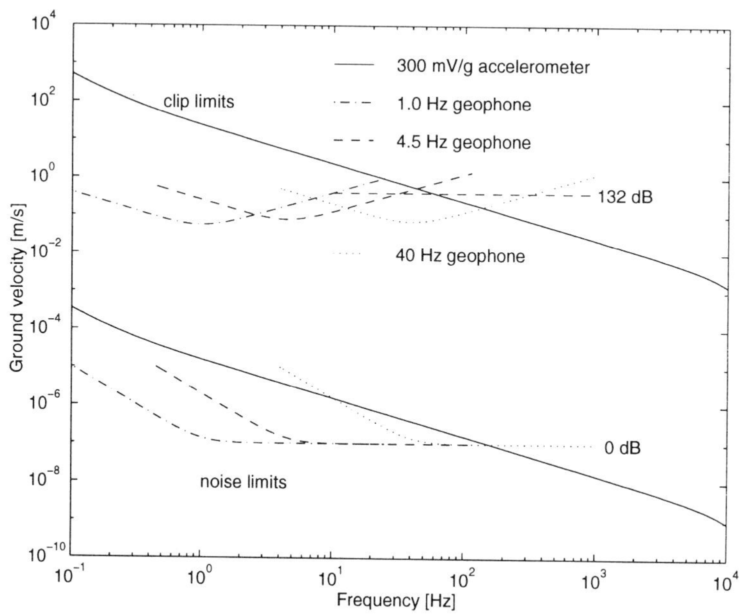


**Fig. 1.7** Sensitivity and dynamic range of sensors commonly used in mine seismic systems. The region between the limits represents the usable range for each instrument. The geophone's greater sensitivity up to several hundred hertz is clearly shown, as is the loss of dynamic range due to displacement clipping below these frequencies. A data acquisition system dynamic range of 132 dB with its quantization noise matching the expected ground noise of $10^{-7}$ m/s is illustrated.

The level at which an acquisition system with a dynamic range of 132 dB would clip is shown. Since the geophone is a passive device, the signal conditioning equipment or the intrinsic ground noise effectively sets the noise level. A noise floor of $10^{-7}$ m/s environmental noise as expected in mines is used in the figure. Geophones exhibit better sensitivity over a relatively narrow band at the lower frequencies of interest, whereas accelerometers cover a relatively wide band, excelling in sensitivity at the higher frequencies and measuring without distortion the large amplitude, low frequency strong ground motions of nearby large events.

### 1.3.1 Geophones

Geophones operate in the seismometer mode, i.e. the most useful bandwidth is above the natural frequency. The coil and magnet which are used to detect the motion of the inertial mass produce an output proportional to velocity. The unit produces a reasonable amount of power at low impedance for driving cables. Damping is determined by the load resistance and each type of instrument specifies the open circuit damping $b_o$ and a relation for determining the current damping, $b_c$ for a given load.

Geophones exhibit a relatively narrow usable bandwidth. Apparently, it is difficult to construct a suspension for the mass which is relatively weak in the axial direction, to produce a low $f_n$, and very stiff in the radial direction to suppress any transverse response. Where manufacturers have attempted to minimize these effects, they generally quote a "clean" or "spurious response free" bandwidth, $f_{cl}$. In a brief survey of data sheets, the highest value of $f_{cl}/f_n$ found was 38, and some were less than 10. As mentioned in Section 1.1, we need at least a factor of 10 to be able to measure seismic moment and energy for an event.

Figure 1.8 (Oh, 1996) shows how geophone sensitivity varies with the angle of excitation. Transverse resonance modes produce a response almost equal to axial at large angles for specific frequencies.

Real geophone behaviour near the natural frequency conforms well with theory. Provided that $f_n$ and $b$ are known, deconvolution techniques may be used to extend the useful bandwidth to lower frequencies. Deconvolution cannot be applied indiscriminately to small or distant events as the signal below $f_n$ may be buried in noise, since the velocity spectrum of events decreases with decreasing frequency, see Fig. 8.8 in Chapter 8 of this book.

The high frequency behaviour, unfortunately, is not so tractable. The spurious responses tend to vary from one unit to another and are difficult to detect on a shake table because of their association with transverse excitation.

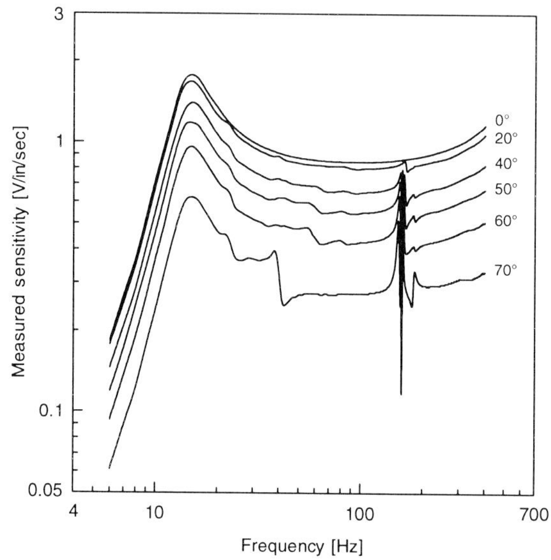


**Fig. 1.8** Variation in geophone sensitivity with angle of excitation. The average sensitivity, as expected, varies as the cosine of the excitation angle (0°=axial, 90°=radial), but the transverse modes can produce almost the full response at a certain frequency even at a large angle. This figure was produced using a 14 Hz geophone. Reproduced from Oh (1996).

We have no choice but to filter out or ignore frequencies which may be affected by spurious responses.

Some examples of manufacturers' specifications for commonly used geophones are reproduced in Table 1.2. The values of maximum velocity at $f_n$ have been calculated from the specified case to coil motion, which has been reduced on the high frequency geophones to compensate for the high tilt angles permitted. To estimate the range of event corner frequencies which a particular sensors covers, we may assume the frequency response is extended down to $f_n/2$ by deconvolution. The range of corner frequencies is then $f_n$ to $f_{cl}/5$. Where $f_{cl}$ is not specified, $f_{cl}=25f_n$ is a good conservative choice.

The corresponding range of moments and stress drops may be seen by drawing the horizontal lines on Fig. 1.1.

Half the case to coil motion may also be compared to the displacement in the seismic source region given in equation (1.7) and illustrated in Fig. 1.2. For example, an event with $m_M = 1$ and $\Delta\sigma = 1$ MPa, will have a displacement of about 1 mm, and a corner frequency of about 50 Hz (from Fig. 1.1) which is above the natural frequency of all the geophones in the table. If one of the miniature geophones is near an event with a larger moment or stress drop, it will clip.

*Miniature geophones* High quality miniature geophones with natural frequencies from 4.5 Hz up to approximately 100 Hz are inexpensive, reliable and commonly available thanks to extensive use by the oil exploration industry. Their strong point is their sensitivity which means that, when used in combination with good amplifiers, the ambient ground noise determines the system noise level of ~$10^{-7}$ m/s at the quietest sites in mines. These frequencies (see clean frequencies in Table 1.2) also propagate through the rock with little attenuation, so the sensor sites may be fairly widely spaced throughout the mine. Miniature geophone output reflects ground velocity, so the kinetic energy may be calculated directly, and only one integration is necessary to obtain displacement for the seismic moment. Geophones are inexpensive and relatively easy to install in boreholes long enough to reach intact rock, as long as some care is taken to ensure they are precisely vertical or horizontal.

A weak point of these geophones is the distortion and clipping introduced at larger displacements produced by nearby large events. Because the geophone output reflects ground velocity, the displacement clipping is not immediately apparent on inspection of the seismograms. The theoretical dynamic range of these sensors is very large (see Fig. 1.6), but as the large events produce low frequencies, we are concerned with the dynamic range at the natural frequency, as listed in Table 1.2. The acceleration limit of 50$g$ is actually a shock limit, so not only will the geophones not reproduce these signals accurately but such high accelerations may lead to permanent damage or change in characteristics.

Due to the low cost of these units, it is reasonable to extend the bandwidth coverage by installing more than one triaxial set at one site. For example, the 4.5 Hz and 40 Hz geophones described in Table 1.2 complement each other well.

*Low frequency geophones* Geophones with natural frequencies between 0.5 Hz and 2 Hz are manufactured for earthquake monitoring and are an order of magnitude more expensive than miniature geophones. Such instruments are essential for measuring moments of events with corner frequencies below 4.5 Hz (see Fig. 1.1).

### 1.3.2 Accelerometers

As described in Section 1.2, accelerometers operate below their natural frequency and measure the force applied to the inertial mass. In principle, the frequency response extends right down to zero (DC) and, even where this is not achieved in practice, a broad band is usable. The stiff spring generally produces no unexpected side effects.

**Table 1.2** Manufacturer's specifications for some commonly used geophones

| *Type* | L-4C[§] | HS-1-1A[¶] | SM-6B[‡] | SM-7B[‡] | SM-15B[‡] | SM-11FT[‡] | GS-20DM[¶] |
|---|---|---|---|---|---|---|---|
| Natural frequency $f_n$ [Hz] | 1 | 4.5 | 4.5 | 8 | 14 | 30 | 40 |
| $f_n$ tolerance | ±5% | ±20% | ±11% | ±6% | ±5% | ±5% | ±5% |
| Tilt for $f_n$ tolerance (vertical only) | 5° | 5° | 5° | 20° | 25° | 180° | 90° |
| Distortion at 18 mm/s | N/S | N/S | <0.3% 12 Hz | <0.2% 12 Hz | <0.2% 14 Hz | <0.2% 30 Hz | <0.2% 40 Hz |
| Tilt for distortion | N/S | N/S | N/S | 15° | 25° | 180° | 90° |
| Clean frequency $f_{cl}$ [Hz] | N/S | N/S | N/S | 310 $=38f_n$ | >500 $>35f_n$ | 500 $=17f_n$ | >850 $>21f_n$ |
| Case to coil motion (peak to peak) [mm] | 6.3 | 7.6 | 4 | 2 | 1 | 1 | 1 |
| Seismic mass $m$ [g] | 1000 | 22.7 | 11.1 | 11 | 11 | 9.2 | 5.6 |
| Max velocity at $f_n$ [mm/s] | 20 | 107 | 56 | 50 | 44 | 42 | 62 |
| dB referred to $10^{-7}$m/s | 106 | 121 | 115 | 114 | 113 | 113 | 116 |
| Coil resistance $R_c$ [Ω] | 5500 | 225 | 375 | 375 | 375 | 360 | 270 |
| Sensitivity (undamped) [V/m/s] | 276 | 18.1 | 28.8 | 28.8 | 28.8 | 30 | 15.1 |
| Open circuit damping $b_o$ | 0.28 | 0.30 | 0.56 | 0.31 | 0.18 | 0.55 | 0.42 |
| Current damping $b_c$ shunt $R_s$ [Ω] | $\frac{1.1R_c}{R_c+R_s}$ | $\frac{254}{R_c+R_s}$ | $\frac{6000}{(R_c+R_s)f_n}$ | $\frac{6000}{(R_c+R_s)f_n}$ | $\frac{6000}{(R_c+R_s)f_n}$ | $\frac{8310}{(R_c+R_s)f_n}$ | $\frac{0.30R_c}{R_c+R_s}$ |
| Operating temperature range [°C] | −29 60 | −29 70 | −40 100 | −40 100 | −40 100 | −40 100 | −45 100 |
| Shock limit [g] | N/S | 50 | N/S | N/S | N/S | N/S | N/S |

Manufacturers: § Mark Products, US, Inc. ¶ Oyo Geospace Corporation ‡ Sensor Nederland BV.

The force balance accelerometer uses electromagnetic or electrostatic force to control the movement of the mass, so an analogue of the restoring force is available in electrical form. Instruments based on mechanical springs generally measure the strain in the spring. This is small and a variety of methods are used to measure it. Nevertheless, noise in the sensing process sets the noise level of the sensors and matching or built-in amplifiers are often part of the package.

The sensitivity of a given accelerometer to ground velocity or radiated seismic energy increases with frequency, allowing smaller events to be detected than by a comparable geophone, assuming that the trigger measures LTA as broadband noise in the time domain and that the higher frequency environmental noise is attenuated by the rock. The relative insensitivity at low frequency reduces the danger of clipping on nearby large, low corner frequency events. The peak acceleration produced by an event depends on the dynamic stress drop, see equation (1.9).

*Piezoelectric accelerometers* Piezoelectric materials produce an electrical charge in response to mechanical strain. As there is always some electrical leakage, these instruments do not measure down to DC. The damping coefficient $b$ is generally 1% to 2% which results in a sharp resonance peak at $f_n$. However, between these extremes there is a wide bandwidth where acceleration response is flat (see examples in Table 1.3). An insufficiently rigid mounting can cause an apparent downward shift of the resonance frequency into the band of interest.

Output may be provided directly from the piezoelectric element for use with a separate charge amplifier, but, more conveniently, a FET amplifier is often included in the sensor package. These units require a constant current supply and modulate the voltage to produce an output. More than 100 m of cable may be driven in this way. All those listed in Table 1.3 operate in this fashion.

In mine monitoring networks, these accelerometers are usually used to detect smaller, higher frequency events than may be achieved with geophones. The units with upper frequency limits of 7000 Hz to 15000 Hz provide a useful increase in the frequency range and high sensitivity within this range, generally allowing coverage down to $m_{\mathrm{M}}$=−3 (see Fig. 1.1). At higher frequencies, for coverage down to $m_{\mathrm{M}}$=−4, even the accelerometers become less sensitive. The 40 kHz sensor listed in Table 1.3 has an rms noise of 0.002 $g$ so the smallest useful signal must have an amplitude of about 0.02 $g$. From Fig. 1.3, we can see that the lower stress drop events must be within 10 m for successful detection. The lower frequency units, while providing uniform coverage over a wide band, do not generally offer sufficient advantage over geophones to justify the added cost.

**Table 1.3** Manufacturer's specifications for some piezoelectric accelerometers

| *Type* | 1030§ | 793L-4¶ | 1136§ | 793¶ | 501‡ |
|---|---|---|---|---|---|
| Sensitivity [mV/*g*] | 7000 | 300<br>±10% | 100<br>±1dB | 100<br>±5% | 10<br>±10% |
| Frequency range ± 3 dB [Hz] | 0.1 – 1000 | 0.2 – 7000 | 1 – 10 000 | 0.5 – 15 000 | 2 – 40 000 |
| Mounted resonance $f_n$ [Hz] | 2400 | 16 000 | 18 000 | 25 000 | 65 000 |
| Max. transverse Sensitivity | 5% | 7% | 7% | 5% | 5% |
| Max. acceleration [*g*] | 0,7 | 15 | 48 | 80 | 212 |
| Nonlinearity | 2% | 1% | 2% | 1% | 1% |
| Noise floor wideband [μ*g* rms] | 3 | 10 | 250 | 600 | 2000 |
| Dynamic range [dB] | 107 | 123 | 106 | 102 | 100 |
| Power supply | 2.2 mA<br>15 V | 2 – 10 mA<br>18 – 30 V | 1 – 4 mA<br>15 – 30 V | 2 – 10 mA<br>18 – 30 V | 0.5 mA<br>12 – 30 V |
| Shock limit [*g*] | N/S | 2500 | 5000 | 2500 | 10000 |

Manufacturers: § Vibra Metrics, Inc. ¶ Wilcoxon Research, Inc. ‡ Vibrometer Corp.

Frequencies of 3 kHz and above are generally heavily attenuated by the rock. This varies from site to site, but even in competent quartzite, source to sensor distances of more than 300 m cause the higher frequency signal to be lost.

*Force balance accelerometers* These instruments replace the mechanical spring with an electronic feedback circuit. They are generally more expensive than 1 Hz geophones, but provide broader bandwidth, e.g. DC to 100 Hz. Generally, acceleration and velocity outputs are available.

## 1.4 Network considerations

High frequency miniature geophones provide reasonable coverage down to $m_M$=0, as shown by comparing clean frequencies from Table 1.2 with corner frequencies in Fig. 1.1, provided that there are five stations within 1 km of each event. Coverage up to $m_M$=3 in the same area can be accomplished by pairing with 4.5 Hz units. A dense network is required to compensate for both the geophones' relative insensitivity and the increasing attenuation of the rock mass to the dominant frequencies of these events. Care must be taken not to accept results from stations near enough to large events to suffer low frequency, large displacement distortion and clipping problems.

**Table 1.4** Typical networks with different sensor densities

| Lower magnitude limit | −3 | 0 | 1 |
|---|---|---|---|
| Network type | Accelerometer at least 5 stations < 300 m | Geophone at least 5 stations < 1 km | Geophone 5th station > 2 km |
| Sensor/ anti-aliasing filter | 10 kHz | 500 Hz | 200 Hz |

Apart from large, low stress drop events, full coverage down to $m_M$=−3 may be provided by a dense network of accelerometers, which have the high frequency sensitivity to reliably detect negative magnitude events and the low frequency insensitivity to measure close, large events. The required density for full coverage is at least five stations within 300 m of any expected source area, so, effectively, a separate network is established around each longwall or other working area. Even so, several times more stations will probably be required to cover a given mine with accelerometers than geophones. Table 1.4 illustrates three typical network configurations.

### 1.4.1 Results of sensor evaluation field trials

In order to better evaluate the performance side of the performance/cost ratio for different sensors, several geophones and piezoelectric accelerometers were installed in a single holder, in a borehole at a deep level gold mine in South Africa.

Three sensors were monitored at a time with one of the geophones always included as a reference. Two examples illustrate the expected superiority of the accelerometer at high frequencies.

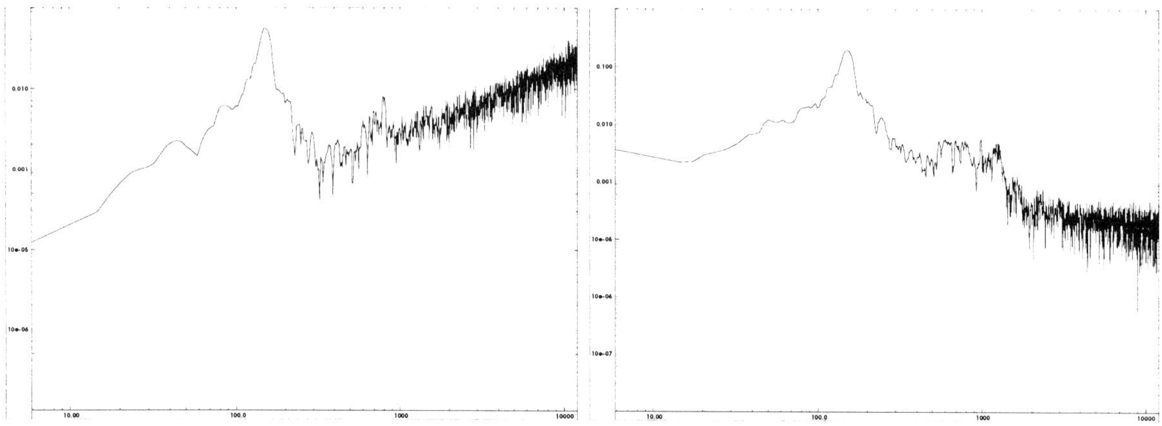

**a** **b**

**Fig. 1.9** Acceleration spectra obtained by differentiating a geophone signal (**a**) and directly from an accelerometer (**b**). The increase in noise on the geophone derived spectrum at higher frequencies is clear. Note that on the accelerometer spectrum, features near 1000 Hz do not appear on the geophone data.

Figure 1.9 shows the spectra of the differentiated geophone output compared to that of an accelerometer. The increase in noise and loss of signal at high frequencies on the differentiated signal are clearly visible. Figure 1.10 shows the recording of a small high frequency event which was recorded by an accelerometer while only noise is visible on the geophone trace. The events recorded during the duration of the experiment were small and within several hundred metres of the sensor, so there was no opportunity for the geophone to demonstrate its low frequency sensitivity, or the accelerometer its strong ground motion abilities.

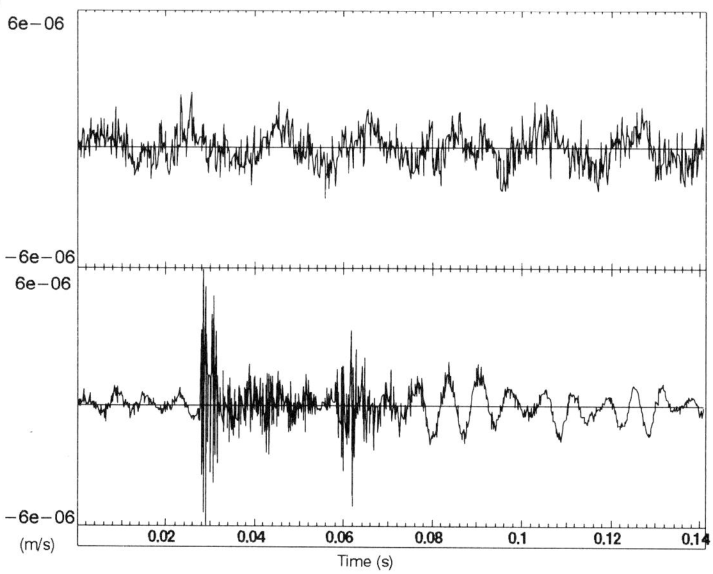


**Fig. 1.10** High frequency event clearly recorded by an accelerometer (bottom) but not visible on geophone output (top). The acceleration data have been arbitrarily scaled so that the 150 Hz signal between 0.08 s and 0.1 s should have the same amplitude as on the geophone data, where it is barely visible.

## 1.5 Sensor orientation

There are two aspects to sensor orientation: firstly, if the lower natural frequency geophones are not installed precisely vertically or horizontally they do not function correctly; secondly, the true directions of ground motion must be found for each event to be described quantitatively. For these reasons, it is advantageous to be able to install sensors accurately with a given orientation, and to be able to estimate the orientation of sensors once installed.

For installation in a borehole, the sensors are usually mounted on to a holder or "boat" which is almost as wide as the borehole and much longer than it is wide, so that it closely assumes the orientation of the borehole, which may be determined by surveying. A keyed rod may be used to rotate the boat within the borehole during installation to allow orientation about the borehole as an axis. For a vertical hole the handle of the keyed rod may be aligned with a survey mark and the boat assumed to be aligned likewise with a few degrees margin of error. For an off-vertical hole some feedback may be obtained as to the orientation of the sensors themselves, either by driving the sensors (if they are geophones) and rotating the boat until a symmetric clipping pattern is obtained, or by including a mercury switch or other tilt sensitive device into the boat. Of course, care must be taken to preserve sensor signal polarity throughout the data acquisition modules.

Once the sensors are operating, any deviation from the assumed orientation may be calculated by comparing the P wave polarization direction with the hypocentral direction for artificial sources or natural events whose locations are known accurately. Given three or more such known locations, it is possible to determine a rotation matrix **A** such that $\hat{\mathbf{x}}_i = \mathbf{A}\hat{\mathbf{y}}_i$, where $\hat{\mathbf{x}}_i$ is the unit direction vector from the sensor to the source for the $i$th event in standard coordinates determined from the relative source and sensor locations and $\hat{\mathbf{y}}_i$ is the same vector in sensor coordinates determined from the P wave polarization. The matrix **A** has nine elements, but can be parametrized in terms of the three Eulerian angles. Since the order in which rotations occur is important the angles are defined as follows: first a rotation of $\phi$ about the $z$-axis; then a rotation of $\theta$ about the new $y$-axis; and, finally, a rotation of $\psi$ about the new $z$-axis. These rotations are illustrated in Fig. 1.11; $\theta$ and $\phi$ correspond to dip and dip direction respectively. Matrix **A** can then be expressed as:

$$\begin{bmatrix} -\sin\phi\sin\psi + \cos\theta\cos\phi\cos\psi & \cos\phi\sin\psi + \cos\theta\sin\phi\cos\psi & -\sin\theta\cos\psi \\ -\sin\phi\cos\psi - \cos\theta\cos\phi\sin\psi & \cos\phi\cos\psi - \cos\theta\sin\phi\sin\psi & \sin\phi\sin\psi \\ \sin\theta\cos\phi & \sin\theta\sin\phi & \cos\theta \end{bmatrix} \quad (1.14)$$

If the direction of one of the sensor axes is known precisely, the unknown rotation about that axis may be determined from a single event. If, for example, the sensor $z$-axis is known to be vertical, $\theta$ and $\psi$ are 0 and matrix **A** simplifies to:

$$\begin{bmatrix} \cos\phi & \sin\phi & 0 \\ -\sin\phi & \cos\phi & 0 \\ 0 & 0 & 1 \end{bmatrix} \tag{1.15}$$

In general, some optimization scheme would be used to determine the best values of $\theta$, $\psi$ and $\phi$ from many events for each sensor.

The rotation matrix **A** may then be used to transform the ground motion from sensor coordinates into standard coordinates for each sensor and thus contribute to determining the source mechanism for each event.

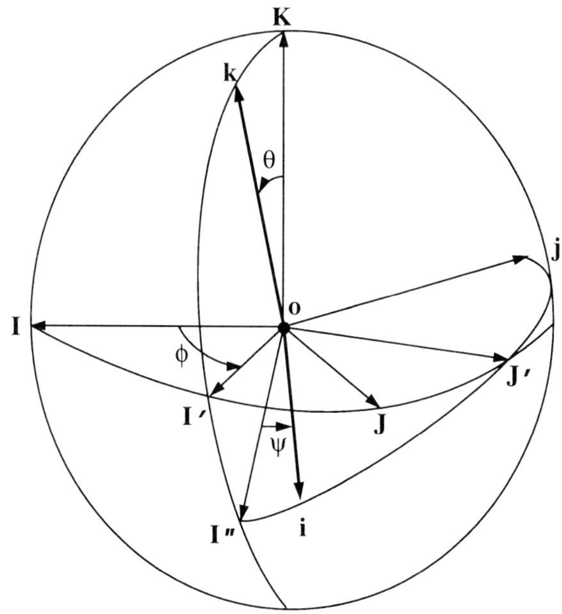


**Fig. 1.11** The Eulerian angles. A rotation in three dimensions may be uniquely defined by a rotation $\phi$ about **K**, followed by a rotation $\theta$ about **J**' and finally a rotation $\psi$ about **k**. The axes **I**,**J**,**K** might represent a mine coordinate system and **ijk** the orientation of the three sensor components.

# 2 Seismic Monitoring Systems

Seismic sensors are distributed around or throughout the volume of interest. The monitoring system must accurately record the amplitude and timing of any significant ground motion over a wide range of amplitudes, frequencies and durations, and assemble the records at a central point for processing, within a reasonably short time so that action may be taken in response and at a high rate for maximum information retrieval.

The seismic data pass through the following stages: monitoring each sensor continuously to decide when the signal becomes significant (triggering); ensuring that the signal represents a seismic event (validation); deciding which records from which sensors represent the same event (association); extracting source and path parameters from the raw ground motion data for each event (seismological processing); and inferring from a history of these parameters the processes which are taking place within the volume being monitored (interpretation). The following chapters of this book are concerned with the details of the last two stages, so here we only note that they are implemented as complex, generally interactive, software processes, and attempt to cover all other aspects of the seismic data acquisition system.

The basic functions and data flow are illustrated in Fig. 2.1. Triggering is not essential, but widely used to reduce the amount of data, thus three types of system may be distinguished: those that do not trigger, but record continuously, where event detection becomes part of the processing stage; those that record data following a trigger, for which the delay provides pre-trigger information; and those that store only the time of each trigger. The most useful mine monitoring systems fall into the second category.

The details of an implementation differ greatly depending on how and when the data are transmitted to the central site, indicated by the vertical lines on Fig. 2.1. If transmission is continuous and immediate, as indicated by the left-most line, then relative timing of the signals at the central site is implicit. For absolute time all data may be time-stamped from a single clock and network triggering and real time association may be performed, where any combination of triggers may be used to control the recording of any or all stations.

However, the properties of available transmission media often dictate that the signal bandwidth be uncoupled from the transmission bandwidth. This leads to an arrangement, the upper central vertical line in Fig. 2.1, where the signals are time-stamped and recorded locally, and transmitted to the central site when the transmission channel is available. In this case each station requires a clock, which must be periodically synchronized with the network

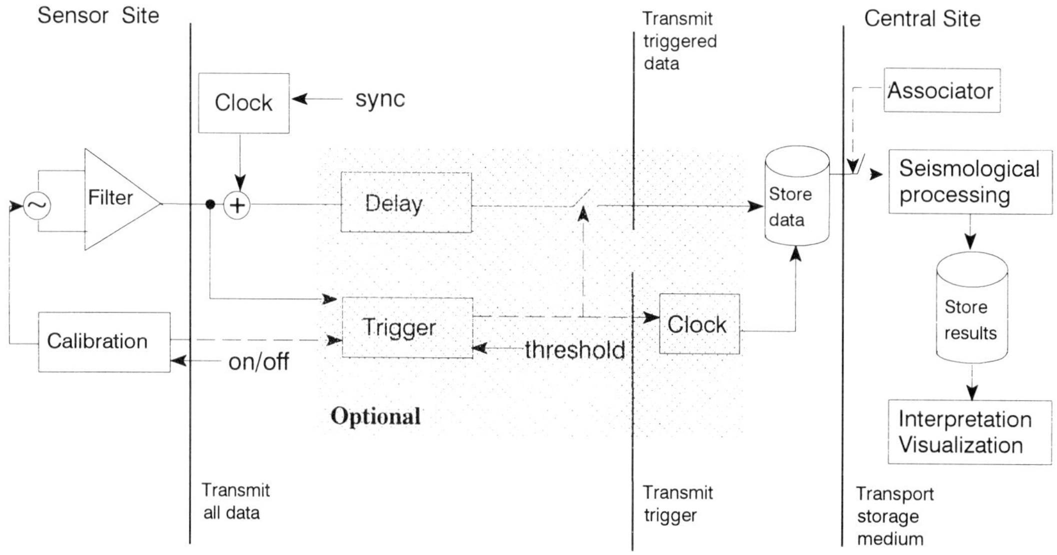


**Fig. 2.1** Basic functions of a seismic system: signal conditioning, represented by filtering and calibration; time-stamping from a clock; storage; and processing. Dashed lines represent control signals. Triggering is optionally used to reduce the amount of data. In an extreme case, only the trigger time is stored. Usually more than one sensor site is incorporated into a seismic network. Vertical lines represent points at which data may be transmitted from multiple sites to a central point for common processing. Those functions to the left of the chosen line are distributed to sensor sites, while those to the right are shared centrally.

clock, and bidirectional communication with the central site. The associator operates on delayed trigger information and may be able to set data transmission priorities, but cannot control the recording process which must be done locally at each sensor.

In that this scenario of distributed control and delayed transmission is most widely applicable, the system functions will be discussed in this context, with alternate possibilities noted where necessary.

The question of system organization is but the most awkward aspect of the more general problem of discussing a system in terms which will not soon be rendered redundant by technological development. Note that Fig. 2.1 illustrates the system functions without specifying analogue to digital conversion, type of data transmission or even digital data storage. Three areas of interest in which change is now under way are the relentless annual increase in the power of digital computers, the increase in communications bandwidth made possible by optical fibre, and the advent of oversampling A/D converters with their unprecedented linearity. Where possible, emphasis will be placed on the physical principles underlying each step in the data processing sequence, rather than the technology.

Following the signal from the sensor through to the central site, we first require signal conditioning to take care of any special sensor needs, injection of a calibration signal, anti-alias filtering, possible reduction in dynamic range by amplitude compression or gain ranging, and digitization. Any direct data transmission also logically belongs with this group. Then we consider the specialized seismic requirements of triggering and record validation. Digital data communication in the distributed system comes next, followed by association and data storage and management.

## 2.1 Signal conditioning

The input to the signal conditioning circuitry must provide for any special requirements of the sensor, a matching damping resistor for geophones and a constant current supply for the accelerometers with built-in FET amplifiers being the most obvious examples. Because this input is connected to the outside world, a certain amount of robustness is desirable.

### 2.1.1 Calibration signal injection

To verify that the system is functioning correctly, it is useful to be able to introduce a known signal as close to the sensor as possible. The sensors designed for earthquake monitoring usually accept a calibration signal. In the case of geophones, this drives a coil which applies an additional force to the mass which must be countered by the spring, effectively causing a displacement of the equilibrium position. For the force balance accelerometer, the calibration signal is added to the error signal.

For miniature geophones, the signal may be applied to the sensing terminals. This interferes with measuring the response, but, again, causes a displacement of the mass. The piezoelectric accelerometers with built-in amplifiers generally do not permit *in situ* calibration.

If the calibration signal is in the form of a pulse which is long enough for the sensor to reach equilibrium at its new position, we will see two instances of the step response of the sensor, one when the pulse is applied and another when it is removed. Note that the displacement step is differentiated by the geophone to give a velocity impulse response, whereas the accelerometer assumes the displacement of the mass is due to acceleration. If the sensor is underdamped, ($b < 1$) then the step response will be oscillatory and we may measure $b$ and $f_n$ from it. See Chapter 1, Section 1.2 and Fig. 1.6, for the theory of sensors.

We may determine $f_n$ independently for a geophone as the frequency at which a sinusoidal calibration signal is in phase with the geophone output. However, this is probably too complex an operation to perform *in situ*.

For piezoelectric accelerometers with built-in amplifiers, and with any sensor to test for crosstalk between channels in the signal conditioning circuitry, it is advisable to be able to inject a broadband signal into one channel at a time, and measure the response of the electronics and the effect on the other channels. Typically, a pseudorandom binary sequence would be used (Ljung, 1987), which is easily generated using a shift-register and has a perfect autocorrelation function. The level of crosstalk is important as it has the same effect as cross axis sensitivity in the sensor and is also a good diagnostic tool.

### 2.1.2 Anti-aliasing filters

Before digitization or other sampling process may be performed on the signal, the bandwidth must be restricted to less than half the sample frequency to prevent aliasing. Since we are generally interested in low frequencies, this is achieved by a low pass filter.

Any signal present at frequency $f$ above half the sampling frequency will be aliased, and corrupt the low frequency data by appearing at frequency

$$f_a = |f - nf_s|$$

where $n$ is an integer such that

$$0 \leq f_a < f_s/2 .$$

In general, we would like to maximize the usable bandwidth for a given sampling frequency. However, realizable analogue filters which cut off very sharply in the frequency domain cause phase distortion even in the signals which are not attenuated.

The characteristics of three different types of filter are plotted in Fig. 2.2. The transfer functions were taken from Baher (1990). A fifth order filter is shown in each case and the Chebyshev has a 0.1 dB ripple in the passband. The abscissa is normalized to the cutoff frequency ($f_c = 1$) where the signal is reduced to half power (−3 dB). This is generally considered the usable bandwidth. To introduce <1% distortion due to aliasing, we must sample at double the frequency at which the attenuation reaches 40 dB.

*Chebyshev filter* The best by this criterion is the Chebyshev characteristic, where we may set $f_s = 4 f_c$. The problem with this approach is illustrated by the group delay, showing that frequencies near to the cutoff are delayed by almost a full cycle compared with lower frequencies. This dispersion is a problem for broadband signals such as seismograms where accurate time differences must be measured.

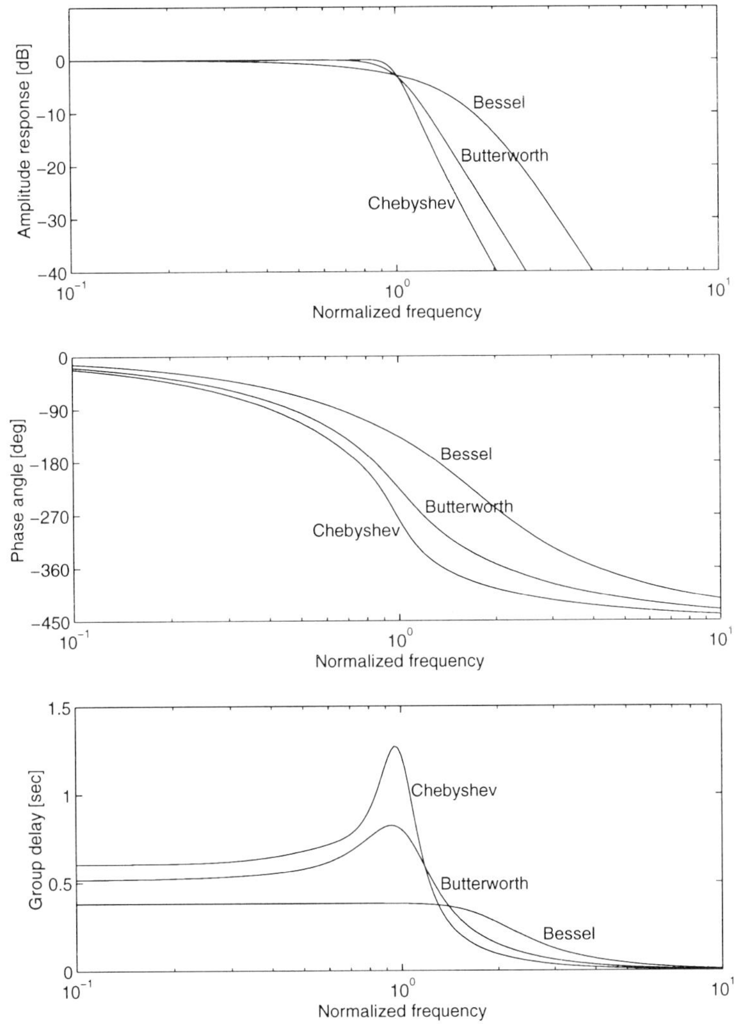


**Fig. 2.2** Characteristics of analogue low pass filters. All filters are fifth order and the Chebyshev has 0.1 dB ripple in the passband. The frequency axis is normalized for 1 Hz half power bandwidth. While the Butterworth and Chebyshev have better attenuation characteristics allowing a lower sample rate for a given signal bandwidth, the dispersion, indicated by the variation of group delay with frequency, makes them unsuitable when accurate picks must be made in the time domain from broadband signals such as seismograms. The transfer functions were taken from Baher (1990).

*Butterworth filter* While not as sharp as the Chebyshev, this still exhibits group delay variations.

*Bessel filter* We are therefore left with the Bessel filter which is designed for constant group delay (linear phase) but which requires a sampling rate $f_s = 8f_c$. Because of its gentle cutoff profile, the Bessel also produces little ringing and overshoot, so that peak amplitudes are also accurate.

*FIR filter* Another filter with the required characteristic is the symmetric finite impulse response (FIR) digital filter. Very high factors of oversampling

(~100) may be used which means that the analogue anti-alias filter may be first order. The low pass FIR filter is applied to the oversampled signal followed by a reduction in sampling rate through decimation. The characteristics of a 64 point FIR filter are plotted in Fig. 2.3. It was designed using a program incorporating the Parks and McClellan optimizing algorithm (McClellan *et al.* 1973). The frequency axis is normalized to the sample rate. The cutoff frequency is above 0.2 $f_s$ and 40 dB attenuation is achieved before 0.25 $f_s$ so the sample rate may be halved after filtering, leaving a signal which utilizes 80% of the available bandwidth.

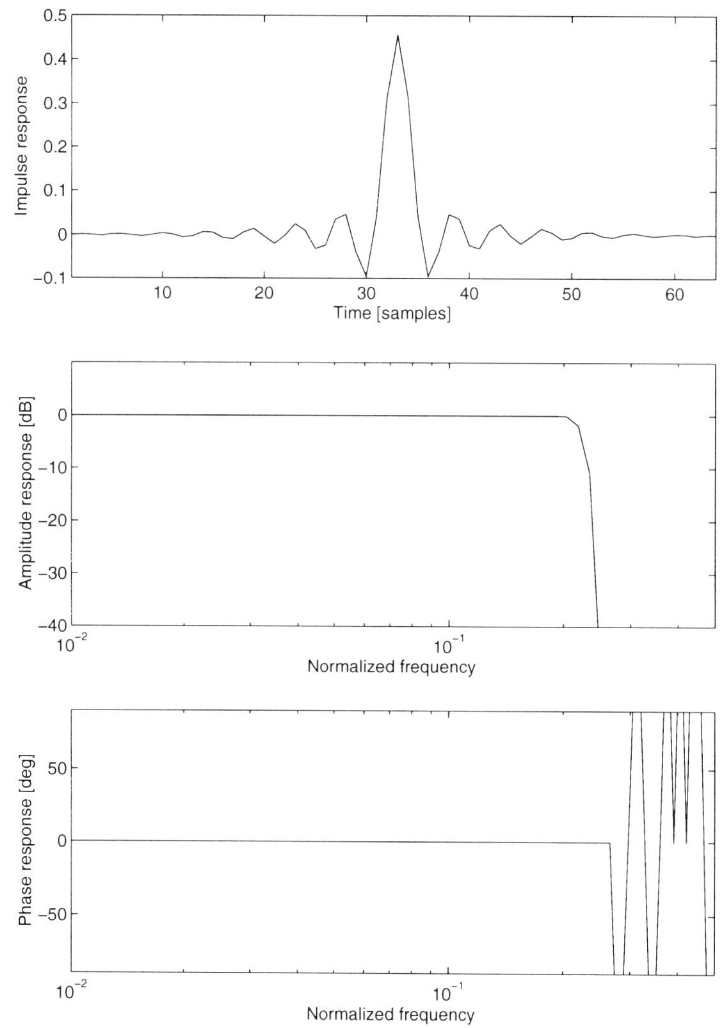


**Fig. 2.3** Characteristics of a 64 point finite impulse response (FIR) digital low pass filter. The frequency axis is normalized to the sample rate. The filter was designed to allow halving of the sampling rate, so reaches the required 40 dB attenuation at 0.25 $f_s$. The passband extends to 0.2 $f_s$, utilizing 80% of the possible bandwidth after decimation by a factor of two. The group delay is a constant 32 samples.

### 2.1.3 Reduction in dynamic range

The signal from the sensor has a very wide dynamic range – the ratio of the maximum signal to the quiescent noise level. Maintaining this dynamic range through the signal conditioning circuitry is difficult as electronic components add noise and limit the maximum amplitude. However, the sensor signal has limited accuracy because of the distortions in the transduction process. Whereas noise for a given circuit combination is fixed, the error due to distortion is proportional to the signal amplitude. If we assume we can resolve successive levels of output signal separated by the noise amplitude, then, when the signal becomes large enough for the distortion to exceed the noise, we have many redundant levels which could all represent the same signal differently affected by the distortion. At these signal amplitudes meaningful levels are separated by the ratio $1 + 1/n$, where the distortion is expressed as 1 part in $n$. For a dynamic range $D$, we have the relation

$$D = n\left(1 + \frac{1}{n}\right)^{(N-n)} \tag{2.1}$$

where $N$ is the total number of significant signal levels. The relation between input signal level and output level number is illustrated in Fig. 2.4. Taking logs and rearranging yields an expression for $N$:

$$N = n + \frac{\log D - \log n}{\log\left(1 + \frac{1}{n}\right)}$$

For example, a dynamic range of $10^6$ (120 dB) with a distortion factor of 0.1% (1 part in 1000) yields 7912 distinct levels which may be represented in 13 bits instead of the 20 bits required to represent $10^6$.

A circuit with the partly linear, partly logarithmic transfer function implied by equation (2.1) is almost impossible to construct and maintain with the required accuracy. Multiple linear circuits with different gains may be used to achieve the same effect. As the signal increases, so a lower gain circuit is selected. This process is called automatic gain ranging, and is illustrated by Fig. 2.4. The switchover criterion is that at least $n$ levels must be distinguishable at the lower gain.

For the above example suppose that 16 000 levels are distinguishable at each gain (14 bits). Then a maximum gain of 64 is required to reach the $10^6$ total dynamic range, the next lower gain could be 8, and followed by a final

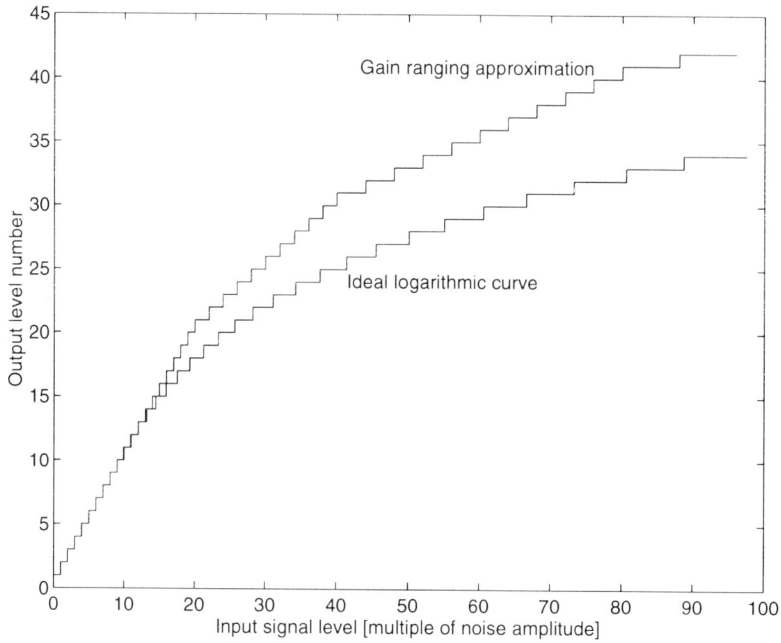


**Fig. 2.4** Quantization levels in the presence of noise and distortion. Ideally, the levels are linearly spaced for small levels where noise predominates, and exponentially spaced at higher levels where distortion produces the greatest uncertainty. The graph was calculated for a dynamic range $D = 100$ (40 dB) and distortion of 10% so that the steps are clearly visible. The piecewise linear transfer function given by selecting one of multiple linear gains is also shown.

gain of 1. Sixteen thousand counts at a gain of 64 corresponds to 2000 at a gain of 8, and similarly for gains 8 and 1, so we easily maintain our maximum distortion criterion.

### 2.1.4 Analogue to digital conversion

A/D converters may be broadly classified into four types: integrating, successive approximation, flash and sigma-delta. In general, integrating converters are too slow for seismic applications, but their high accuracy has been inherited by sigma-delta converters.

Successive approximation converters approach the solution one bit at a time, from the most significant bit, subtracting the current approximation from the input signal before determining the next bit. Flash converters take the parallel approach, having a comparator for each possible input level and encoding the result for output.

Subranging converters are appearing which may be regarded as successive approximation converters which determine more than one bit at a time or pipelined flash converters.

Sigma-delta (or delta-sigma) converters use noise-shaping, oversampling and digital filtering techniques to provide high accuracy and relatively high speed.

The sigma-delta modulator uses an integrator and comparator to produce a high rate bitstream which, when averaged by a digital filter and decimated, produces a very accurate representation of the signal. The accuracy is attributed to the stability of the integrator and the use of a single comparator. The technique is still evolving, with high order multibit modulators and a wide variety of digital filters appearing recently.

At present, 12 or 14 bit successive approximation or subranging converters combined with the gain ranging dynamic range reduction option, described above, provide a cost effective and flexible solution. One converter may be shared between several channels or gain ranges by means of an analogue multiplexer. Synchronous sampling of all channels is necessary for polarization measurements on triaxial sensors or beam forming on arrays. Where an A/D converter is shared, this may be achieved by digital interpolation or using a track and hold amplifier per signal after dynamic range reduction.

Sigma-delta converters offer a wider dynamic range, but at present (Analog Devices, 1994) the limit for filter cutoff frequencies of 30 Hz and greater appears to be about 18.5 bits (111 dB) even from devices which produce 24 bit output, thus requiring some amplitude compression or two gain ranges per sensor. Some devices have very sharp filters so that more than 90% of the Nyquist frequency is available for use. In general, multiplexing a single converter between channels or gain ranges defeats the noise reducing properties of the filter and more than one converter is often included in a single package.

*A/D specification* The accuracy of a converter may be expressed in the static case as a nonlinearity error or in dynamic terms as the ratio of signal to total harmonic distortion plus noise. Usually, these specifications will be better, i.e. smaller, than the distortion figure of the sensor. The important specification is then the dynamic range, which is the ratio of maximum possible signal to noise in the absence of signal. The noise may be specified as a voltage, a histogram of output values for a constant input or as the floor of a fast Fourier transform (FFT) plot.

### 2.1.5 Data transmission

The signal produced by the sensor typically has a noise power level of the order of $10^{-14}$ W. A high quality screened twisted pair cable has crosstalk attenuation of about 120 dB/km, so the proximity of a cable carrying 1 kW of power for even 100 m would induce $10^{-10}$ W of interference into the signal cable, increasing the noise level and severely limiting the dynamic range. Simply amplifying the signal improves the situation somewhat, but an amplifier's dynamic range is inversely proportional to its gain. Thus this direct baseband method of data transmission is only useful over short distances and far from any sources of interference.

Reducing the dynamic range (see Section 2.1.3 above) requires either compression hardware or multiple signal cables carrying the same signal at different gains.

Under less benign conditions some type of modulation is required. The rate at which a given communication channel can carry information is given by the Hartley–Shannon law (Connor, 1972):

$$R = B\log_2(1 + S/N)$$

where $B$ is the bandwidth, $S$ is the maximum signal power and $N$ is the noise power introduced in the channel. $R$ is the data rate expressed in bits per second. In this case the required rate is fixed by the sensor, so the law basically states that we may trade increased bandwidth for poorer signal to noise ratio.

One alternative is frequency modulation, where the sensor signal amplitude is used to vary the frequency of a carrier signal. This broadens the bandwidth of the signal which, as expected, makes it more tolerant to noise and also shifts it away from the frequency of mains power which is the chief source of interference and may now be filtered out without altering the signal. However, the modulation and demodulation circuits introduce their own distortions and noise.

Frequency modulation has the practical advantage that the carrier signal is always present, so faults which cause a loss of signal may be automatically detected. Frequency division multiplexing is also possible, where signals with different carrier frequencies are transmitted over a single cable.

The next step in improving data transmission is quantization (Taub and Schilling, 1971). If the transmitted signal can assume only certain discrete values which are separated by more than the amplitude of the noise, then it may be recovered exactly at the receiver by restoring it to the nearest permitted value. This is of vital importance when transmitting over sufficiently long distances to require many repeaters which would otherwise amplify any induced noise with the signal, but is also useful when the transmission medium is nonlinear.

In the extreme case the signal is encoded as a sequence of binary digits (bits) so only two possible levels need to be distinguished by the receiver. This naturally requires the maximum bandwidth. Once the data are encoded in binary form, an arbitrary number of bits may be grouped together, depending on how many levels may be distinguished in the prevailing noise conditions, and sent as one symbol, allowing a tradeoff between bandwidth and signal to noise ratio. Hence we distinguish between the baud rate, which is the number of symbols per second and limited to less than double the bandwidth, and the data rate, which is the number of bits per second.

When dealing with a Gaussian noise distribution, for example, there is a finite probability of encountering an arbitrarily large error, so some symbols will be misinterpreted. Thus distortion of the signal has been replaced by an error rate. By adding some redundant bits, errors may be detected at the receiver, and adding still more makes correction possible.

Since the signal will eventually be digitized for processing on a computer, digitizing as close to the sensor as possible and using digital data transmission offers the optimal solution.

## 2.2 Triggering and validation

Generally automatic detection of the presence of a seismic signal is desirable. When this detection takes place in real time and is used to initiate further action from the system such as recording and association, it is referred to as triggering.

For low frequencies and distant, long duration events, continuous recording may be justified, but in the mining environment where the relatively high frequencies of interest necessitate a correspondingly high sampling rate and the nearby events produce ground motion of short duration, triggering is necessary to reduce the amount of data recorded and to initiate the process which culminates in producing a report. In situations which might defeat a simple trigger algorithm, and especially where further processing must proceed automatically because of the volume of data, it is useful to have a further validation phase which verifies that the entire segment of recorded data conforms to a single processable seismic event.

### 2.2.1 Event detection

In the mining environment the prime concern has historically been the quantification of damaging events, which naturally have a very high signal to noise ratio. With the shift in emphasis to ever more sensitive networks in the search for precursive activity, no shortage of events with good signal to noise ratio has been experienced, hence the processing routines are still designed to function under these circumstances, and the triggering algorithms need only look for a sudden increase in amplitude.

*Fixed threshold* The simplest trigger is simply a fixed threshold and a trigger is declared if a single sample exceeds this value. This does not work well where the noise is environmental, as the level tends to change with mining activity.

*STA/LTA ratio* The noise level may be represented by a long term average (LTA) of some estimate of the instantaneous signal amplitude. This is then compared with a short term average (STA) of the amplitude, the trigger criterion being

$$\mathrm{STA/LTA} > R_t$$

where $R_t$ is the trigger ratio. There are several parameters which characterize this algorithm. The period over which the LTA is taken represents a cutoff between the shorter period signal and the longer period noise envelope variations, the period of the STA could be seen as the minimum duration of a valid event, and the trigger ratio is the minimum signal to noise ratio for a valid recording.

Many sources of interference are impulsive in nature, e.g. lightning, switching transients in power lines, single bit digital errors. Although of very short duration, these can give rise to very large amplitudes, which may, even when averaged over the STA period, be sufficient to push the STA over the threshold. One common variation on this algorithm is thus to use the median, rather than the mean, of samples within the STA window.

The signal envelope $E$ may be calculated by combining with a copy of the signal $S$ which has been delayed by a 90° phase shift, performed by a Hilbert transformer

$$E = \sqrt{(S^2 + S_H^2)}$$

where $S_H$ is the output of the Hilbert transformer. A useful approximation is to add a fraction of the derivative to the original signal. This has the effect of smoothing the envelope compared with simply taking the absolute value, and also responding to frequency changes (Allen, 1978).

For a triaxial set of sensors to be equally sensitive in all directions, the trigger algorithm should be performed on the absolute value of the vector sum of the $x$, $y$ and $z$ components

$$|S| = \sqrt{(x^2 + y^2 + z^2)}$$

For practical reasons it is often better to trigger on each component separately while recording all three, which leads to a maximum of $\sqrt{3}$ variation in sensitivity with angle of arrival.

### 2.2.2 Pre-trigger data and end of event

When performing a trigger algorithm in real time, the confirmation of a trigger is only available once the event has actually begun. In addition, more confidence can usually be given to the phase picks if some pre-event data are available for estimation of signal to noise ratios. A temporary store of recent data is therefore held while performing the trigger algorithm, which is recorded permanently if the trigger succeeds. The same algorithms used to determine the onset of an event may be used to determine that it has ended.

### 2.2.3 Validation

The recorded seismic data are sometimes not suitable for processing, whether because of an erroneous trigger or some form of corruption. It is therefore useful to be able to distinguish between such "bad buffers", and "good buffers" which do contain data suitable for processing. A backpropagation neural network (Cichocki and Unbehauen, 1993) is used to perform this operation. The network consists of 200 neurons in the input layer, ten in the hidden layer and two in the output layer. The squared amplitudes of the recorded signal are used as input, averaged over 0.5% of the length recorded. A sigmoidal activation function is used, so the output values should be (>0.5, <0.5) for a good buffer and (<0.5, >0.5) for a bad buffer. The network is trained using 500 buffers which have been visually classified as good or bad, employing a typical gradient teaching method. The result of the training is two weighting matrices, one 200 × 10 in size, the other 10 × 2. These are stored on the system in operation. A success rate of better than 90% has been observed, which is almost the performance of a trained operator.

## 2.3 Digital data communications

In the simplest case one of the data transmission methods described in Section 2.1.5, above, will suffice for immediate and continuous transmission from the sensor to the central processing site. In practice this is hardly ever the case: the interference or attenuation becomes too severe or the communications bandwidth is restricted (usually by a radio link) to the point where the triggered data must be stored and further transmission to the processing site can take place only as and when the communication channel allows; or the data acquisition and processing functions may be allocated to different machines for strategic reasons.

At this point we are dealing essentially with computer to computer communication, which is often discussed in terms of the protocol hierarchy of the International Organization for Standardization's reference model for Open Systems Interconnection (ISO-OSI) shown in Table 2.1 (Tugal and Tugal, 1989). Level 5 and higher are generally regarded as part of the application and in this case would include message types for the activation of an acquisition unit and communication of triggers and seismic data. Level 4 and below perform the actual data transport in an application independent and data transparent manner. If there is an established infrastructure such as a local or wide area network (LAN or WAN) then these levels are defined by the network hardware and operating system software, and care must only be taken to maximize use of the available resources. However, often the communications system is dedicated to and considered part of the seismic system, in which case some details of the low level functions are of interest.

**Table 2.1** Protocol hierarchy in the ISO-OSI reference model

| Level | Name | Scope | Examples |
|---|---|---|---|
| 1 | physical | link | connectors, voltages, interface control |
| 2 | data link | link | addressing, error detection, flow control |
| 3 | network | link | network control, multiplexing |
| 4 | transport | end-to-end | message reconstruction |
| 5 | session | end-to-end | login, logout |
| 6 | presentation | end-to-end | ASCII characters |
| 7 | application | end-to-end | file transfer |

### 2.3.1 Maximizing the information rate

When dealing with anti-aliasing under signal conditioning in Section 2.1 above it became clear that while theoretically a signal need only be sampled at twice the highest frequency present, in practice this is not achieved for a variety of reasons. The first method for maximizing the amount of information in a given number of samples is thus to ensure that the Nyquist bandwidth is being utilized as much as possible, by using the sharpest anti-aliasing filter practical.

Since we are aware of the nature of the seismic data and the processing steps which will be performed on them, further reductions may be made in the number of samples necessary to represent an event. In Section 2.2.2 it was suggested that a recording be terminated as soon as the amplitude had fallen below some multiple of the noise level in an inverted trigger test. Seismic parameter extraction requires frequencies up to only five times higher than the dominant frequency, so the higher frequencies may be filtered out and the sampling rate reduced to conform to the new Nyquist. By analogy with the automatic gain ranging which enables a system to cope with a wide dynamic range in amplitude, finding the end of an event and applying adaptive decimation may be said to yield a wide dynamic range for duration and frequency content.

The wavelet transform, which uses finite orthogonal functions as opposed to the infinite sine waves of the Fourier transform, promises a better type of decimation or lossy data compression, where some high frequency information, and hence the accuracy of phase picks, may be retained while representing the signal with far fewer samples than would otherwise be required.

Finally, we may use some lossless data compression scheme such as Huffman coding (Connor, 1972) to ensure that each sample is represented in the minimum number of bits possible. While the primary concern here is making maximum use of scarce communication bandwidth, a useful reduction in the amount of data storage required also results from the use of these techniques.

### 2.3.2 Low level protocols

When the communication system is part of the seismic system, various principles may be applied to enhance the utility of the communications. Discrete messages are exchanged between the computers, rather than the continuous data streams produced by the sensors.

Levels 4 and below of the OSI model are concerned with the transport of messages from one computer system to another, in this case the data acquisition unit and the central processing site. Levels 3 and below are link level protocols, so that if the path covers multiple physical links, then these levels apply separately to each link while level 4 supplies the end-to-end transport.

The error detection, correction and flow control functions are performed at levels 2 and 3, in hardware and software respectively, because the details depend very much on the link capabilities. For slow or error prone links, the maximum message size may be restricted so that individual messages have more chance of being transmitted correctly and retransmission for error correction is quicker. Level 4 then reassembles the original messages for the application. Other than this, the lower levels should be data transparent, i.e. the content of the messages should not be restricted. Often hardware support is used to allow link specific error detection and control information to be added to each message while maintaining data transparency.

For example, an acquisition unit might send data via cable at 1200 b/s to a multiplexer which forwards data from many stations over a radio link at 9600 b/s to a second multiplexer with an Ethernet link to the central site. The cable link, while slow, should be reliable, so a simple checksum might be included for error checking, and any timeouts, always the last resort for error detection, could be long. The radio link might add a significant amount of information for detecting burst errors. If the radio frequency is shared, polling flow control could be used, where the central site requests data from each site in turn. The serial links are point to point, so the low level addressing consists of directing the data to the correct port. Ethernet, by contrast, is a bussed medium, so each message has hardware source and destination addresses added.

It is, of course, necessary that the systems at each end of the link agree on the interpretation of the signals exchanged. Standards which are widely used in inter-computer communication have been drawn up by the Electronic Industries Association (EIA), the International Telegraph and Telephone Consultative Committee (CCITT) of the United Nations International Telecommunications Union, and the Institute of Electrical and Electronic Engineers (IEEE). These standards include the EIA's RS232C and RS422 for level 1 protocol between data terminal equipment (DTE), usually a computer, and data communications equipment (DCE), usually a modem. These are mirrored and extended by the CCITT's V series of standards. IEEE 802 covers levels 1 and 2 for a wide variety of LANs, and other IEEE standards cover more recent high speed low level protocols which are often given impetus as computer to peripheral communication methods.

## 2.4 Association

The associator must decide which triggers from individual stations correspond to the same event. For association between two triggers, the difference in times must be less than or equal to the seismic wave travel time $\Delta t_{ij}$ between the two stations:

$$|T_i - T_j| \leq \Delta t_{ij} = \frac{\sqrt{(x_i - x_j)^2 + (y_i - y_j)^2 + (z_i - z_j)^2}}{V_p} \tag{2.2}$$

where $T_i$ is the trigger time of the $i$th station at position $(x_i,y_i,z_i)$ and $V_p$ is the wave propagation velocity for the phase on which it is assumed the stations triggered.

Equation (2.2) only describes a test for pairs of triggers, and further rules are required to find all triggers from an event, especially when spurious triggers may occur which result in a situation where trigger A associates with trigger B and trigger B with trigger C, but trigger A does not associate with trigger C. A popular strategy, especially in systems with direct transmission, is to open a time window upon receipt of the earliest trigger, with a duration equal to the longest travel time between any two stations in the network, and accept any triggers occurring within this window as part of the same event (Lee and Stewart, 1981). If fewer than some minimum number of triggers are received within this window, the event and at least its initial trigger are discarded, and the process repeated for the succeeding trigger. This method is sometimes referred to as a "network trigger", rather than association, by analogy with individual station triggers. An event defined in this way may still contain spurious triggers, and it is then worthwhile to apply equation (2.2) exhaustively to all pairs to define the largest subset of these triggers which is mutually consistent, before proceeding with location or further processing (Lawrence, 1984).

In order to unambiguously separate the triggers caused by sequential events, there must be no triggers for a period equal to the travel time between the most widely separated stations under consideration. The associator may use various heuristics to improve its performance when this criterion is not met, such as including each station only once in each event, but any sustained activity which violates this principle will, in the presence of spurious triggers, result in triggers from different events being associated or triggers from the same event being separated.

There is thus a relationship between the diameter of a group of stations and the maximum rate at which events on which they trigger can occur and be reliably associated. The word "group" is deliberately used in this instance as it may comprise a subset of the stations in a network. The defining property is that only members of the group be considered for mutual association. If the maximum travel time between any pair of stations within the group is $\Delta t_{max}$, then events must be separated by at least $2\Delta t_{max}$ so that the last trigger of the first event is separated by $\Delta t_{max}$ from the first trigger of the second event. If we assume a Poisson (random) distribution of events with time, then the probabilty $P$ of no events occurring during the period $2\Delta t_{max}$ is related to the average rate of occurence of events $r$ by (Mood *et al.*, 1974):

$$P = e^{-r\,2\Delta t_{max}} \tag{2.3}$$

if we use a probability of failure $F = 1 - P$, to express the probability that our criterion will not be met, take natural logs on both side of equation (2.3), and use the approximation $\ln(1+x) \approx x$ for small $x$, then

$$F = r \cdot 2\Delta t_{max} \tag{2.4}$$

so

$$r \approx \frac{f}{2\Delta t_{max}} << \frac{1}{2\Delta t_{max}}$$

i.e. for a low probability of interference between events, the average event rate must be much less than the reciprocal of the minimum separation time. This problem is compounded by the fact that mining often does not proceed continuously, but in discrete steps caused by periodic blasts. The event rate following such a blast greatly exceeds the daily average for a long time compared with the travel times.

For example, consider a mine network with a longest travel time of 0.5 s which gathers 1000 events per day. The average time between events over this period is then 86.4 s. If we make the common assumption that 90% of events occur in 10% of the time after a blast, the average interval during this period reduces to 10.4 s, and from equation (2.4) the probability of two events occurring within the 1 s minimum spacing is 10%. Considering that each such overlap could result in two erroneously associated sets of triggers, this network is already a candidate for introducing groups for association purposes. The true situation is worse, as the event rate is not constant for two hours after the blast, but decreasing, and immediately after the blast would be far higher than the one per 10.4 s rate used above. Also limits on system throughput might result in many events being lost during this time and not reflected in the 1000 recorded events per day which was the initial assumption.

This analysis assumes that the only information available to the associator is the arrival time of a single seismic phase. This is generally the case because the associator (or network trigger) decides whether a station trigger should be recorded for further processing or discarded. If resources are available for the temporary storage of, and further parameter extraction from, a potentially spurious waveform, then maximum amplitude, dominant frequency, polarization parameters, the arrival of a second phase, or a full location, from a single station or two or more closely spaced stations, may be used to improve the associator performance when events occur almost simultaneously in different parts of the network.

## 2.5 Central processing site

The central site must collect all the seismic data and support association, seismological processing, interpretation and permanent data storage. It is convenient, and generally possible in the mining environment, for the stations to be largely controlled from the central site.

It is probably safe to assume that all these operations will be computer based. Given the range of computer operating systems available today, it is still necessary to stress that these functions must all be able to proceed concurrently, and so the system must be able to perform multiple tasks at once, whether through use of a single multitasking machine, or multiple machines communicating via network. If a trend may be discerned, it is towards the multiple machine approach, as the price/performance ratio of computer hardware continually improves and the ability of diverse machines to communicate with one another over networks becomes almost universal.

Change is continuous in the computer environment. The data acquisition function of the central site is the most critical to maintaining continuous coverage. In many ways it is the most mature and best defined part, and yet also one of the most difficult to engineer because of the realtime constraints and explicit parallelism involved in collecting and evaluating data from multiple stations at once. Thus in the multiple machine environment it is attractive to embed this function, i.e. to deploy it on a dedicated machine with specialized hardware and operating system software of which the user may remain essentially ignorant, seeing it simply as a seismic data server, which is more reliable and stable because of its specialized nature. The interactive processing steps may then be carried out on a multipurpose desktop machine, accessing the data in a standard format via the network.

## 2.6 System performance

This chapter has concentrated on individual aspects of the seismic system, especially where there are physical limits to its performance, and it must be stressed that the performance of the system as a whole is only as good as its weakest link. Each functional component – the transducer, signal conditioning, triggering, data communication, association and central site – has the potential to distort or discard data to the point where the system's usefulness as a tool is questionable. It is only by maintaining technical excellence in all phases of system deployment from design and installation through operation and maintenance, that the potential results may be achieved.

The quality of the data collected may be assured by frequent calibration of all system components, preferably on an automatic basis. The quantity of data is also important in ensuring that every bit of information broadcast by the seismic waves about the state of the rock mass is captured.

Seismic system parameters differ with the size of the volume to be covered. Table 2.2 shows three target network sizes differing by an order of magnitude in each case, and the changes which are necessary in the specification of the system.

In the case of the system limited by communications bandwidth there are three distinct data rates: the burst rate, which is determined by the number of events that can be stored locally at a sensor site; the sustained rate, which is the rate at which data from events may be continuously transmitted to the central site; and the daily or weekly rate, which takes into account periodic patterns in the seismicity rate caused by mining methods.

The final rate which is important is that at which the data are processed, interpreted and fed back to the management of the mining operation. No matter how fast and accurate the system is, it only serves its purpose when it contributes to the operation of the mine.

**Table 2.2** Seismic network parameter variation with size

| | Regional | Local | Micro |
|---|---|---|---|
| $m_{min}$ | 1 to 0 | 0 to −1 | −3 to −4 |
| $m_{max}$ | 4 to 5 | 4 | 3 |
| Average volume [km] | 30×30×5 | 3×3×3 | 0.3×0.3×0.3 |
| Events/day | 100 | 1000 | 10 000 |
| Sensors | 1 Hz; 4.5 Hz geo | 4.5 Hz; 28 Hz geo | 10 kHz acc |
| Minimum density [km] | 5th station > 2 | 5 stations < 1 | 5 stations < 0.3 |
| Useful frequency band [Hz] | 0.5 to 300 | 2 to 1000 | 3 to 10 000 |
| Comms [kbps] | 1.2 | 9.6 | 115 |
| Storage [GB] | 0.2 | 2 | 20 |
| CPU [Mflops] | 1 | 8 | 80 |

# 3 Deconvolution, Polarization and Wavelet Transform of Seismic Signals

Signals recorded by any seismic system usually pass through a number of linear causal systems (such as different filters) and each of them introduces a certain amount of distortion in the original signal. Therefore, prior to any processing, it is very important to recover, at least to some extent, the original signal using all available information. This will be discussed in Section 3.1 below.

Another very important aspect of signal processing is the polarization. Different signals have different types of polarization and this difference can be used for discrimination between various signal types. Body waves, such as P and S waves, are usually linearly polarized, whilst microseismic noise usually is not. Therefore, employing various polarization analysis techniques, described in Section 3.2, it is possible to discern different types of signals. This feature is usually widely employed in automatic arrival pickers.

The last subject which will be discussed in the current chapter is the wavelet transform. This recently introduced technique can be used for various purposes – signal filtering, data compression and time–frequency analysis.

## 3.1 Deconvolution

The recovery of a segment of the input signal (true ground motion in terms of displacement, velocity or acceleration) from the related segment of the output signal of a seismic system is known as the restitution problem in seismology. If this problem is considered from the point of view of communication theory, it is the deconvolution problem of a linear causal system. In order to get the true ground motion it is therefore necessary to construct and apply the inverse digital filter. In order to remove possible problems of inverse filter instability the time domain approach rather than the spectral domain is used.

### 3.1.1 Deconvolution filters for seismic systems

The signal distortion of a seismic system is mainly due to the phase biasing characteristic of its transfer function introduced by the seismic system combinations or by the anti-aliasing filter in digital systems.

Usually, seismic systems can be considered to be a conventional seismometer in series with the Butterworth or Bessel anti-aliasing filter. We will not consider here the design of an inverse filter to correct for the anti-aliasing filter, although it can be easily done following almost the same method as we will describe later in case of the Butterworth filter. The transfer function of the seismometer is given by (Ferber, 1989):

$$H_0(s) = g \frac{s^n}{s^2 + 2h_0\omega_0 s + \omega_0^2} \tag{3.1}$$

$g$ gain factor measured in V s/m;
$\omega_0$ angular cutoff frequencies of a system;
$h_0$ damping factor.

The exponent $n$ determines the output to input relation as follows: output to displacement input ($n$ = 3), to velocity input ($n$ = 2), and to acceleration input ($n$ = 1).

Therefore, the transfer function of the overall seismic system can be considered as a cascade of second order Butterworth high cut filters in series with $n$ differentiators.

As follows from equation (3.1) the deconvolution problem can be split into two parts: the deconvolution of second order Butterworth high cut filters and a deconvolution of integrators and differentiators.

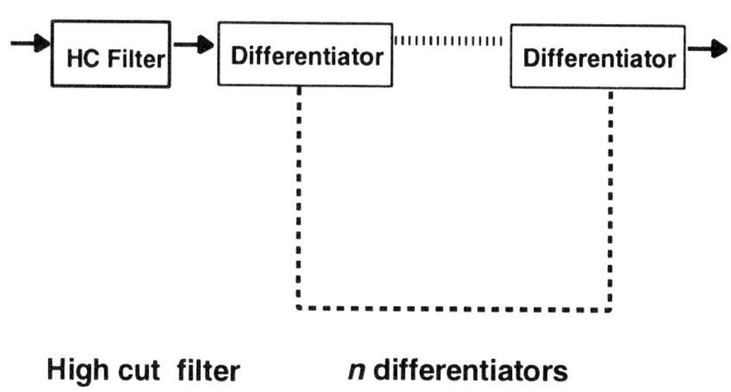


**Fig. 3.1** The transfer function of the overall seismic system can be represented as a cascade of second order high cut filters in series with $n$ differentiators.

### 3.1.2 Inverse digital filters for second order Butterworth high cut filters

In this section the inverse digital filter construction for a second order Butterworth high cut filter is briefly described. Using a bilinear $z$-transform and frequency pre-warping, the $z$-transform of a digital filter simulating a second order Butterworth high cut filter can be derived:

$$F(z) = a_0 \frac{1 + 2z^{-1} + z^{-2}}{1 + a_1 z^{-1} + a_2 z^{-2}} \tag{3.2}$$

where the following abbreviations are used:

$$a_0 = \frac{\omega'^2}{1 + k + \omega'^2} \tag{3.3}$$

$$a_1 = \frac{2\,(\omega^{-1} - 1)}{1 + k + \omega'^2} \tag{3.4}$$

$$a_2 = \frac{1 - k + \omega'^2}{1 + k + \omega'^2} \tag{3.5}$$

$$\omega' = \tan\left(\frac{\Delta t}{2}\omega\right) \tag{3.6}$$

$$k = 2h\omega' \tag{3.7}$$

in which $\omega$ is the natural frequency of the geophone, and $h$ is a damping factor.

It is necessary to point out that the simple inverse filter design approach of interchanging the numerator and denominator will not result in a stable filter, due to the pole on the complex unit circle. Although various more stable approaches are available, the best is to design some shaping filter having a minimal phase or zero phase property. Stabilization may also be achieved by shifting that pole into the stability region. Using this technique (shifting poles and applying a shaping filter), a minimum phasing inverse digital filter is given by the following one-sided recursive filter:

$$F^{-1}_{mp,\varepsilon}(z) = \frac{1}{a_0} \frac{1 + 2z^{-1} + z^{-2}}{1 + 2\varepsilon z^{-1} + \varepsilon^2 z^{-2}}, \qquad \varepsilon < 1 \tag{3.8}$$

while a zero phasing inverse digital filter is given by the following two-sided recursive filter:

$$F^{-1}_{zp,\varepsilon}(z) = \frac{1}{a_0}\,\frac{z + a_1 + a_2 z^{-1}}{(1 + \varepsilon z^{-1})(1 + \varepsilon z)}, \qquad \varepsilon < 1 \tag{3.9}$$

This filter is two-sided because the subfilter $1/(1 + \varepsilon z^{-1})$ is used in a forward as well as backward recursion, yielding a non-causal impulse response.

Another approach which enables easier interpretation of the stabilization in physical terms is to design a shaping filter with an eigenfrequency $v$ near to the Nyquist frequency and a damping constant $d$ near to 1 to replace the given Butterworth filter. Let $b_0$, $b_1$, $b_2$ denote the coefficients of a digital filter simulating the Butterworth filter on to which the given Butterworth filter should be shaped; thus, these coefficients can be computed similar to the coefficients $a_0$, $a_1$, and $a_2$ using $v$ and $d$ instead of $\omega$ and $h$.

Using this approach, a minimum phasing shaping filter is given by:

$$F(z) = \frac{b_0}{a_0}\,\frac{1 + a_1 z^{-1} + a_2 z^{-2}}{1 + b_1 z^{-1} + b_2 z^{-2}} \tag{3.10}$$

### 3.1.3 Inverse digital filters of integrators and differentiators

The differentiator of a seismometer–galvanometer system is given by

$$D(s) = \frac{s^3}{\omega^2} \tag{3.11}$$

Using the bilinear $z$-transform with frequency pre-warping, a digital filter simulating this differentiator is given by:

$$D(z) = \frac{2}{\Delta t\ \omega'^2}\left(\frac{z - 1}{z + 1}\right) \tag{3.12}$$

$$D(z) = \alpha\,\frac{1}{Q(z)^3}\,P(z)^3 \tag{3.13}$$

where

$$\alpha = \frac{2}{\Delta t\, \omega'^2}$$
$$Q(z) = 1 + z^{-1}$$
$$P(z) = 1 - z^{-1}$$

$P(z)$ describes a notch filter with a broad notch at a zero frequency (Shanks, 1967), which also distorts the rest of the amplitude spectrum from zero to the Nyquist frequency. An inverse digital filter for this notch filter may be designed by locating a pole near to zero frequency to sharpen the notch, resulting in less amplitude distortion for frequencies greater than zero. A one-sided recursive inverse filter is given by

$$P^{-1}_{os,\epsilon}(z) = \frac{1}{1 - \epsilon z^{-1}} \tag{3.14}$$

while

$$P^{-1}_{ts,\epsilon}(z) = \frac{1 - z}{(1 - \epsilon z^{-1})\,(1 - \epsilon z)} \tag{3.15}$$

denotes the two-sided recursive inverse filter, from which the minimum phasing

$$D^{-1}_{mp,\epsilon}(z) = \frac{1}{\alpha}\, Q(z)^3\, (P^{-1}_{os}(z))^3 \tag{3.16}$$

and zero phasing

$$D^{-1}_{zp,\epsilon}(z) = \frac{1}{\alpha}\, Q(z)^3 \left(P^{-1}_{ts}(z)\right)^3 \tag{3.17}$$

inverse digital filters can be derived.

Another approach to the problem of designing an inverse digital filter of integrators and differentiators is to design a finite impulse response (FIR) filter which will satisfy the necessary imposed conditions. There are a variety of algorithms which allow for the design of FIR filters for that purpose but probably the most widely used is the FIR filter design method based on Chebyshev optimization and the Remez exchange algorithm (Rabiner and Gold, 1975). This algorithm allows one to specify the desired magnitude response in the frequency band of interest, e.g. instead of designing a separate FIR filter for the integrator and high pass filter, one can specify the magnitude response for the integrator in the filter pass band. The resulting filter will combine the properties of both filters.

Therefore, as was pointed out earlier, deconvolution is split into two separate parts: deconvolution of Butterworth anti-aliasing filters and deconvolution of integrator/differentiator systems. The seismogram of a seismic event recorded by the Welkom Seismic Network (Fig. 3.2a), was processed with the integrator (Fig. 3.2b), and differentiator (Fig. 3.2c). In both cases the FIR filter design with a full band specified was used for a differentiator and 0.08–0.5 (relative units) pass band for the integrator.

This approach to the problem of restoring the true ground motion is the most flexible one, allowing for anti-aliasing filters, applied prior to digitization, to be corrected separately and for integrator/differentiator systems, enabling the researcher to obtain displacement, velocity or acceleration proportional recordings using the same ideology of processing.

In the following we will briefly discuss some properties of Bessel filters which can be used as anti-aliasing filters.

*Bessel filters* The Bessel filter has a transfer function described by following equation:

$$F(f) = \frac{d}{B_n(f)} \tag{3.18}$$

where $B_n(f)$ is the $n$th order Bessel polynomial and $d$ is a normalizing constant. The Bessel polynomials result from the recursion relation:

$$B_n(f) = (2n - 1)B_{n-1}(f) + f^2 B_{n-2}(f) \tag{3.19}$$

with initial conditions

$$B_0(f) = 1; \; B_1(f) = if + 1$$

Bessel low pass filters are characterized by the property that the group delay is maximally flat. The step response of Bessel filters exhibits extremely low overshoot, typically less than 1%. A common and useful approach to the synthesis of filters is to perform a suitable digitization of the continuous filters. Note, that bilinear transformation (which is usually used for design of the infinite impulse response (IIR) filters for many filter types, such as Butterworth, Chebyshev, etc.) cannot be used here, because it fails to preserve the maximally flat group delay property of the Bessel filters (Baher, 1990; Thiran, 1971).

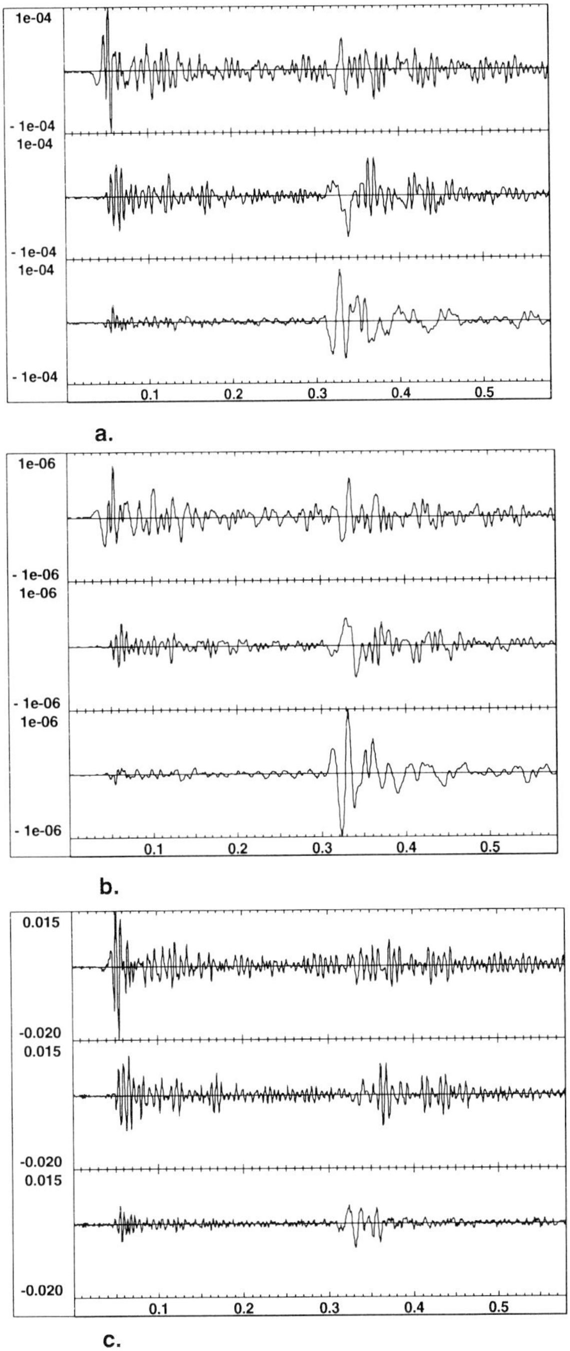


**Fig. 3.2**
**a**. Three component seismogram of a seismic event recorded by Welkom Seismic Network.
**b**. Results of processing seismograms of (a) with an integrator with pass-band 0.08–0.5 (relative units).
**c**. Results of processing seismograms of (a) with a full band differentiator.
Vertical axis units are m, m/s and m/s$^2$ respectively, while the horizontal axis units are seconds.

It can be shown that the nonlinear factor which gives the analogue prototype its maximally flat group delay property is not usually preserved by the bilinear transformation.

Baher (1990) and Thiran (1971) derived the digital counterpart of the analogue low pass Bessel filter. However, since all the available degrees of freedom have been used to shape the phase, the resulting amplitude response is rather poor. For the design of such filters the combination of two parameters has to be found experimentally to meet the specifications of the analogue filter. It should be noted that tabular values of the amplitude and phase responses of analogue Bessel filters are available.

### 3.1.4 An iterative technique for the deconvolution of seismograms

Another computationally more simple approach to the problem of deconvolution of seismic recordings and restoration of true ground motion was proposed by Alessandrini and Perazolo (1987). This method makes use of an iterative technique by Brigham *et al.* (1968) which also performs that task in the time domain. The seismogram $y(t)$ is the convolution of the impulse response of a seismic system, $h(t)$, and the ground motion, $x(t)$:

$$y(t) = h(t) \circledast x(t) \tag{3.20}$$

As the first approximation to $x(t)$ we may assume that:

$$x_0(t) = y(t),$$

with the error being equal to

$$e_1(t) = y(t) - h(t) \circledast x_0(t) \tag{3.21}$$

A better approximation $x_1(t)$ can be obtained by adding $e_1(t)$ to $x_0(t)$, and so on until

$$x_n(t) = y(t) + e_1(t) + \ldots + e_n(t) \tag{3.22}$$

$$e_n(t) = \{\delta(t) - h(t)\} \circledast^n y(t) \tag{3.23}$$

where the symbol $\circledast^n$ indicates $n$ time convolutions, and $\delta(t)$ the delta function.

It can be shown that the iterative procedure is absolutely convergent if $|1 - H(\omega)| < 1$, where $H(\omega)$ is the Fourier transform of $h(t)$. If this condition does not hold we can achieve convergence by means of an artifice, i.e. by applying to $y(t)$ a system whose impulse response is $h(-t)$, where $h(-t)$ is defined as $h(t)$ but for $t < 0$.

Although the iterative technique described above appears to be a more stable method for obtaining the inverse filter in the case of a seismograph it has the serious shortcoming that it may take quite a few iterations before convergence is achieved.

## 3.2 Polarization

The polarization properties of vector wavefields have been studied extensively in various areas from electromagnetics to physical oceanography and seismology. Special emphasis has been placed on combining polarization methods with other methods for event locations, backazimuth estimation, phase identification, detection, etc. The basic principles which are used in polarization properties estimation are reviewed below.

### 3.2.1 Three axis principal components method

Historically, one of the first methods which made use of polarization properties of seismic signals to discriminate between different types of signals was proposed by Flynn (1965). This method is based on the three axis principal component method. The covariance matrix for a time window of $N$ points can be represented as

$$\mathbf{C} = \begin{Vmatrix} \sigma_{nn} & \sigma_{ne} & \sigma_{nz} \\ \sigma_{ne} & \sigma_{ee} & \sigma_{ez} \\ \sigma_{nz} & \sigma_{ez} & \sigma_{zz} \end{Vmatrix} \tag{3.24}$$

where $\sigma_{xy}$ represents the zero lag cross power between the signals on the $x$ and $y$ axes

$$\sigma_{xy} = \frac{1}{N} \sum_{i=1}^{N} x_i \, y_i \tag{3.25}$$

If necessary, the same tapering can be used (Hamming, Hanning, etc.). It can be shown that **C** is a real symmetric, positive semidefinite matrix (positive definite if there is some uncorrelated noise on all three axis). Flynn (1965) used principal components analysis to characterize the polarization of seismic waves. This method involves a three dimensional coordinate rotation into the principal directions of **C**. The principal directions lie along the eigenvectors of **C**, and the rotated covariance matrix $\mathbf{C}'$ has the form

$$\mathbf{C}' = \begin{Vmatrix} \lambda_1 & 0 & 0 \\ 0 & \lambda_2 & 0 \\ 0 & 0 & \lambda_3 \end{Vmatrix} \tag{3.26}$$

where the diagonal elements are the real eigenvalues of **C**, with $\lambda_1 \geq \lambda_2 \geq \lambda_3$. The eigenvectors define a polarization ellipsoid with the best least squares fit to the data samples. This ellipsoid may have an arbitrary orientation relative to the original *n, e, z* system. For P waves the dominant eigenvector, associated with $\lambda_1$, is assumed to lie along the direction of propagation. To resolve the 180° ambiguity we can use the fact that the signal arrives from below the station. The projection of this vector on to the horizontal plane will give the radial axis $+r$ of the new *r, t, z* system and the estimated backazimuth to the source is $-r$. It is necessary to point out that in general two rotations are needed to convert from computed principal coordinates to this *r, t, z* system: first about $\lambda_1$'s eigenvector until one axis is horizontal, then about this axis (tangential axis, *t*) into *r, t, z*. After these two rotations none of the off-diagonal elements of the covariance matrix is necessarily zero. The relative sizes of the three eigenvectors indicate the degree of linear polarization, and the wave type most appropriate therefore. They also can be used to estimate the backazimuth estimation errors (Walck and Chael, 1991). These calculations are done for each point of the seismic trace in a moving time window.

Polarization filtering can be implemented by weighting the rotated seismograms by some measure of rectilinearity of motion, in the preferred propagation direction.

Each of the three eigenvalues $\lambda_i$ is proportional to the energy associated with the waves having the polarization of its corresponding eigenvector. The degree of linear polarization or rectilinearity can be estimated in a variety of ways. For example, according to Samson and Olson (1980):

$$P^2 = \frac{(\lambda_1 - \lambda_2)^2 + (\lambda_1 - \lambda_3)^2 + (\lambda_2 - \lambda_3)^2}{2\,(\lambda_1 + \lambda_2 + \lambda_3)^2} \tag{3.27}$$

The polarized P wave arrival should stretch the polarization ellipsoid along the proper direction, making $\lambda_1$ dominant. If the background noise recorded at a station is not isotropic, then the polarization ellipsoid will be biased towards the dominant noise direction.

Numerous refinements and improvements have been proposed since the original paper by Flynn, e.g. Montalbetti and Kanasewich (1970) suggested the weighting function to be used as polarization filter weights must be smoothed over a time window approximately half of the original window's length, in order to smooth out spikes which may be present due to ambient noise presence.

In Fig. 3.3a we show three component recording of seismic event recorded by Welkom Seismic Network in South Africa. In Fig. 3.3b different rectilinearity functions calculated from eigenvalues for the same event are shown.

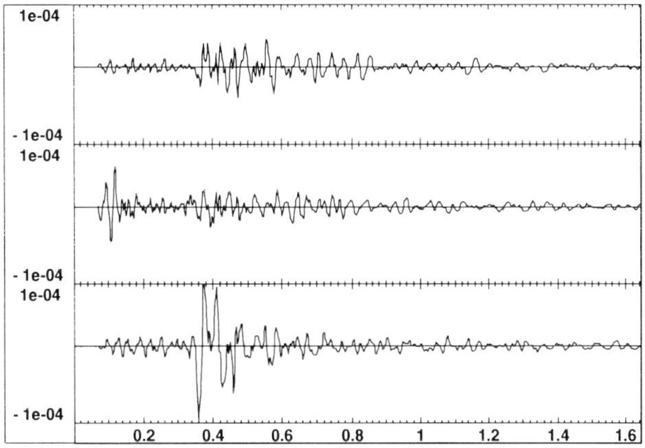


**Fig. 3.3a** Three component seismogram, recorded by Welkom Seismic Network, South Africa.
The vertical scale units are m/s, while the horizontal units are seconds.

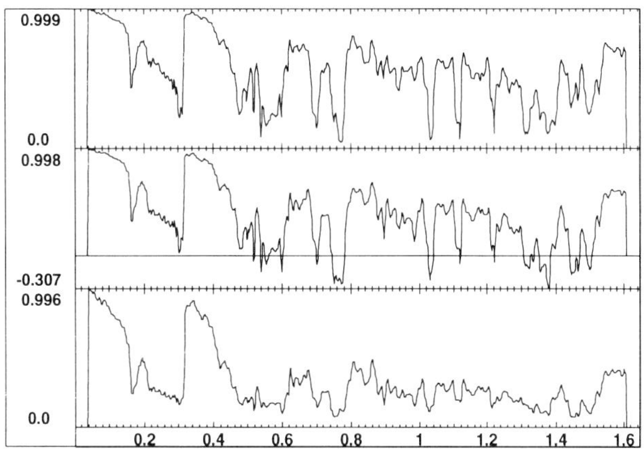


**Fig. 3.3b** Different rectilinearity motion estimate functions for the seismogram shown in Fig. 3.3a.
The vertical scale units are relative units, while the horizontal units are seconds.

Polarization estimates obtained by this method may be employed for S wave arrival pickers (Cichowicz *et al.* 1988). For example, in Mendecki (1993) S wave arrival time is determined by the function operating only on the SH and SV components in a sliding window with some predefined length $T(A_{max})/2$, where $T(A_{max})$ is the dominant period, based on the width of the largest peak $A_{max}$. The polarization estimate may then be constructed in the following form:

$$F(t) = DPS \ \sin(a) \ SNRS \ E_{sp} E_{sp0},$$

where:

*DPS* the polarization of the S wave which is defined as (see equation (3.27)):

$$DPS = \frac{(\lambda_1 - \lambda_2)^2 + (\lambda_2 - \lambda_3)^2 + (\lambda_3 - \lambda_1)^2}{2(\lambda_1 + \lambda_2 + \lambda_3)^2}$$

$a$ angle between the directions of the P and S wave's plane;
*SNRS* signal to noise ratio (normalized to range 0–1);
$E_{sp}$ energy of S wave to the energy of P wave (normalized to range 0–1);
$E_{sp0}$ energy of S wave to the energy of the first pulse of the P wave (normalized to range 0–1).

*Two axis principal components method* This method was proposed by Magotra *et al.* (1987). The main difference between this method and the original three axis principal components method is that for determining the orientation of the radial axis $r$ only horizontal components are used. The vertical component is used only to resolve the 180° ambiguity in the direction of $+r$. It must be pointed out that this is in fact the two axial principal component method, which can be described as a rotation about the $z$-axis in the direction of the dominant eigenvector of the two dimensional covariance matrix obtained from the horizontal components. Along this direction the horizontal signals ($r$ and $t$) must be uncorrelated. The rotated three dimensional covariance matrix resulting from that method is

$$\mathbf{C} = \begin{Vmatrix} \sigma_{rr} & 0 & \sigma_{rz} \\ 0 & \sigma_{tt} & \sigma_{tz} \\ \sigma_{rz} & \sigma_{tz} & \sigma_{zz} \end{Vmatrix} \quad (3.28)$$

and using the constraints $\sigma_{rr} > \sigma_{tt}$ and $\sigma_{rz} > 0$ we can define the direction $+r$. In fact this method is just the first of the two steps, described above, required to get a fully rotated three dimensional covariance matrix. In this case, the noise model assumed is less restrictive than for the previous one. The noise should be uncorrelated and have equal power on both horizontal components, but there are no restrictions on the noise on the vertical component. Using this method the backazimuth can be easily obtained from the north and east elements of the original matrix **C** using the following equation:

$$\theta_0 = \frac{1}{2} \arctan \left[ \frac{2\,\sigma_{ne}}{\sigma_{nn} - \sigma_{ee}} \right] \tag{3.29}$$

and then inspecting the sign of the resulting $\sigma_{rz}$ to resolve the ambiguity $\theta = \theta_0$ or $\theta = \theta_0 + 180°$ to determine backazimuth.

### 3.2.2 Complex polarization filters

The complex polarization analysis of particle motion, which is able to identify the problem of multiple arrivals and handle elliptical polarization, has been proposed by Vidale (1986), with further improvements by Bataille and Chiu (1991). Let us denote $r_r(t)$ for the radial ($x$) component of motion, which is positive outwards, $t_r(t)$ for the tangential ($y$) component, which is positive in the clockwise direction, and $z_r(t)$ for the vertical component, which is positive upwards. Vidale (1986) proposed a conversion to the analytical signal

$$\begin{aligned} r(t) &= r_r(t) + iH\{r_r(t)\} \\ t(t) &= t_r(t) + iH\{t_r(t)\} \\ z(t) &= z_r(t) + iH\{z_r(t)\} \end{aligned} \tag{3.30}$$

where $H$ represents the Hilbert transform and $i = \sqrt{-1}$. This analytical signal is used to compute the covariance matrix

$$\mathbf{C} = \begin{Vmatrix} rr^* & rt^* & rz^* \\ tr^* & tt^* & tz^* \\ zr^* & zt^* & zz^* \end{Vmatrix} \tag{3.31}$$

where asterisks denote complex conjugation. The three eigenvalues $\lambda_i$ and eigenvectors $(x_i, y_i, z_i)$ of the 3 x 3 complex covariance matrix may be computed analytically for each time point.

$$\begin{Vmatrix} X_i \\ Y_i \\ Z_i \end{Vmatrix} [\mathbf{C} - \lambda_i \mathbf{I}] = 0; \quad i = 1,2,3 \tag{3.32}$$

The analytic signal is complex, the covariance matrix is Hermitian and therefore has real positive eigenvalues, and the eigenvectors are in general complex.

The eigenvector $(x_1, y_1, z_1)$ associated with the largest eigenvalue $\lambda_1$ points in the direction of the largest amount of polarization present. However, the phase in the complex plane will initially be arbitrary. First the eigenvector is normalized to have length equal to 1. Then by rotating the eigenvector, corresponding to the largest eigenvalue, by 0° to 180° in the complex plane, the rotation which results in the largest real component can be found. This can be done by searching over $\alpha = 0°$ to $180°$ to maximize the following function:

$$X = \sqrt{Re(x_1\ cis\ \alpha))^2 + (Re(y_1\ cis\,\alpha))^2 + (Re(z^1\ cis\,\alpha))^2} \tag{3.33}$$

where $cis\ \alpha = \cos\alpha + i\sin\alpha$ and $Re(x)$ is the real part of $x$. The vector $(x_1, y_1, z_1)$ is then rotated by angle $\alpha$, and the elliptical component of polarization can be estimated using the ratio of the imaginary part of the eigenvector to the real part, obtained in equation (3.33):

$$P_E = \frac{\sqrt{1 - X^2}}{X} \tag{3.34}$$

For the linearly polarized motion the ratio $P_E$ should be close to 1. The measures of strength of polarization and degree of planar polarization can be constructed in the same way as before:

$$DPS = 1 - \frac{\lambda_2 + \lambda_3}{\lambda_1} \tag{3.35}$$

$$DPP = 1 - \frac{\lambda_2}{\lambda_1} \tag{3.36}$$

This method has some distinct advantages over other methods of polarization parameter estimation. The polarization can be measured from covariance at any point in the seismogram, while in other methods it is necessary to average the covariance over some portion of the wavelength, before polarization parameters may be estimated. The main advantage of including the Hilbert transform is that the eigenvalues of the covariance matrix are invariant to constant phase shifts in the components. The main drawback of this algorithm is that it requires at least four times the computation that is required for real covariance matrices.

Further improvement to Vidale's algorithm has been suggested by Bataille and Chiu (1991). Similarly to Vidale's method, the analytical signals are constructed according to equations (3.30). Then the imaginary part of the covariance matrix **C** (equation (3.31)) contains information only about the noise. Hence, only the real part of the covariance matrix need be considered. After the principal components decomposition, measures of the degree of linear polarization can be constructed the same way as before, for example using equation (3.36). Bataille and Chiu used as polarization filter weights the following function:

$$F = \left[ \frac{\mathbf{p}^{1T} \cdot \mathbf{C} \cdot \mathbf{p}^{1}}{Tr\ \mathbf{C}} \right]^{n} \tag{3.37}$$

where $\mathbf{p}^1$ is the main eigenvector. The function $F$ can be described as the ratio of energy in the direction of the main eigenvector to the total energy. The parameter $n$ can control the bandpass of the directional bandpass filter, described by equation (3.37).

It must be noted that for constructing the measure of the degree of linear polarization of the signal it is not necessary to diagonalize the covariance matrix. Samson (1977) has shown that

$$P^2 = \frac{1}{2(n-1)(Tr\ \mathbf{C})^2} \sum_{j,k=1}^{n} (\lambda_j - \lambda_k)^2 = \frac{n\ Tr(\mathbf{C}^2) - (Tr\ \mathbf{C})^2}{(n-1)\ (Tr\ \mathbf{C})^2} \tag{3.38}$$

Therefore, if it is necessary to estimate the degree of linear polarization, then it can be done without diagonalization of the covariance matrix, but using the covariance matrix only.

All the methods described above deal with seismograms recorded by a single three component station and no spectral analysis has been used.

Further improvements can be achieved using time–frequency decomposition of seismograms and applying polarization parameter estimation methods for each frequency.

Different schemes can be used to decompose the signal in a moving time window prior to polarization parameter estimation: ordinary Fourier transform and multitapering spectral analysis techniques have been applied (Menke *et al.*, 1990) to analyse the polarization of compressional waves from seismograms of chemical explosions.

The most attractive in the computational requirement sense is the polarization analysis technique combined with short time spectral analysis, Samson's method (Samson 1977), because it does not require diagonalization of the spectral matrix. Figure 3.4 shows three component seismograms of two events recorded by the Welkom Seismic Network and its sonogram with spectral values weighted for the degree of linear polarization of signal spectral components in the vicinity of a particular frequency. Spectral analysis was performed in a sliding window of 0.02 s with overlapping equal to half the sliding window size, 0.01 s. Various arrivals are seen in P and S waves coda. This type of analysis, combined with pattern recognition methods, is extensively used for the purpose of quarry blast and earthquake discrimination.

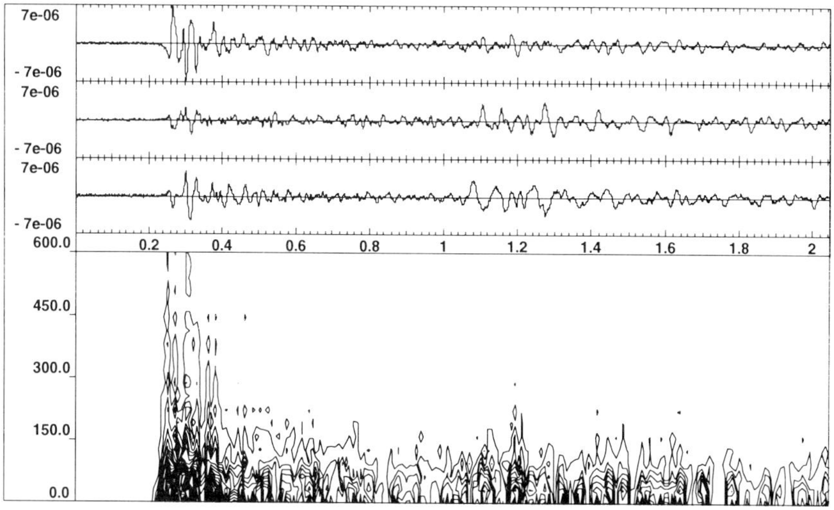


**Fig. 3.4a** Three component recordings of seismic events recorded by Welkom Seismic Network and corresponding sonogram (time versus frequency distribution). Spectral power for each frequency was calculated as energy of all three components within a time window of 0.02 s, with 0.015 s overlapping, and weighted for the linear polarity of that particular frequency, and using equation (3.38).

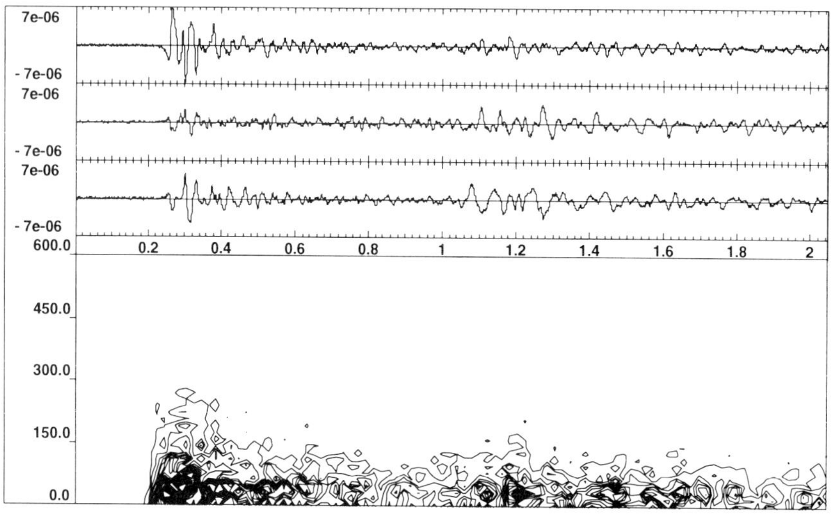


**Fig. 3.4b** Same as Fig. 3.4a, but sliding window length was set to 0.04 s with 0.035 s overlapping. Resolution on the time axis is relatively poor but frequency resolution has been increased compared to Fig. 3.4a.
In both cases, the seismogram vertical scale units are m/s, while the sonogram horizontal scale units are seconds.

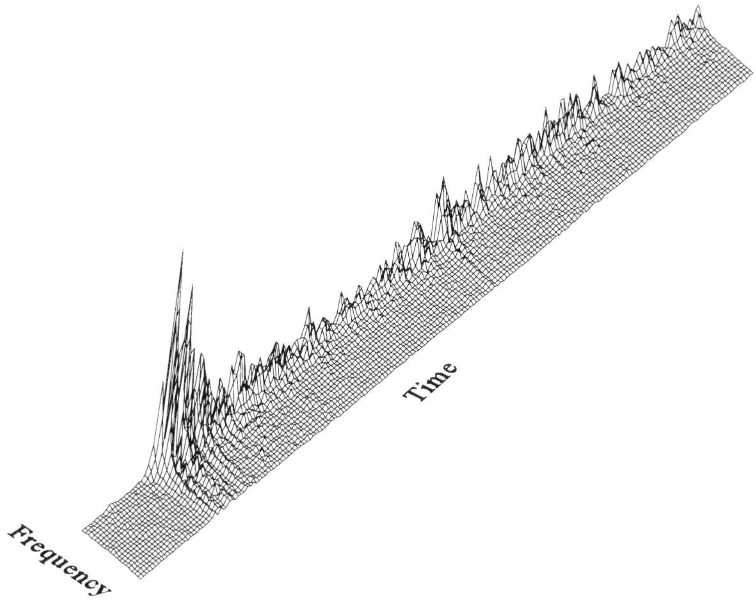


**Fig. 3.4c** Sonogram of event shown in Fig. 3.4a, represented as mesh.

Application of time–frequency decomposition followed by polarization analysis can yield the frequency dependent estimates of polarization parameters and help to get further insight into the fine structure of the seismic wavefield.

Sonogram representation can be used for various purposes: arrival detection, allowing focus on particular frequency bands of interest; and discrimination of quarry blasts and earthquakes.

The most difficult problem is to find the best tradeoff between time and frequency resolutions. Another problem is the heavy computational requirements of these calculations. It is more appealing to use a transform other than Fourier, e.g. for realtime detection purposes, the Walsh transform; for discrimination purposes, the wavelet transform instead of the Fourier transform with varying parameters.

## 3.3 Wavelet transform

During past decade the wavelet transform has found a wide range of applications to various problems in digital signal processing, image processing and compression, harmonic analysis, etc. Although the concept of the wavelet transform originated in seismology and it is now routinely used in diverse areas it has not yet become popular with seismologists. In this Section, we will give a brief description of the wavelet transform and show some examples of its application to seismological problems.

Although the Fourier transform (FT) is widely used in digital signal processing for frequency analysis of both stationary and non-stationary types of signals, it has severe limitations in the latter case. In the case of transient signals the characteristics of the signals to be analysed are strongly time dependent. This imposes restrictions on the use of the FT since the recordings of earthquakes and blasts do not necessarily fulfil stationarity conditions. Thus alternative methods of signal analysis must be used. One of the alternatives is the so-called short time Fourier transform (STFT). In this method, the input signal is FT analysed in some short time window which slides along the input signal length, possibly with some overlay. Various methods and various types of analysing window are described in geophysical literature, which use this type of FT for adaptive filtering, adaptive time–frequency dependent polarization estimation, etc. One of the most difficult problems met when using the STFT is the optimal choice of the analysing time window to achieve the best compromise between frequency and time resolution. Usually this problem is overcome using several time windows and then adopting results which satisfy some criteria. Some examples of STFT are shown in the previous section. The basic idea behind the wavelet transform (WT) is to decompose the input signal using a special function called the *basic* or *mother wavelet*. This function $\psi$ is translated in time (to select the part of signal to be analysed) and either dilated or contracted using a scale parameter $a$ (in order to control the size of the time window to be used). The WT can be expressed by the following relationship:

$$S(a,b) = |a|^{\frac{1}{2}} \int s(t) \psi\left(\frac{t-b}{a}\right) dt \qquad (a \neq 0) \tag{3.39}$$

where $S(a,b)$ is the WT of the signal $s(t)$, and $b$ is the translation in time. It can be seen immediately that we are no longer dealing here with frequency in the FT sense, but with scale. As is shown elsewhere (e.g. Dorize and Gram-Hansen, 1992) the proper choice of an analysing wavelet may overcome this problem, making the output of the WT to be similar to the frequency–time distribution obtained with the STFT. Therefore, the output of the WT is a time–scale plot. The advantages are obvious: there are no assumptions about the stationarity of input signal and no problems with finding the proper length and shape of time window to apply. In the WT, higher frequency components provide better time resolution relative to lower frequency components. In low frequency regions, the time resolution is poor, which prevents us from making incorrect estimates. We thus see that the WT provides an analysis tool which gives different resolution depending on which part we focus our attention when comparing to the STFT where the resolution is fixed and controlled only by the change in the analysing time window length.

The choice of the basic wavelet is neither unique nor arbitrary, but depends on the information one hopes to gain from using the WT, since there are many functions which can be used for the purposes of the WT. We will briefly outline the admissibility conditions (Kumar and Foufoula-Georgiu, 1994):

- Compact support, or sufficiently fast decay – this condition is necessary to get localization in space.
- Zero mean, which is expressed as $\int_{-\infty}^{\infty} \psi(t)dt = 0$.

These two conditions ensure that the wavelet has the wiggle shape and is wavelike.

A typical example of a wavelet fulfilling the admissibility condition is shown in Fig. 3.5. This wavelet is given by:

$$\psi(t) = \frac{2}{\sqrt{3}} \pi^{-1/4}(1 - t^2)e^{-t^2/2} \tag{3.40}$$

where the constant is chosen in such a way that $\|\psi\|^2 = 1$. This wavelet is the second derivative of the Gaussian smoothing function (Kumar and Foufoula-Georgiu, 1994).

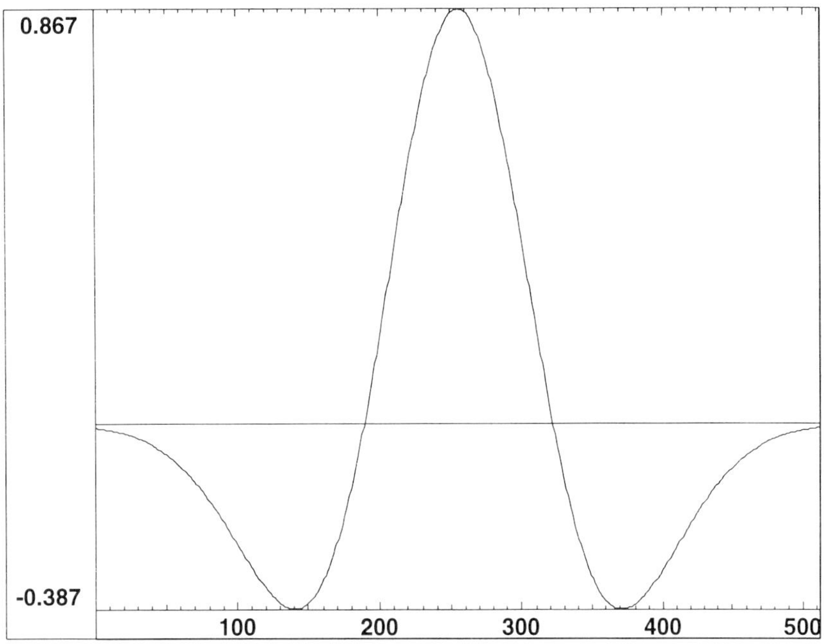


**Fig. 3.5** Mexican hat wavelet. Vertical axis units are relative units, horizontal axis units are number of samples.

Equation (3.39) is applicable to a continuous wavelet, therefore this transform is called the *continuous wavelet transform* (CWT). When the parameters $a$ and $b$ take discrete values then it is called the *discrete wavelet transform* (DWT). There is a very efficient DWT computation scheme analogous to the fast Fourier transform (FFT) known as the fast wavelet transform (FWT). Daubechies (1992) proposed the implementation of a wavelet transform with a specially designed pair of finite impulse response (FIR) filters called a quadrature mirror filter (QMF). The most distinguished feature of QMF is that the frequency responses of the two FIR filters separate the low frequency and the high frequency components of the input signal. The dividing point usually is halfway in between zero frequency and Nyquist frequency. Then implementation of the DWT is obvious:

- Pre-calculate QMF coefficients using basic wavelet coefficients.
- Pass the input signal through both low pass and high pass QMFs.
- Decimate the resulting low and high pass filter output by a factor of 2.
- Store the high frequency component obtained at the previous step and, using the low frequency component as input signal, repeat steps 2 to 4 as many times as necessary. This is schematically shown in Fig. 3.6.

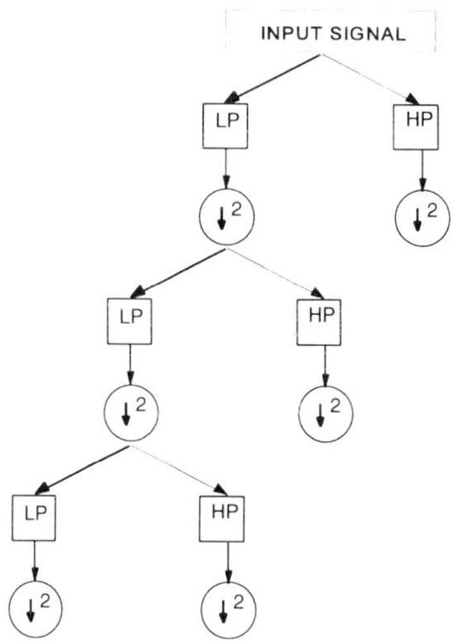


**Fig. 3.6** Flowchart for FWT.

The output of the low pass filter at every step is called the *approximation*, the output of the high pass filter is called the *detail* of the input signal. Thus signal is fully represented by approximation at the last step and all intermediate details. There is no redundancy in the FWT - the total length of approximation at the last step and the length of all details is equal to the length of the original signal. In Fig. 3.7 the chirp signal is shown with approximations in Fig. 3.7a,c,e,g, and detail in Fig. 3.7b,d,f,h, on levels 1, 2, 3, 4 respectively.

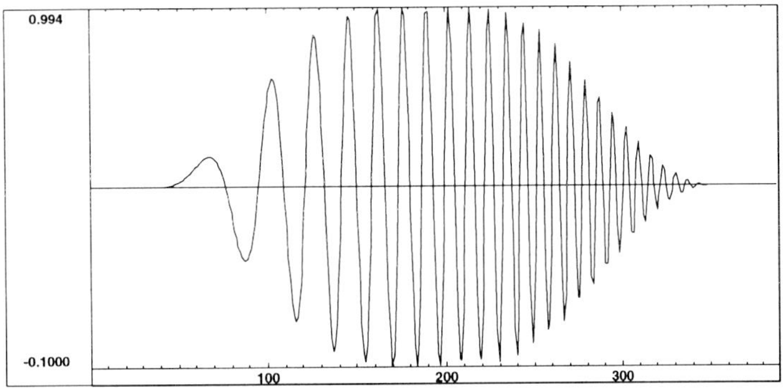


**Fig. 3.7** Chirp signal - sinusoid with linearly increasing frequency and modulated amplitude (Top).
**a.** Approximation on level 1.
**b.** Detail on level 1.
**c.** Approximation on level 2.
**d.** Detail on level 2.
**e.** Approximation on level 3.
**f.** Detail on level 3.
**g.** Approximation on level 4.
**h.** Detail on level 4.

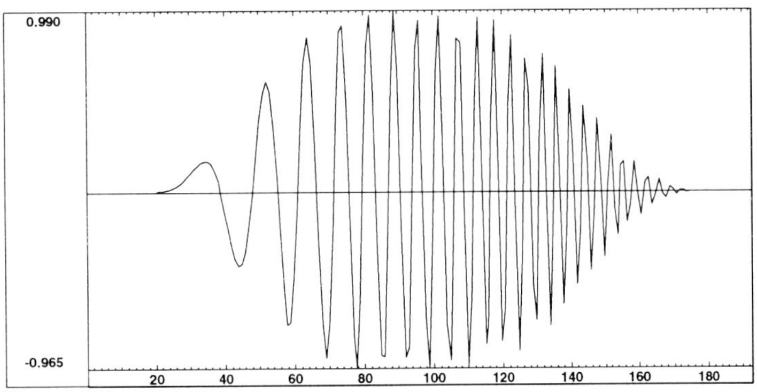


**a.**

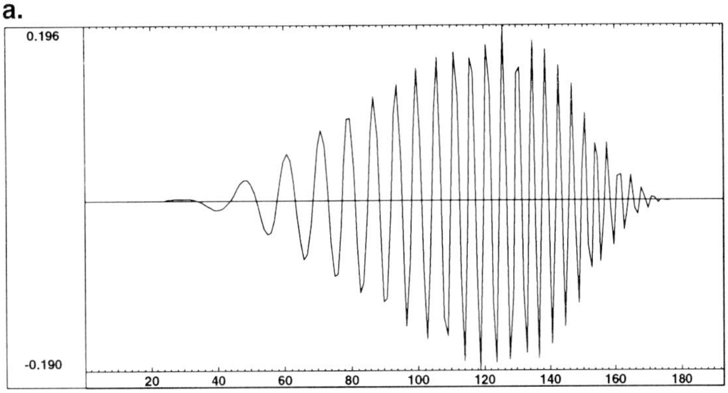


**b.**

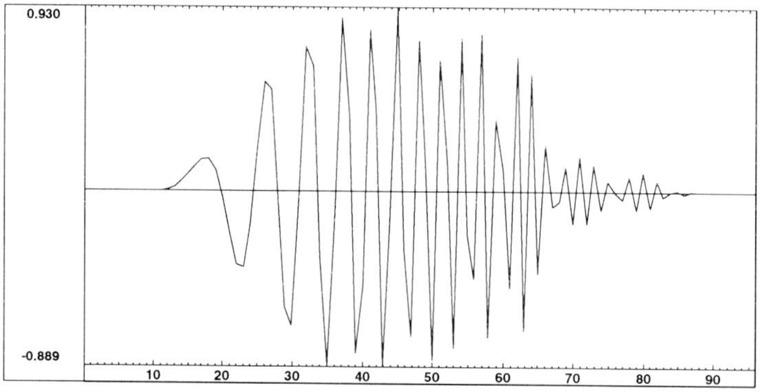
0.930
-0.889
10
20
30
40
50
60
70
80
90

c.

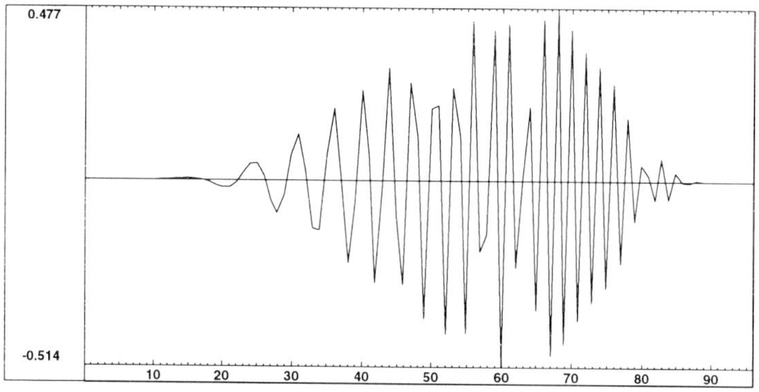
0.477
-0.514
10
20
30
40
50
60
70
80
90

d.

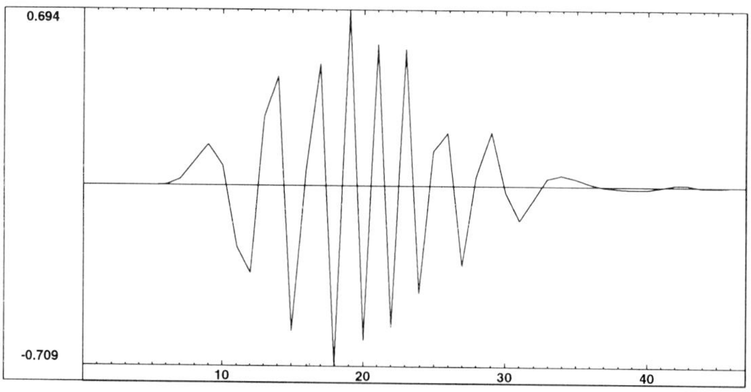
0.694
-0.709
10
20
30
40

e.

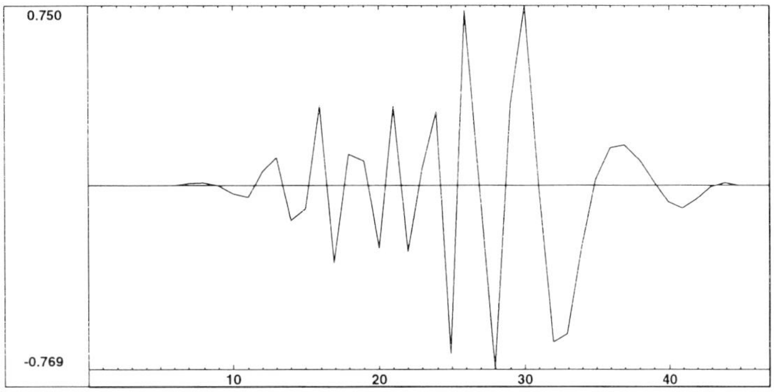


f.

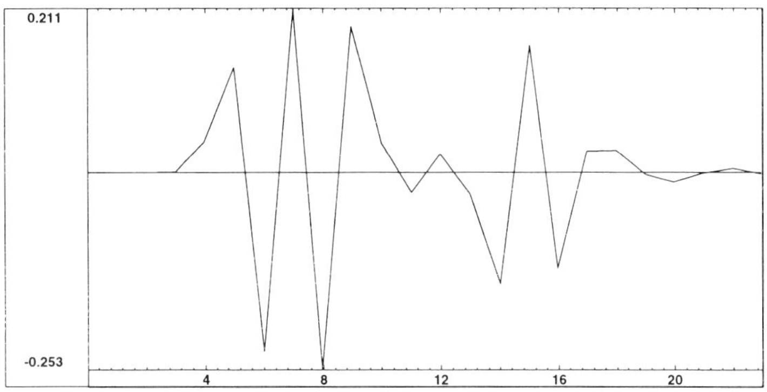


g.

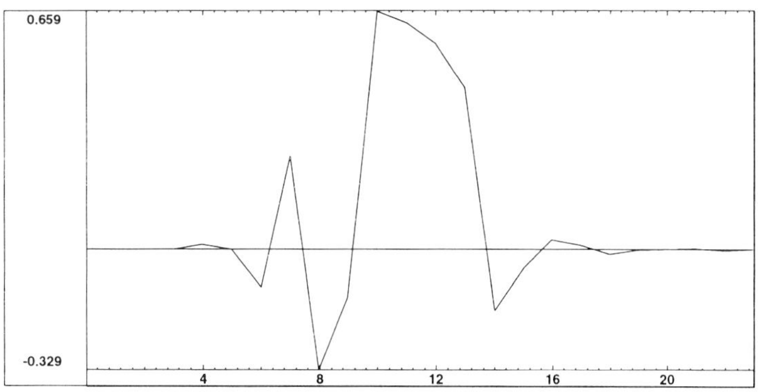


h.

In Fig. 3.8, the magnitude of the continuous wavelet transform is shown for the chirp signal. Linearly increasing frequencies of the chirp are clearly seen. In order to get the same results using the STFT we would have to compute the STFT of the original chirp signal using several sliding time windows with different lengths, to focus on different resolutions of the time–spectral representation. The wavelet transform allows us to obtain the complete behaviour of scale (frequency) versus time of the initial time series in one step, giving different resolutions in different parts of the plot, e.g. the resolution in the regions of high frequency (scale) is better in the time domain, but worse in scale (frequency) domain, and *vice versa* which is why the wavelet transform is often called *multiresolution analysis*.

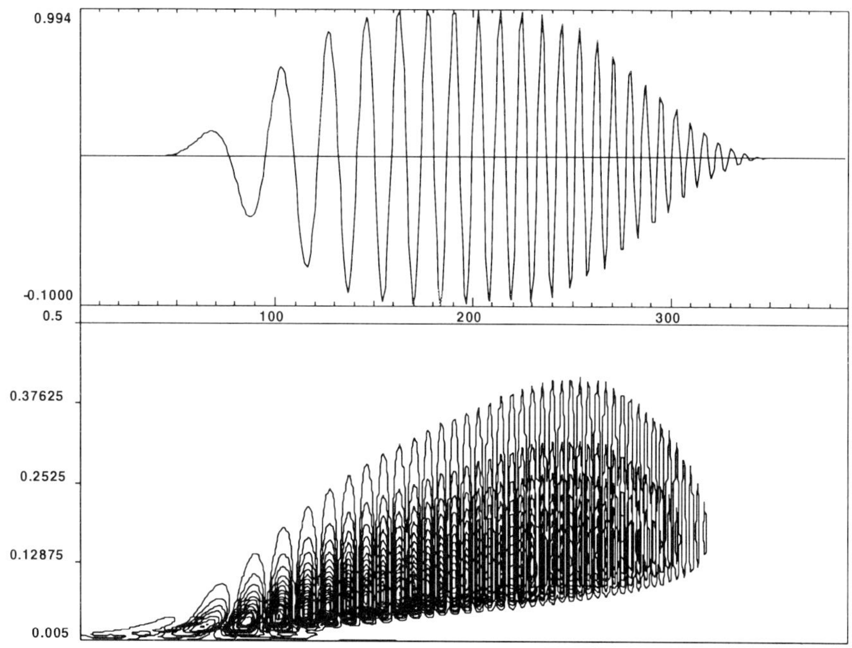


**Fig. 3.8** Continuous wavelet transform of chirp signal shown in Fig. 3.7, using Mexican hat wavelet. Note the linear distribution of maxima in scale time domain (spectral time domain) in the STFT.

We will not discuss here various applications of the wavelet transform as there is a vast amount of published work on the subject, but mention should be made that recently the WT has been applied by Lilly and Park (1995). They successfully applied the concept of CWT to polarization analysis of seismic records in a manner not unlike the multitapering spectral estimation method (Thomson, 1982) applied by Park *et al.* (1987a, b) to polarization analysis.

We will now show some examples of the application of the WT to seismograms for the purposes of compression and decimation. In Fig. 3.9 a one component recording of a seismic event recorded by the Welkom Seismic Network in South

Africa is shown along with its CWT. As mentioned above, the most distinguished feature of such plots is multiresolution. In order to obtain similar results using the STFT, one should run the STFT with different window lengths and combine results. This feature of the WT has been exploited by Lilly and Park (1995) for polarization analysis using different scales of CWT.

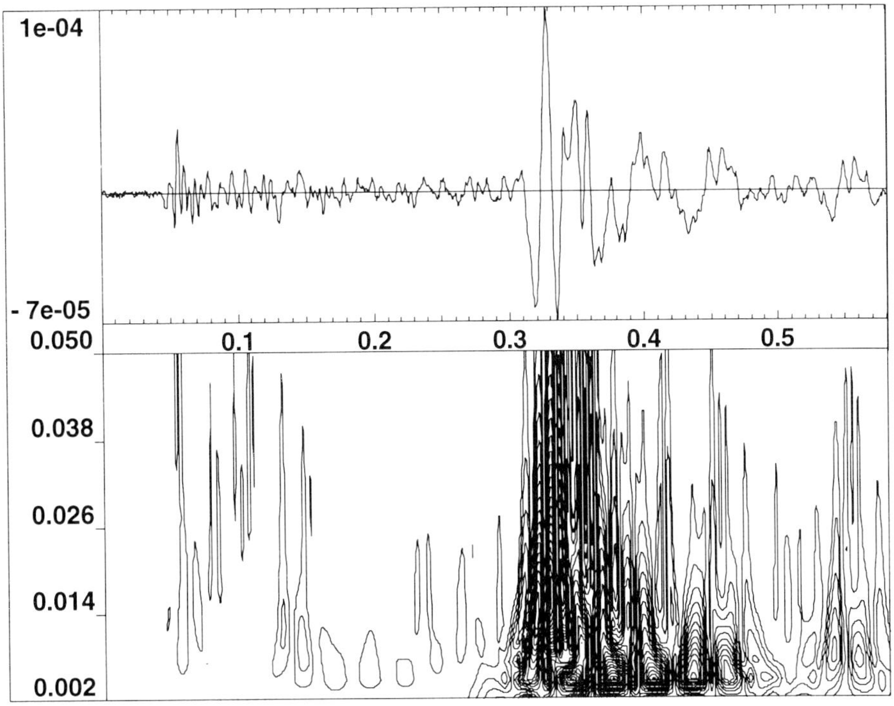


**Fig. 3.9** Seismogram recorded by Welkom Seismic Network in South Africa and its CWT. Units shown on the right-hand side of the CWT plot denote relative units (0.5 Nyquist frequency). Note the varying resolution of the CWT – the lower the frequency the better the resolution in the frequency domain and the worse the resolution in the time domain and vice versa. Horizontal axis units are seconds, while vertical axis units for seismogram plot are m/s, and for the sonogram plot relative units.

The WT can be used for decimation purposes. The basic concept here is to remove some of the signal details and then reconstruct the signal. With the proper choice of mother wavelet, reconstruction of the signal can be stopped at some level. The signal obtained then contains 'redundant' information, that is, signal levels which can be ignored. It should be pointed out that some of the signal processing may be applied directly to signal decomposition, e.g. arrival picking, polarization estimation, etc., thus giving the 'scale' (frequency) dependent information. In Fig. 3.10 we show the original seismogram and the seismogram obtained by first decomposing the initial signal into five levels, and then reconstructing it up to the third level. The number of samples in the decimated seismogram is one-eighth the number in the original seismogram.

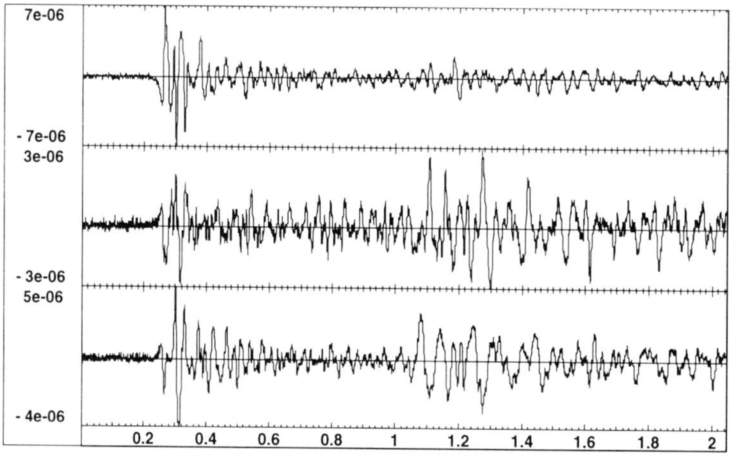


**Fig. 3.10a**
Original seismogram.

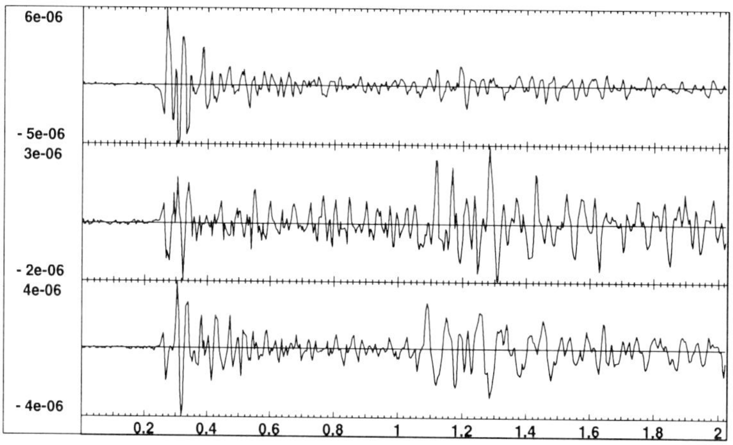


**Fig. 3.10b** Seismogram obtained by removing levels 1,2,3, resulting in a seismogram containing one-eighth of the samples of the original. Horizontal axis units are seconds, while vertical axis units are m/s.

# 4 Seismic Raytracing

The success rate with the location of seismic events and the inversion of seismic source parameters depends to a great degree on the accuracy of reconstructing the ray paths which seismic waves follow in a given geological medium. In particular, tangents to the seismic rays at the source and at the stations are of importance if routine inversions of seismic moment tensors are to be performed.

Propagation of seismic waves can be calculated in a variety of ways. More complex media require more computationally expensive and tricky schemes to find the transit time, trajectory and associated amplitude and phase shift. If the medium is homogeneous, the path of seismic waves is a straight line connecting source and receiver and the travel path is easy to calculate. When the medium is horizontally layered, the path along which the energy flows may be derived using the invariance of the ray parameter. The determination of travel times in media which vary both vertically and laterally requires some sort of raytracing.

All raytracing procedures can be said to consist of two steps: kinematic and dynamic raytracing.

Kinematic raytracing includes ray trajectory and travel time determination, but does not imply any estimation of dynamical parameters.

Dynamic raytracing additionally estimates amplitudes, phase shift and a ray-spreading Jacobian. Usually, kinematic raytracing precedes dynamic raytracing, but some methods (e.g. wavefront construction methods) estimate both kinematic and dynamic parameters simultaneously. Using the output of dynamic raytracing, it is possible to compute synthetic seismograms provided that the source function is known.

The basic concept underlying raytracing is that seismic energy of infinitely high frequency follows the trajectory determined by the raytracing equations. Physically, these equations describe how the energy continues in the same direction until it is refracted or reflected by velocity variations.

Raytracing methods can be divided into several categories:

- shooting and bending methods;
- point-to-curve methods;
- finite difference methods;
- wavefront construction methods.

## 4.1 Shooting and bending

In shooting methods of raytracing, a fan of rays is shot from one point in the general direction of the other point. The correct path and travel time to connect the two points may then be approached by more accurate estimation procedures (e.g. Julian and Gubbins, 1977; Červený *et al.* 1977).

Bending methods of raytracing start with an initial guess for the ray path. The ray path is then bent by a perturbation method until it satisfies a minimum time criterion. Other names for the bending method are *point-to-point raytracing* and *two-point raytracing*. The problem is then defined by specifying the ray congruence (in terms of a source and the list of reflections and mode conversions at all interfaces intersected by the ray) as well as a receiver. Bending methods were introduced in seismology by Wesson (1971), Chander (1975) and Julian and Gubbins (1977). In the original method of Julian and Gubbins (1977), the most detailed description of which can be found in Pereyra *et al.* (1980), the raytracing equations

$$\frac{d}{ds}\left(\frac{1}{c(\boldsymbol{r})}\,d\frac{\boldsymbol{r}}{ds}\right) - \nabla\left(\frac{1}{c(\boldsymbol{r})}\right) = 0 \tag{4.1}$$

for a ray $r$, parametrized by the arc length $s$, in a velocity field $c(\boldsymbol{r})$, are approximated by discretizing the ray path into $k + 1$ points $(\boldsymbol{r}_0, \boldsymbol{r}_1,...\boldsymbol{r}_k)$ and by using finite differences for the derivatives. The system of equations obtained can then be linearized and solved to find the perturbation that must be applied to some initial guess of the ray in order to reduce the right-hand side of the ray equations to zero. Since the linearized system is approximate, several iterations will usually be sufficient for convergence.

The second type of bending method (Um and Thurber, 1987, Prothero *et al.* 1988) tries to minimize directly the travel time as a function of the ray curve $\gamma$.

$$T(\gamma) = \int_{\gamma} \frac{ds}{c} \rightarrow \min \tag{4.2}$$

The advantages of this subgroup of bending methods over methods described above are obvious. The precision of the finite difference approach is restricted by the precision of numerical differentiation, which is inherently more unstable than numerical integration. The linearization of the system of equations described above introduces small errors. Taking into account the ill conditioning in the presence of a complicated velocity model, these small errors may prevent the solution from convergence to the desired accuracy, or even cause divergence.

The difficulties with raytracing using the above described shooting and bending methods fall into several categories:

- For strongly varying velocity fields there can be many paths connecting two points of interest and, taking that into account, it is easy to miss the ray path with minimum travel time. This raises the important issue of how to generate all the rays in the given congruence which join the source and the given receiver.
- If travel paths to many points are needed, the computing time necessary for it may render the raytracing impractical.
- Even in a smooth medium, there may be a shadow zone, where pairs of points will be connected only by rays that have a very small geometric amplitude because a small change in the takeoff angle results in a large change in the ray path. Shooting methods of raytracing often have trouble finding the correct ray in a shadow zone.

Bending methods do give an answer in shadow zones; however, in both bending and shooting methods, it is possible that the answer is only a local minimum and therefore the global minimum travel time and corresponding ray path remain unknown (Hanyga, 1991).

## 4.2 Point-to-curve

Most of the problems described above can be overcome by the *point-to-curve raytracing algorithms* (Hanyga, 1991). These algorithms solve a somewhat relaxed problem: the receiver is not fixed but rather allowed to move along the curve in the model space. The algorithm determines successive receivers and associated rays. The output is a curve in space whose points define the rays as well as the associated receivers. This new problem is well defined at the caustics where the solution curve folds over the receiver curve, returning to the receivers already visited along a new solution branch. By adjusting the ray parameters, the ray end point usually marches backwards and forwards along the receiver curve, thereby visiting the receivers once or several times or not at all and thus generating different arrivals. This technique is based on so-called 'continuation methods' (Allgower and Georg, 1990).

Continuation methods amount mathematically to solving an undetermined system of equations or equivalently forcing the solution to follow an implicitly defined curve. All the branches connected into a single curve are successively found, essentially resolving the problem of tracing all the rays at every receiver. Although these methods are much more powerful than *point-to-point* raytracing methods they are not guaranteed to find all arrivals. This can be attributed to the fact that the implicit curve possibly consists of several disconnected branches.

Therefore, after following one branch, it is not certain that there is no other branch and that all arrivals have been found. Caution needs to be exercised to avoid endlessly generating the same arrivals, which may happen if a branch is a closed loop. Another problem which may be encountered is receiver curve definition. Some of the solutions may not be found depending on the actual curve definition and model box size.

Let us now briefly formulate the *point-to-curve* algorithm. Rays satisfy the ray differential equations:

$$\frac{d\boldsymbol{x}}{d\sigma} = \boldsymbol{p}$$

$$\frac{d\boldsymbol{p}}{d\sigma} = \frac{1}{2}\nabla u^2 \tag{4.3}$$

$$\frac{dT}{d\sigma} = u^2$$

where $\boldsymbol{x}$ denotes the ray location, $\boldsymbol{p}$ the slowness vector, $u$ the slowness (a twice continuously differentiable function of $\boldsymbol{x}$), $T$ is the cumulative travel time along the ray, and $\sigma$ a parameter along the ray defined by equation (4.2). Let us suppose that a ray in three dimensions is characterized by two ray parameters $\gamma_1$ and $\gamma_2$, for instance two initial ray angles in three dimensional space. Then the ray field is described by $\boldsymbol{x}$ $(\gamma_1,\gamma_2,\sigma)$ and for a fixed receiver $\boldsymbol{x}^R$ the *two point* raytracing problem is given by:

$$\boldsymbol{x}(\gamma_1,\gamma_2,\sigma) - \boldsymbol{x}^R = 0 \tag{4.4}$$

which is a system of three nonlinear equations with three unknowns $\gamma_1$, $\gamma_2$ and $\sigma$. Such a system has, generally speaking, a finite number of isolated solutions which correspond to the finite number of arrivals at the receiver. The main idea behind the point-to-curve raytracing method is to replace the fixed receiver location $\boldsymbol{x}^R$ by a receiver curve. Then equation (4.4) can be rewritten as:

$$\boldsymbol{x}(\gamma_1,\gamma_2,\sigma) - \boldsymbol{x}^R(\lambda) = 0 \tag{4.5}$$

which is a system of three nonlinear equations with four unknowns or an implicit description of a curve in a four dimensional space. That curve can be explicitly defined by:

$$(\gamma_1(s),\gamma_2(s),\sigma(s)\lambda(s)) \tag{4.6}$$

where $s$ is the arc length along the curve. Then the four dimensional curve whose definition is given by (4.6) represents the solutions of a sequence of point-to-point problems. In general, the methods for implicit curve tracing can be divided in two groups: *predictor-corrector* methods and *piecewise-linear* methods. For detailed descriptions of these methods, the reader is referred to Allgower and Georg (1990).

Let us now consider example of the application of point-to-curve raytracing in some model geometry, containing caustic points, i.e. points where the geometrical spreading matrixis approaching zero. The receiver is located inside the caustic cusp of the source. In following discussion, we will use MODEL8, described in Hanyga (1988). The model consists of two homogeneous blocks. The rays are contained in the upper block and are reflected without conversion. The propagation speed in the upper block is assumed to be 3.5 km/s.

| | |
|---|---|
| Model box: | $-10 \leq x \leq 20,\ 0 \leq y \leq 20,\ 0 \leq z \leq 20;$ |
| Interface: | $z = 0.01\,(x-10)^3 + 0.05\,(x-10)^2 + 5$ |
| Source: | (6, 10, 16) |
| Receiver end: | (0, 10, 19.5) |
| Receiver line direction: | (1, 0, 0). |

The receiver curve has been defined as the line passing through the receiver end point with the direction cosines shown above. This is basically the so-called 2.5D model. The ray distribution found using the point-to-curve algorithm is shown in Fig. 4.1.

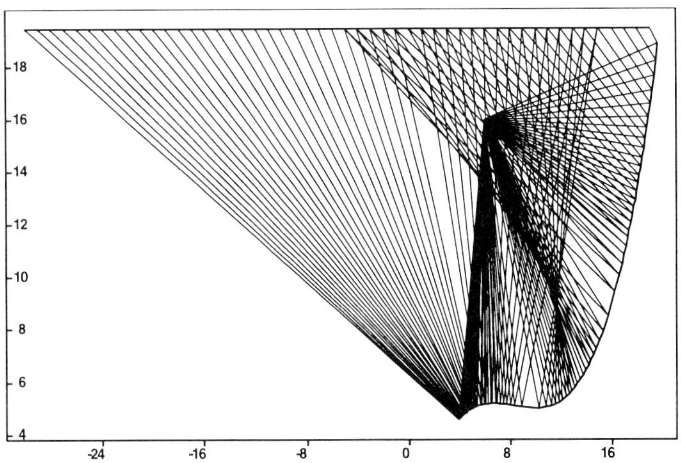


**Fig. 4.1** Ray distribution for MODEL8 (Hanyga, 1988) calculated for source (6,10,16) and receiver curve defined by the receiver end point (0,10,19.5) and direction cosines (1,0,0) using the point-to-curve raytracing algorithm. For the travel time plot, see Fig. 4.2. Units on both axes are km.

This example demonstrates the ability of point-to-curve algorithms to find multiple solutions – triplications, bifurcation points. The travel time plot for the above mentioned model and definition of receiver curve is shown in Fig. 4.2. In order to check the stability of the point-to-curve raytracing method for different receiver curve definitions, we calculated the solutions of the above formulated problem for two different receiver line directions.

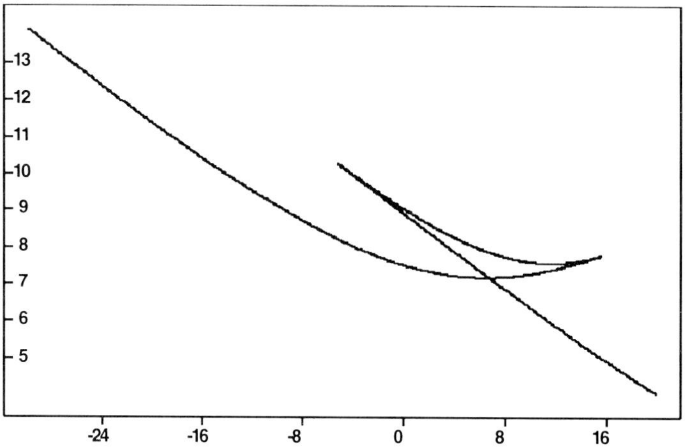


**Fig. 4.2** Travel time plot for MODEL8. Horizontal axis is in km, while vertical axis is in seconds.

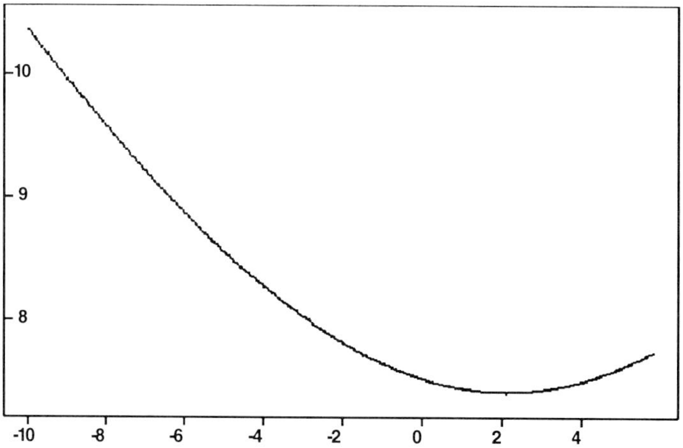


**Fig. 4.3** Travel time plot for MODEL8, with a receiver line passing through the receiver end point (0,10,19.5) and with direction angles (60°,30°,0°).

In Fig. 4.3 the travel time distribution is shown for a receiver line passing through the point (0, 10, 19.5) and making a 60° angle with the *X* axis (direction angles are 60°, 30°, 0°). No triplications are found. Figure 4.4 shows the travel time plot for a receiver line passing through the same receiver end point (0, 10, 19.5) but with direction angles now (30°, 60°, 0°). Triplications are present in the solution. The main reason explaining why the triplications of travel time curve are not found in Fig. 4.3, is the restriction on the receiver curve crossing the greater model boundary. Once the ray curve is outside the greater model boundary (which must be supplied as user defined input) the solution stops. It should be pointed out that in order to overcome that problem the greater model boundary can be defined to change adaptively during the calculation process, correspondingly increasing the computational requirements.

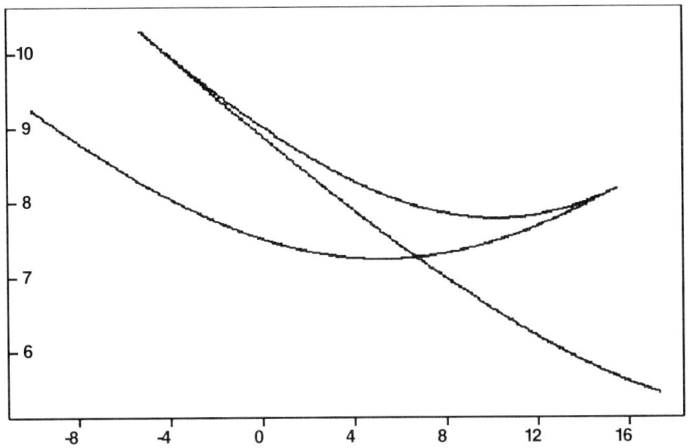


**Fig. 4.4** Travel time plot for MODEL8, with a receiver line passing through the receiver end point (0,10,19.5) and with direction angles (30°,60°,0°).

## 4.3 Finite difference

Finite difference methods also overcome most of the shortcomings of the shooting and bending methods, but are less accurate.

Vidale (1988) described a finite difference method that extrapolated travel times away from the source. Vidale's algorithm was initially based on a two dimensional earth model consisting of square cells with nodes positioned at the cell corners. Energy is propagated from the source throughout the grid in expanding (square) rings of nodes centred around the source. Let us consider Fig 4.5. The source is assumed to be at grid point $A$ and the timing process is initiated by assigning point $A$ the travel time of zero. Then, through the use of the finite difference solution of the eikonal equation, all four points adjacent to point $A$, labelled $B_1$ through $B_4$ in Fig. 4.5, can be timed as follows:

$$t_i = \frac{h}{2}(S_{B_i} + S_A) \tag{4.7}$$

where $h$ is the mesh spacing, $S_A$ is the slowness at the point $A$, and $S_{B_i}$ is the slowness at the grid point $B_i$ being timed.

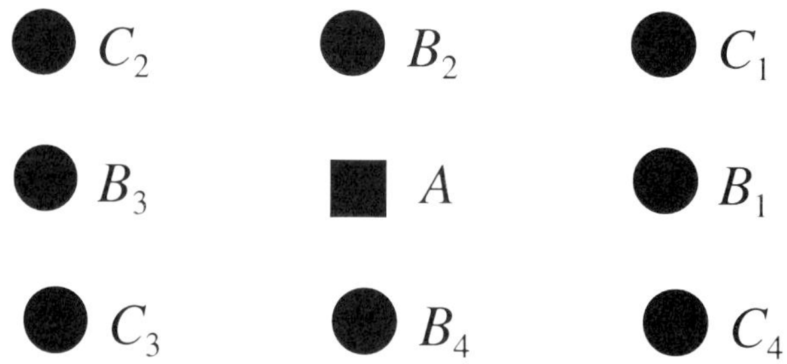


**Fig. 4.5** The source grid point $A$ and the grid points in the ring surrounding the source point.

The travel times at the four corners are then found by one of two methods, as follows:

a. Plane-wave approximation

$$t_{C_1} = t_A + \sqrt{2(hs)^2 - (t_{B_2} - t_{B_1})^2} \tag{4.8}$$

This equation gives the travel time to point $C_1$ using the travel times from the source to points $A$, $B_1$ and $B_2$ in a plane wave approximation, The point $A$ does not need to be a source point for this equation to apply.

b. Assuming locally circular wavefronts

The second approach assumes locally circular wavefronts. A circular wavefront can be specified by the three parameters: the $x$ coordinate of the virtual source point – $x_s$; the $z$ coordinate of the virtual source point – $z_s$; and the origin time of the virtual source $t_s$. The virtual source must lie at the centre of the curvature of the circular wavefronts. Assuming that the origin of this coordinate system is placed at point $A$ and grid points $B_1$, $B_2$ and $C_1$ have coordinates $(h,0)$, $(0,h)$ and $(h,h)$ respectively, the travel times to points $A_1$, $B_1$ and $B_2$ can be expressed by the following system of equations:

$$
\begin{aligned}
t_0 &= t_s + S\sqrt{x_s^2 + z_s^2} \\
t_1 &= t_s + S\sqrt{(x_s + h)^2 + z_s^2} \\
t_2 &= t_s + S\sqrt{x_s^2 + (z_s + h)^2}
\end{aligned}
\tag{4.9}
$$

This is a system of three unknowns which can be solved to produce the values of $x_s$, $t_s$, $z_s$, with which $t_3$ can be calculated as follows:

$$
t_3 = t_s + S\sqrt{(x_s + h)^2 + (z_s + h)^2} \tag{4.10}
$$

Analysis of the accuracy of both solutions has shown that for wavefronts with very little curvature, equation (4.8) produces the quicker and more accurate answer. For strongly curved wavefronts, equation (4.10) is a better approximation. For optimal accuracy, a 'mixed' scheme must be used.

Using this method to find the travel times at the fourth corner of a square, the travel times may be found throughout the grid. The order of solution is very important. Firstly, points on the square around the source are timed, then, starting with the minimum travel time on each side of the initial square, travel times are calculated on the next larger square. The solution proceeds the same way until all points in the medium are timed (Fig. 4.6). Vidale (1990) later extended his method to the three dimensional case and proposed a new extrapolation scheme. Vidale used an implicit assumption that the travel time at a node on a ring can be computed from locally adjacent nodes on the prior inner ring.

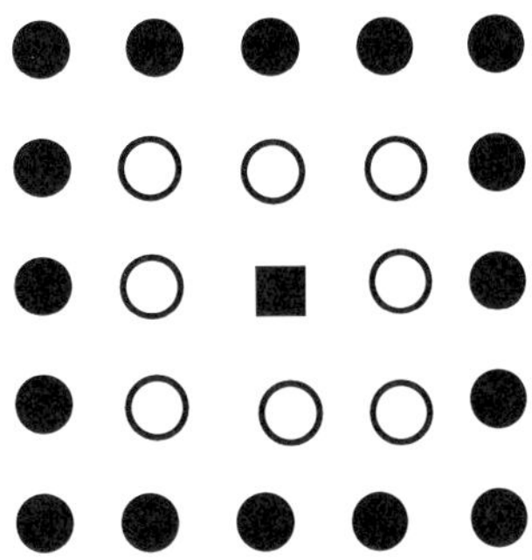

**Fig. 4.6** Progress of solution of 2D case. The ring of points shown as filled circles are to be timed at this stage. The filled rectangle at the centre of the grid is the source. The hollow circles show the position of the ring timed at the previous stage.

However, in some cases, when there are large velocity variations of the media, the first arrival energy may actually follow the circuitous route and approach the node from the outer ring. To illustrate this point, consider the two layer model in Fig. 4.7, where the source is located in a low velocity layer. The actual arrival at point $D$ is the head wave transmitted along the path $A$ $B$ $C$ $D$. However, the expanding square algorithm will time the point $D$ before the edge of the square reaches the interface. Therefore, the influence of the high velocity layer will not be included and will lead to incorrect timing of this and many other points.

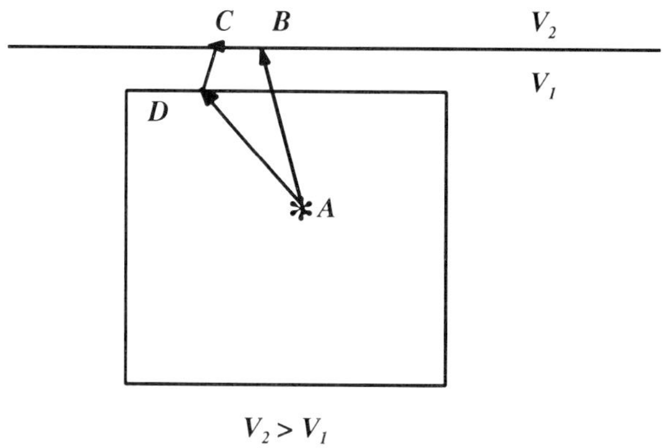


**Fig. 4.7** A two layer model where Vidale's (1988) finite difference scheme will fail to produce the correct solution. $A$ is the source point, $D$ the point to be timed. $V_1$ and $V_2$ are the velocities of two layers and $V_2 > V_1$ (Qin *et al.* 1992).

Another alternative approach which avoids that problem, but which requires large computational cost, has been proposed by Qin *et al.* (1992). In this scheme, the solution region progresses outwards from an 'expanding wavefront' rather than an 'expanding square' and therefore respects causality. On every step perimeter the travel time array is constructed and sorted in order of increasing travel time. Then the solution proceeds, starting with the global minimum travel time point, updating the perimeter array until the global minimum travel time is found. Therefore, the solution is found along the expanding wavefront and causality is preserved, although computational requirements are increased significantly.

All the above described finite difference methods make use of the traditional cell-centred node scheme. An alternative approach has been used by Cao and Greengalgh (1994), which is based on a new corner node velocity model discretiziation scheme. They have shown that this method offers accuracy advantages. A substantial efficiency improvement is achieved by the incorporation of the wavefront tracking algorithm based on the construction of a minimum travel time tree.

A characteristic feature of the finite difference methods is that they only calculate the travel times for the first arrivals at each point on a regular grid. As such, they cannot be, in general, considered as an alternative to other raytracing methods, although they represent interesting complementary techniques, mainly due to their efficiency, speed, ease of use and ability to deal with peculiar waves such as headwaves or diffracted waves.

As has been mentioned, the finite difference methods described above provide solutions only for the first arrival travel times, which may be sufficient for some problems, but the first arrivals method can be used to obtain, indirectly, the later arrivals as well (Podvin and Lecomte, 1991). A more systematic approach to calculating later arrivals with a modified version of that algorithm has been presented in Podvin (1992).

Once travel times are known, rays can be traced back to the source using the direction of steepest descent from the grid cell closest to the receiver.

The expanding wavefront method of Qin *et al.* (1992) has been applied to a raytracing problem for Vaal Reefs Gold Mine, South Africa.

Geological information available consists of three layers, where the velocity of the upper layer is 6834 m/s, of the second 6116 m/s and the underlying layer has a velocity of 5779 m/s. Each layer of information has been defined in a very sparse grid (1 km grid spacing in $X$ direction, 2 km spacing in $Y$ direction).

Obviously, initial data could not be used without prior interpolation and model space definition. Each layer's initial grid has been used to interpolate on a fine, uniformly spaced grid with 100 m grid spacing, using B-splines. Perspective mesh plots of each layer's bottom are shown on the left in Fig. 4.8. Contour line representation of the same information is shown on the right.

In each case, the model space used is:

$$\begin{aligned} 2000\text{ m} &\le X \le 17\,000\text{ m};\\ 70\,000\text{ m} &\le Y \le 75\,000\text{m};\\ 0\text{ m} &\le Z \le 5000\text{ m} \end{aligned}$$

The grid dimensions therefore are 50 x 50 x 50. The source location used was (12 500 m, 72 500 m, 4250 m).

In Fig. 4.9 isochron distribution is shown on different model slices by planes parallel to the horizontal plane at various depths: $Z = 0$ m, 500 m, 1000 m, 1500 m, 2000 m, 2500 m.

Ray trajectories for different source positions and receivers situated along the model boundaries and in a plane passing through the source and parallel to the $XZ$ plane are shown in Fig. 4.10. Rays have been smoothed prior to display. Comparing the ray distribution in Fig. 4.10, one can see caustics and shadow zones present.

Isochrons on slices by planes parallel to the $YZ$ plane are shown in Fig. 4.11, for $X = 70\,000$ m, 71 000 m, 72 000 m, 73 000 m, 74 000 m, 75 000 m. It should be noted that in this case up to a depth of about $h = 2500$ m wavefronts are almost spherical. When the wavefronts reach another interface with sharp velocity changes, and after depth of $h = 2000$ m, the wavefronts are quite different from spherical. There are various sharply pronounced inhomogeneities which are reflected in the isochron distribution in Fig. 4.9.

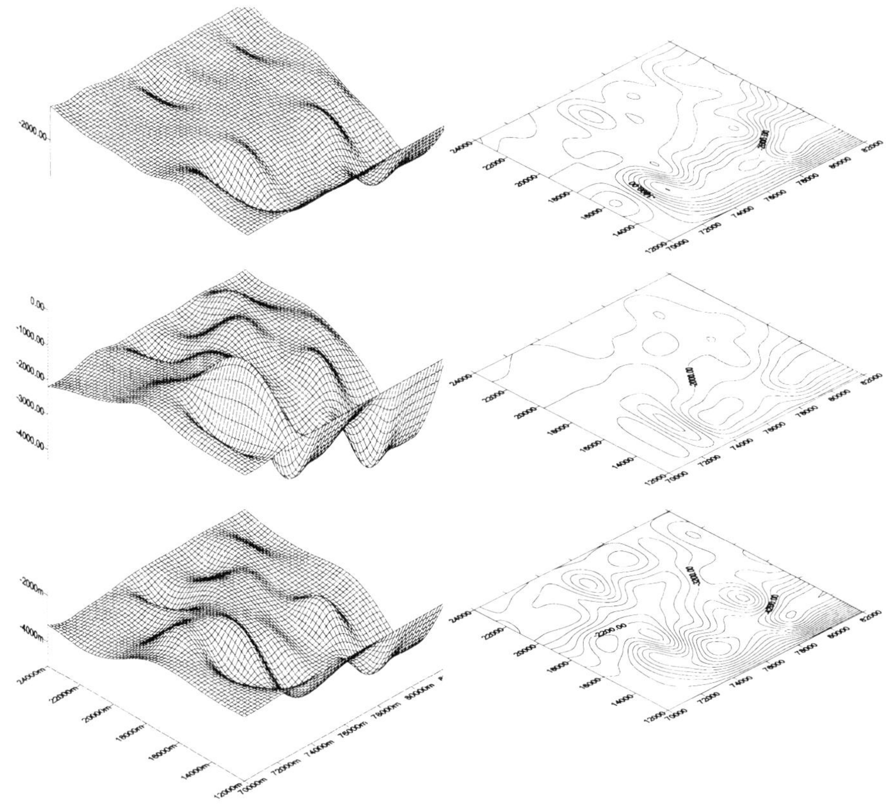

**Fig. 4.8** Geometry of interfaces for the Vaal Reefs Gold Mine region represented as a 3D mesh plot for model space: 70 000 m $\leq X \leq$ 82 000 m; 12 000 m $\leq Y \leq$ 24 000 m; 0 m $\leq Z \leq$ 5000 m. The depth is shown on the left hand axis. The fine mesh grid has been obtained by B-spline interpolation of the initial sparse grid.

In order to avoid such sharp velocity variations another approach to model creation should be adopted, that is, an initial sparse grid with velocity distribution must be created, and then this newly created grid must be interpolated on a fine three dimensional grid using three dimensional B-spline interpolation. Thus all sharp variations of velocity at intefaces will be smoothed out, which will lead to more stable calculations.

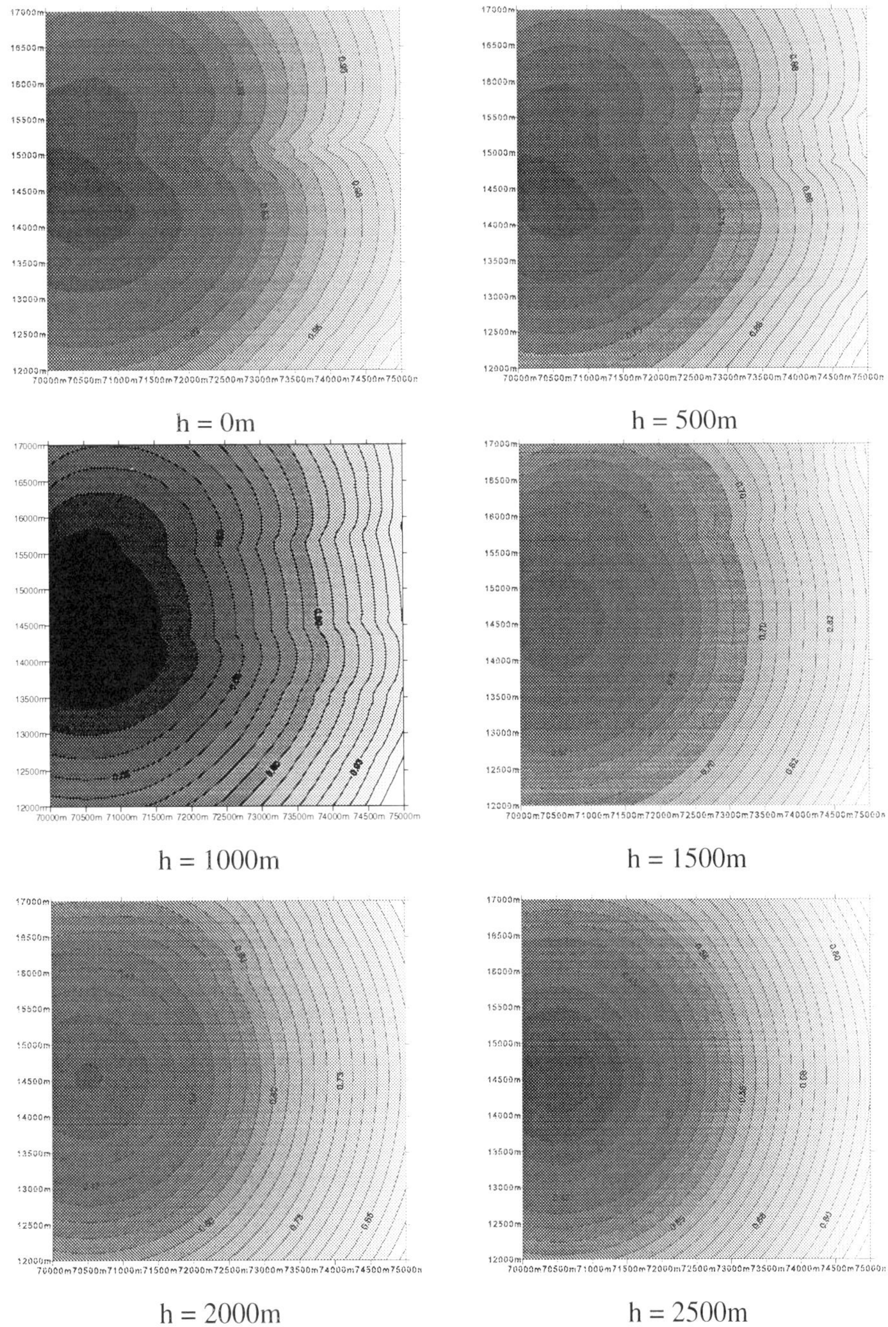


**Fig. 4.9** Isochrons on slices by planes parallel to XY plane at different depths.

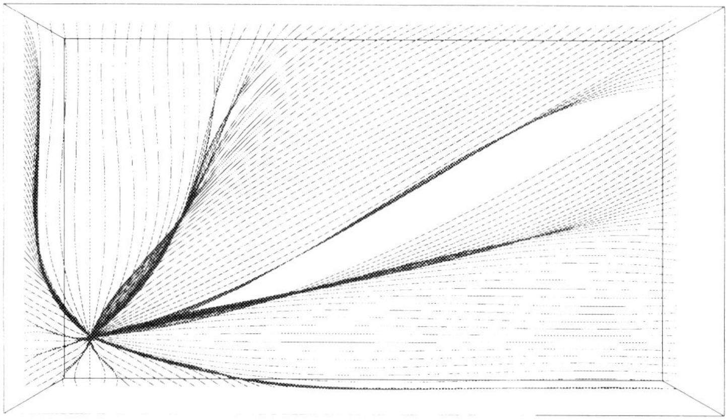

(12 500 m, 72 500 m, 4250 m)

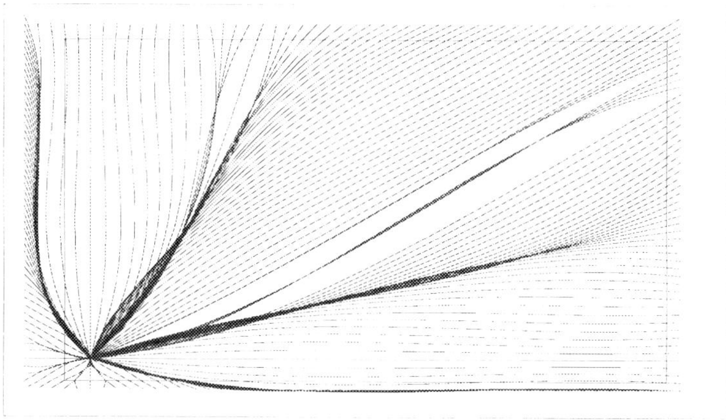

(12 500 m, 72 500 m, 4500 m)

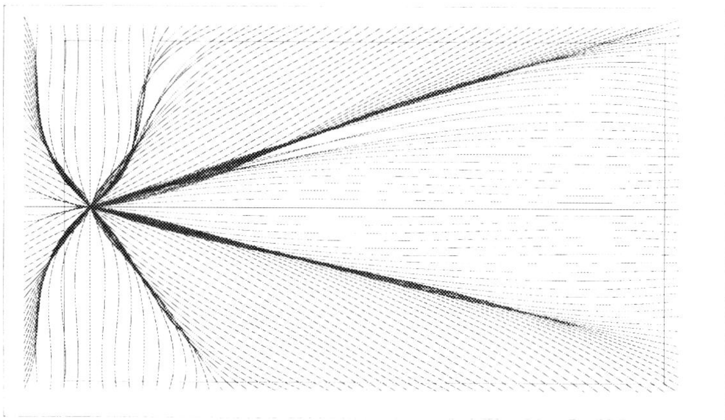

(12 500 m, 72 500 m, 2500 m)

**Fig. 4.10** Ray plots for receivers situated along the model boundaries and lying in the plane passing through the source and parallel to the $XZ$ plane for source locations as indicated.
The model space used is 12 000 m $\leq X \leq$ 17 000 m; 70 000 m $\leq Y \leq$ 75 000 m; 0m $\leq Z \leq$ 5000 m. The grid dimension is 50x50x50.
Rays have been smoothed prior to display.

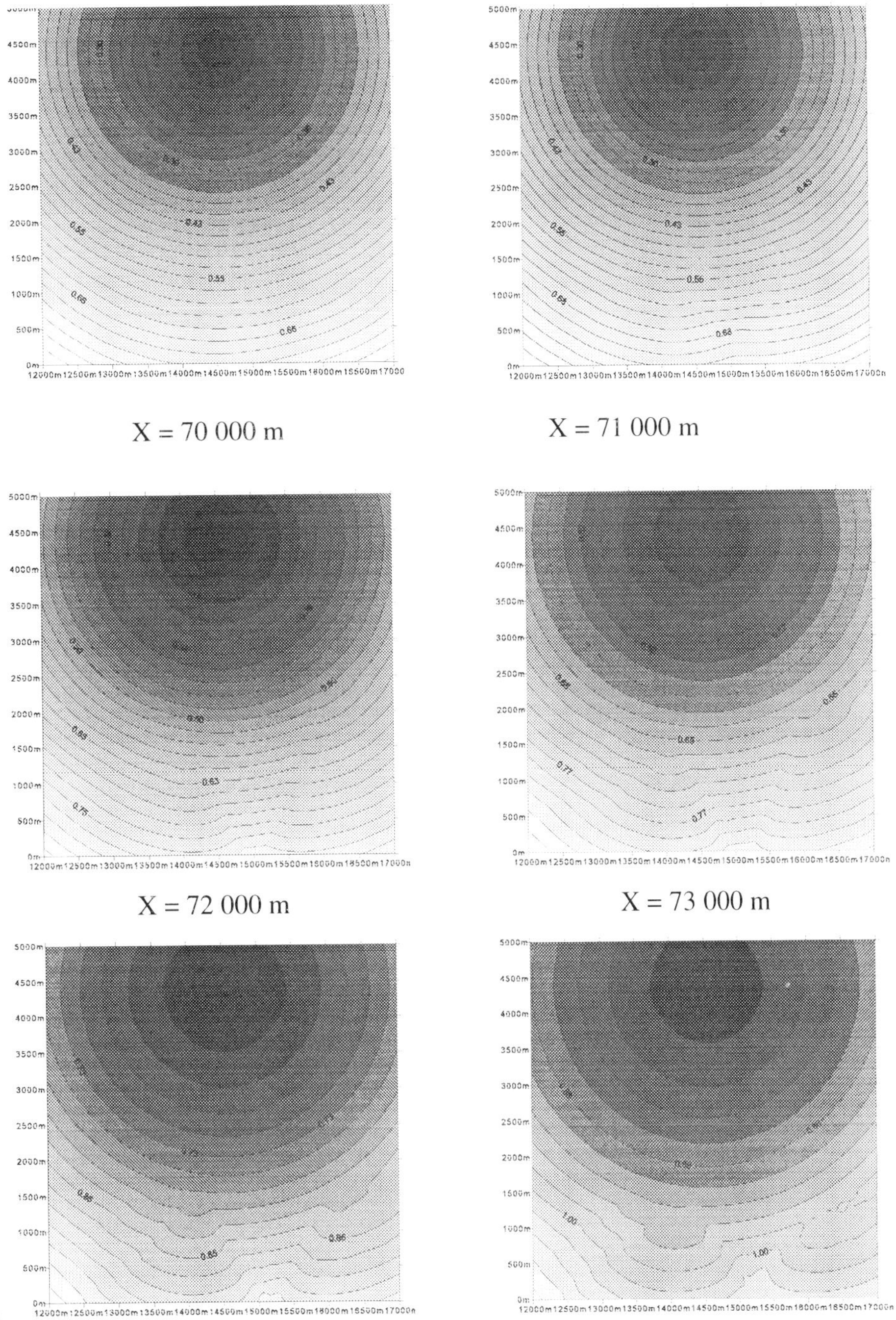


X = 70 000 m

X = 71 000 m

X = 72 000 m

X = 73 000 m

X = 74 000 m

X = 75 000 m

**Fig. 4.11** Isochrons on slices by planes parallel to YZ plane

## 4.4 Wavefront construction methods

This new group of travel time and amplitude estimation methods has been pioneered by Vinje *et al.* (1993). These methods are based on raytracing with the same robustness (in smooth media) as the grid methods, e.g. finite difference methods, but have the ability to provide the later arrivals and amplitudes. These methods are related to the method of graphical wavefront construction described by Thornburgh (1930) and Riznichenko (1946).

This section briefly outlines the main ideas on which these methods are based.

Consider the wavefront shown in Fig. 4.12 (Vinje *et al.* 1993). The front is closed and the starting points of this initial wavefront are numbered from 1 to 7. These points will be used as the initial positions of rays which will propagate the wavefront from a position at time $\tau$ (the position of the wavefront at that time is shown as a solid line) to time $\tau + \Delta\tau$ (dashed line). Different raytracing procedures may be used, but the most straightforward approach is to use the standard Runge–Kutta method or simple Taylor expansion for the purpose of wavefront extrapolation from the old to the new position.

In Fig. 4.12, the end points of the rays are marked with filled circles. The distance between each new point is calculated and checked against some predefined distance $\Delta S_{max}$. If the distance between two neighbouring points on the wavefront exceeds $\Delta S_{max}$, a new start position is interpolated on the old wavefront. This is done between points 1 to 7 in Fig. 4.12.

The following parameters are interpolated (Vinje *et al.* 1993):

a. start position of the ray;
b. unit ray tangent vector;
c. geometric spreading;
d. amplitude coefficient; and
e. ray entrance coefficient at the source.

In this way, new rays are created between all points on the old wavefront, as shown in Fig. 4.12, and the end points of the interpolated rays are marked with open circles. As a result of this procedure, a new wavefront is obtained which consists of new points with the distance between the neighbouring points not exceeding the predefined maximum distance $\Delta S_{max}$.

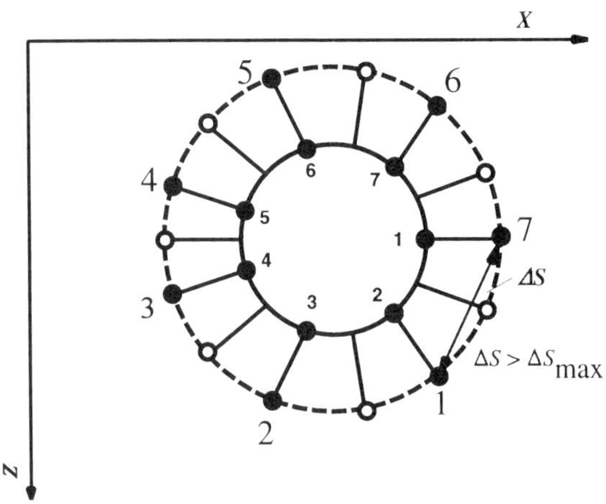


**Fig. 4.12** Wavefront construction scheme. A new wavefront (dashed line) is constructed from the previous wavefront (solid line). New rays are interpolated between each of the points from 1 to 7 in the old front since $\Delta S$ exceeds $\Delta S_{max}$. Vinje et al (1993).

The main advantage of this method is that it is easy to trace not only first arrivals but also almost all others. The method can easily be adapted to simplify and significantly speed up the computation required if only first arrivals are to be traced. It is easy to modify the method to include reflection and transmission at discontinuous interfaces. While conventional raytracing may give no arrivals in areas of large geometrical spreading, and it is difficult to cover the entire model space with a bundle of rays from a source point, the above method solves this problem. Vinje *et al.* have shown that this method has very high accuracy compared with finite difference methods. In a complicated medium, however, there is the problem of undersampling of the ray field at caustics (Lucio *et al.* 1995). This problem is due to the improper ray sampling criterion, as the ray density has to be controlled in order to ensure accuracy as well as computational efficiency of the algorithm (Lucio *et al.* 1995). The ray density can be controlled at some predefined places like a sequence of flat horizons (Lambare *et al.* 1994). In order to keep the ray density above some minimum level, it is necessary to insert additional rays as the ray field spreads out from the source. An additional ray is created using interpolation on the previous wavefront position and subsequent tracing of the ray.

Several criteria for the density control have been suggested, among them the metric distance between adjacent rays (Vinje *et al.* 1993); their angular distance (Sun, 1992); and the uniform ray density criterion (Lucio *et al.* 1995). The latter criterion seems to be most accurate and robust, even in regions with strong caustics and multiple arrivals.

Consider now an example of the application of the wavefront construction method to the model consisting of a single isotropic block. The propagation velocity is defined by the following:

$$c(z) = \frac{1}{2}\left(2.5z + 2.5 + \sqrt{(1.5z - 7.5)^2 + 0.1}\right) \tag{4.11}$$

The source position is (0.0, 0.0), the grid size 301 x 301 points with grid step equal to 0.1 km, the maximum distance between rays 0.1 km, and the time increment between successive isochrons 0.1 s. In this model, the caustic cusp generated by the velocity variation in the medium is clearly seen. The velocity distribution model, along with its derivatives used in wavefront construction for this model, has been modelled by cubic B-splines. The results of applying this wavefront construction method, shown in Fig. 4.13, are in good agreement with the results obtained by extended asymptotic ray theory (Hanyga, 1988).

**Fig. 4.13** Wavefront distribution for a velocity model given by equation (4.11). Model dimension 301 x 301, grid size 0.1 km.
The wavefronts are plotted in 0.1 s increments, the source position is (0.0,0.0), and the model box dimension is $-5.0 \le X \le 25.0$; $-10.0 \le Z \le 20$.

**Fig. 4.14** Results of application of wavefront construction method to a two-layered model. Model dimensions $0.0 \leq X \leq 2$ km; $0.0 \leq Z \leq 2.0$ km; grid dimensions 201 x 201, grid step size 0.01 km, velocity in upper layer 1.5 km/s, in lower 1.9 km/s, source position (0.25, 0.75) km. The second layer starts at 1.0 km depth. Wavefronts calculated with 0.01 km increment, $DS_{max} = 0.02$.

The results of the application of the wavefront construction method to a two-layered model are shown in Fig. 4.14. The model size is 201 x 201 points and the grid size 0.01 km. The velocity in the first layer is 1.5 km/s, and 1.9 km/s in the second. The second layer starts at depth 1.1 km. The velocity distribution is represented by B-splines. The source position is (0.25, 0.75) km. Headwaves generated at the interface are clearly observed.

The velocity model for the cross-section of one of the gold mines in South Africa is shown in Fig. 4.15. The results of applying the wavefront construction method are shown in Fig. 4.16.

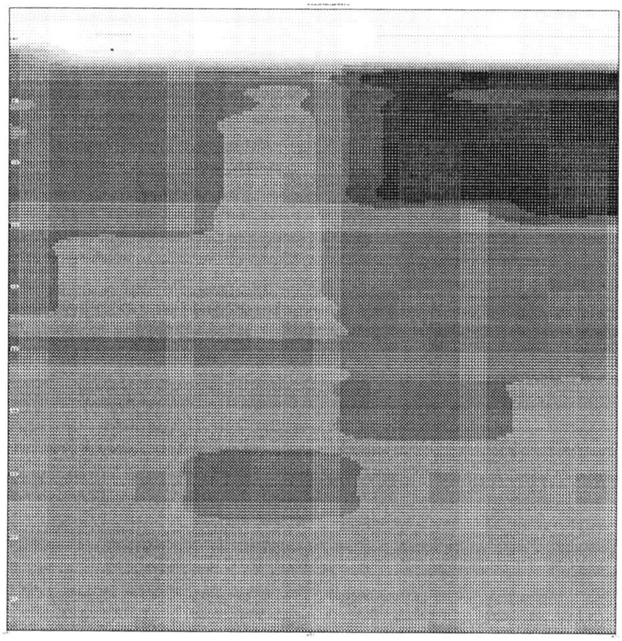

**Fig. 4.15** Velocity model for the cross section of one of the gold mines in South Africa. Darker shades correspond to higher velocity values. The model is represented by cubic B-splines, using 201x201 points with 10 m spacing. Model box dimension: 2 km x 2 km.

**Fig. 4.16** Wavefronts distribution for a velocity model shown in Fig.4.15. Wavefronts have been calculated with 0.01 second increments, $\Delta S_{max}$ = 10 m. Source location is (300.0, 1000.0) m. Numerous caustics are observed.

# 5 Location of Seismic Events

The location of seismic events is of fundamental importance, since all subsequent seismological processing depends, to some degree, upon the event position and the distances to the stations. Finding the location of a seismic event amounts to solving an inverse problem – a set of unknowns has to be retrieved from a set of data. The four unknowns consist of the location and origin time vector, $\mathbf{x} = (\mathbf{h}, t_0)^{\mathrm{T}} = (x_0, y_0, z_0, t_0)^{\mathrm{T}}$. In the case of a mine seismic network, the unknowns are to be retrieved from data consisting of P and S wave arrival times, the direction of the wavefronts, both of which are derived from waveforms, the velocity model and the station coordinates. The data have associated errors which can produce location errors. The location errors also depend on the spatial distribution of stations with respect to the position of the source, discussed in Section 5.4 below, and on the physics of the rupture process. If slow rupture starts at a certain point, the closest station(s) may record waves radiated from that very point while others may only record waves generated later in the rupture process by a high stress drop patch of the same source. In this case, care has to be taken in determining the arrival times if the location of rupture initiation is sought, otherwise the location will be a statistical average of different locations within the source. It is helpful therefore to think of a seismic source as a spatial event with characteristic size of the order of a metre for a moment magnitude $m_{\mathrm{M}}$=-4 event, up to a few hundred metres for an event of $m_{\mathrm{M}}$>4.

## 5.1 Location by arrival times and/or directions or azimuths

One can write the following arrival time equation for the *j*th seismic station that recorded the P or S wave:

$$LOC_j(\mathbf{x}) = t_j - t_0 - T_j(\mathbf{h}) \qquad (5.1)$$

where:

| | |
|---|---|
| $t_j$ | the observed P or S wave arrival time at the *j*th station; |
| $t_0$ | the unknown origin time of the event; |
| $T_j(\mathbf{h})$ | the unknown travel time of the P or S wave to the *j*th station; |
| $LOC_j(\mathbf{x})$ | the residual i.e. the difference between the observed $[t_j]$ and the calculated $[t_0 + T_j(\mathbf{h})]$ arrival times. |

Since equation (5.1) is nonlinear, one needs a well behaved system consisting of at least five equations to be able to find a unique solution. Due to the sparseness of and the large errors in the data and/or poor configuration of stations, the location problem can be unstable. A practical way to stabilize the solution is to eliminate the origin time, $t_0$, by centring, i.e. subtracting the average arrival time equation from the arrival time equations for each individual station. After centring and rescaling the residuals to metres and multiplying equation (5.1) by the average velocity of the P or S wave along the respective ray paths, it becomes

$$LOC_j(\mathbf{h}) = \bar{V}_j\{t_j - \bar{t} - [T_j(\mathbf{h}) - \bar{T}(\mathbf{h})]\} \tag{5.2}$$

where $\bar{t}$ and $\bar{T}$ are the average arrival time and average travel time respectively, and the origin time is found from $t_0 = \bar{t} - \bar{T}$.

To constrain the location problem, one can add the direction of the wavefront, defined by the principal eigenvector of the covariance matrix of the initial P pulse recorded by the triaxial sensor at the *j*th station (Flynn, 1965; Mendecki, 1993) In this case, equation (5.2) becomes

$$LOC_j(\mathbf{h}) = \bar{V}_j\{t_j - \bar{t} - [T_j(\mathbf{h}) - \bar{T}(\mathbf{h})]\} + Dis_j(\mathbf{h}) \tag{5.3}$$

where $Dis_j(\mathbf{h})$ is the orthogonal distance between the current estimate of **h** and the direction, or vertical plane through the direction (azimuth), at the *j*th station. Directions will constrain the location problem only if the straight line is a fair approximation of the ray path between the source and the station, which is typically true for the nearby stations and/or if the medium can be considered homogeneous between the event and the station. The azimuth, rather than direction, should be used in the case of a layered geological structure with no significant lateral inhomogeneities. If the medium is very inhomogeneous, the travel times have to be calculated using a raytracing method rather than an arrival time equation based on a straight ray path.

In Fig. 5.1, an event recorded by the Vaal Reefs seismic network was located using the assumptions that the ray path is straight and that the medium is homogeneous. Also shown is the location based on the raytracing as discussed in Chapter 4. The velocity model used for the raytracing is shown in Fig. 4.8 and is based on the complex geology in this region. The difference in location is 4.2% of the average hypocentral distance (183 m) and, in this case, the straight ray path assumption is not adequate.

The classical way to solve a system of equations, each of type (5.3), is to linearize it with a Taylor expansion around a trial location and invert the resultant linear system, usually by a least squares method. This means that the

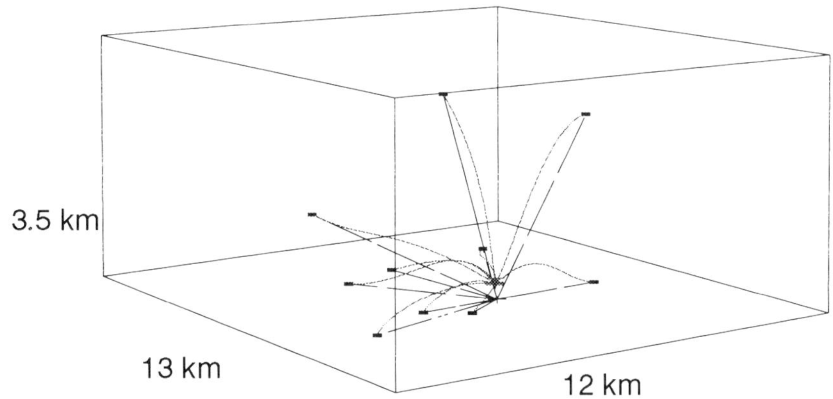


**Fig. 5.1** Event location by finite difference raytracing method (81 900, 19 830, 2770), is shown using curved lines, whereas location using straight ray approximation (81 945, 19 822, 2948), is shown in straight lines. The average hypocentral distance for the location based on raytracing is 4693 m, compared with the average distance for straight ray approximation of 4359 m.

$L_2$ norm (Euclidian norm) of the residual vector is minimized. The use of least squares procedures requires the assumption that the distribution of the residuals is of Gaussian nature. Jeffreys (1932) recognized that the time residuals do not follow the Gaussian distribution and occasional large residuals strongly affect the location of earthquakes. It is therefore preferable to use the $L_1$ norm that minimizes the absolute values of the residuals to solve the system of nonlinear equations since it is less sensitive to outliers. The introduction of automatic arrival time detectors, characteristic of mine seismic systems, increases the probability of picking up noise spikes, or confusing P and S arrivals on a particular channel, thus creating larger outliers.

The following misfit function based on equation (5.3) can be constructed in terms of the $L_1$ norm

$$LOC(\mathbf{h}) = \sum_{j=1}^{na} w_j \left| \bar{V}_j \left\{ t_j - \bar{t} - \left[ T_j(\mathbf{h}) - \bar{T}(\mathbf{h}) \right] \right\} \right| + \sum_{k=1}^{nd} w_k \left| Dis_k(\mathbf{h}) \right|, \tag{5.4}$$

where:

| | |
|---|---|
| $na$ | the number of P and S arrivals; |
| $nd$ | the number of P wavefront directions; |
| $w_j$ | the weighting factor of the $j$th arrival time; |
| $w_k$ | the weighting factor of the $k$th direction. |

The weighting factors are necessary, specifically when automatic processing is performed, to quantify the measure of reliability of the arrival times and directions. The weighting factor for P arrival times can be defined (Mendecki, 1993):

$$w_j(\text{P wave}) = SNRP \cdot DPP \tag{5.5}$$

where:

*SNRP* the normalized signal to noise ratio of the P arrival;
*DPP* the normalized degree of polarization of the P arrival.

*SNRP* and *DPP* can be calculated as follows:

$$SNRP = \frac{\bar{A}_1[T(A_{max})/4] - \bar{A}_2[T(A_{max})/4]}{\bar{A}_1[T(A_{max})/4] + \bar{A}_2[T(A_{max})/4]}$$

$$DPP = 1 - \frac{\lambda_2}{\lambda_1} \tag{5.6}$$

where:

$A_{max}$ the maximum amplitude in the waveform;
$T(A_{max})$ the period associated with $A_{max}$;
$\bar{A}_1[T(A_{max})/4]$ the median average amplitude after the P pick over $T(A_{max})/4$;
$\bar{A}_2[T(A_{max})/4]$ the median average amplitude before the P pick over $T(A_{max})/4$;
$\lambda_1, \lambda_2$ the largest and second largest eigenvalues, respectively, of the covariance matrix after the P pick over $T(A_{max})/4$.

Similarly, Mendecki (1993) defines the weighting factor for S arrivals as:

$$w_j(\text{S wave}) = SNRS \cdot DPS \cdot DA \tag{5.7}$$

where:

*SNRS* the normalized signal to noise ratio of the S arrival, which is the product of the normalized signal to noise ratios of the SH wave and SV wave respectively;
*DPS* the normalized degree of polarization of the S arrival;
*DA* the absolute value of the deflection angle cosine.

*SNRS*, *DPS* and *DA* can be calculated as follows:

$$
\begin{aligned}
SNRS &= SNRS\_SH \cdot SNRS\_SV \\
SNRS\_SH &= \frac{\bar{A}_{1SH}[T(A_{\max})/4] - \bar{A}_{2SH}[T(A_{\max})/4]}{\bar{A}_{1SH}[T(A_{\max})/4] + \bar{A}_{2SH}[T(A_{\max})/4]} \\
SNRS\_SV &= \frac{\bar{A}_{1SV}[T(A_{\max})/4] - \bar{A}_{2SV}[T(A_{\max})/4]}{\bar{A}_{2SV}[T(A_{\max})/4] + \bar{A}_{1SV}[T(A_{\max})/4]} \\
DPS &= \frac{(\lambda_1 - \lambda_2)^2 + (\lambda_1 - \lambda_3)^2 + (\lambda_2 - \lambda_3)^2}{2(\lambda_1 + \lambda_2 + \lambda_3)^2} \\
DA &= |\cos\theta|
\end{aligned}
\tag{5.8}
$$

where:

- $\theta$ the angle between the tangent of the P wave ray at the station and the normal to the plane of the S wave motion,
- $\bar{A}_{1SH}$ the median average amplitude after the S pick over $T(A_{\max})/4$ for the SH waveform,
- $\bar{A}_{2SH}$ the median average amplitude before the S pick over $T(A_{\max})/4$ for the SH waveform,
- $\bar{A}_{1SV}$ the median average amplitude after the S pick over $T(A_{\max})/4$ for the SV waveform,
- $\bar{A}_{2SV}$ the median average amplitude before the S pick over $T(A_{\max})/4$ for the SV waveform.

The weighting factor for direction is as follows:

$$
w_k = |\cos\{\psi(E_1[T(A_{\max})/4], E_1[T(A_{\max})/2])\}|\, w_j(\text{P wave}) \tag{5.9}
$$

where:

- $E_1[T(A_{\max})/4]$ the eigenvector of the covariance matrix after the P pick over $T(A_{\max})/4$
- $E_1[T(A_{\max})/2]$ the eigenvector of the covariance matrix after the P pick over $T(A_{\max})/2$
- $\psi$ the angle between $E_1[T(A_{\max})/4]$ and $E_1[T(A_{\max})/2]$.

For the location of seismic events in mines, the $L_1$ norm and centring were introduced by Prugger and Gendzwill (1988) and used at Saskatchewan potash mines, Canada, and are also implemented in the location procedures used in South African gold mines (Mendecki, 1993).

The solution of equation (5.4) can be found by the Nelder–Mead simplex minimization procedure (Press *et al.*, 1989). Equation (5.4) may have local minima and to find the global minimum one needs to start the iterative procedure from appropriately chosen starting points. Mendecki (1993) proposes:

- the coordinates of the nearest station;
- the average coordinates of specified number of nearest stations;
- the location based on the direction from a specified number of nearest stations.

Other nonlinear minimization techniques which can be used to solve equation (5.4) include genetic algorithms and simulated annealing (e.g. Sambridge and Gallagher, 1993).

The question now arises whether the $L_1$ norm always handles the problem of arbitrarily erratic data the best. Finding an optimal value of $p$ for the norm $L_p$, where $p$ is not necessarily equal to 1, might lead to better estimation of the unknowns.

The concept of the adaptive $L_p$ norm and its application to location of seismic events in mines was introduced by Kijko (1993; 1994), based on work by Rice and White (1964) and Ekblom and Henriksson (1969). They found that the value of $p$ depends critically on the distribution of residuals. Harter (1972) related the $p$ norm value to the kurtosis of the residual distribution. The kurtosis is an inverse measure of the property of sharpness of the frequency distribution peak and is defined as the ratio of the fourth order central moment to the square of the second order central moment. The unbiased estimates of the second and fourth order central sample moments are given by Gonin and Money (1987). Money *et al.* (1982) suggested

$$p = \frac{9}{K^2 + 1} \tag{5.10}$$

for $1 \leq p < \infty$, where $K$ is the kurtosis. A simple adaptive algorithm is used to find $p$ (Gonin and Money, 1985a,b; 1987):

- Solve equation (5.4) using some norm (e.g. $L_1$) and calculate $K$ from the resulting distribution of residuals. Using equation (5.10) to calculate $p$, equation (5.4) can now be solved using the $L_p$ norm. The process is continued until it converges to some specific value of $p$.

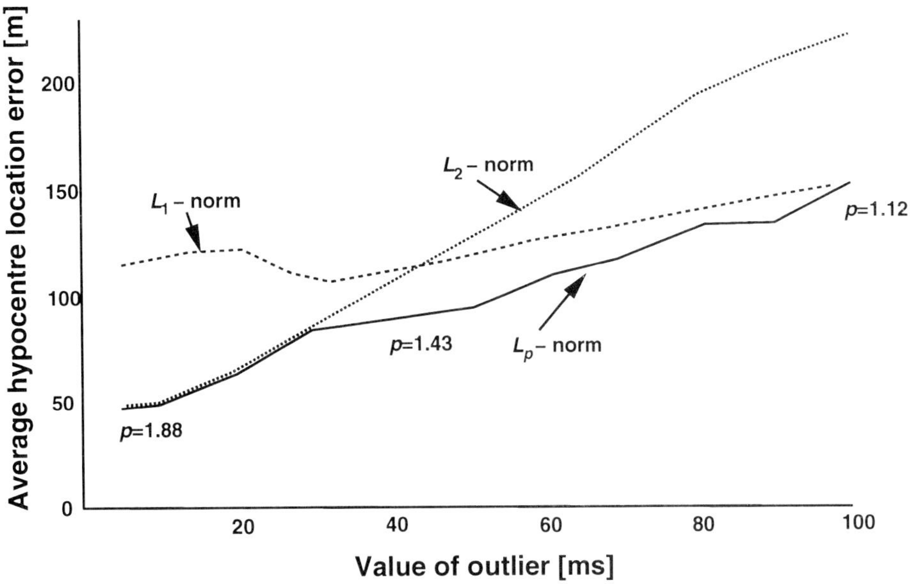


**Fig. 5.2** Comparison of seismic event location procedures based on the $L_1$, $L_2$ and adaptive $L_p$ norms in the presence of one outlying arrival time in the range of 5 to 100 ms. (Figure reconstructed from Kijko 1993.)

Figure 5.2 demonstrates the performance of the $L_1$, $L_2$ and adaptive $L_p$ norms in the presence of one outlying time residual. For each value of the outlying time residual present at the one station, one thousand seismic events were generated. For each event, P arrival times recorded by 11 other stations were perturbed by normally distributed random numbers with a standard deviation equal to 5 ms.

Figure 5.2 shows the average error of the hypocentre location computed by employing the $L_1$, $L_2$ and adaptive $L_p$ norm procedures for outlying residuals in the range from 5 to 100 ms. For large outliers values of $p$ close to 1 are predicted.

The results indicate that up to a certain value of an outlier the employment of the adaptive $L_p$ norm provides a substantial improvement over the $L_1$ norm; for larger outliers the $L_1$ and $L_p$ norm techniques are basically equivalent. When the outliers are small, the least squares technique and the adaptive $L_p$ norm are essentially the same, and both of them are better than $L_1$. As the value of the outlier increases, the least squares approach becomes progressively less efficient. Further numerical experiments show that the superiority of $L_p$ norm becomes less effective as the number of outlying observations increases, leading to the equivalence of $L_1$ and $L_p$ norm procedures.

## 5.2 Relative location and similarity of waveforms

In the case of the velocity model not being known adequately, or if velocities change significantly with time, one can attempt to improve the location by the arrival time difference (ATD) method, also known as 'master event' location or relative location (e.g. Spence, 1980). This procedure requires an accurately located master event (e.g. blast) in the proximity of the event to be relocated (called here the target event), that has reliable arrival times at stations used in the relocation procedure. One must stress the importance of the accurate position of the master event, since all relocations are performed relative to it. It is inherently assumed here that the velocities of the seismic waves from the master event to the stations and those from the target event are the same. Since this is not always the case it is important that the two events should be close to each other; less than 10% of average hypocentral distance would be a good rule of thumb. One can write the following arrival time equations for the $j$th station:

$$\begin{aligned} LOC_j(\mathbf{x}_\mathrm{M}) &= t_{\mathrm{M}j} - t_{\mathrm{M}0} - T_{\mathrm{M}j}(\mathbf{h}_\mathrm{M}) \\ LOC_j(\mathbf{x}) &= t_j - t_0 - T_j(\mathbf{h}) \end{aligned} \tag{5.11}$$

where:

- $\mathbf{x}$ $= (\mathbf{h}, t_0)^\mathrm{T} = (x_0, y_0, z_0, t_0)^\mathrm{T}$, the unknown origin time and hypocentre vector of the target event;
- $t_j$ the arrival time of the seismic wave from the target event at the $j$th station;
- $T_j$ the travel time of the seismic wave from the target event to the $j$th station;
- $\mathbf{x}_\mathrm{M}$ $= (\mathbf{h}_\mathrm{M}, t_{\mathrm{M}0})^\mathrm{T} = (x_{\mathrm{M}0}, y_{\mathrm{M}0}, z_{\mathrm{M}0}, t_{\mathrm{M}0})^\mathrm{T}$, the known origin time and hypocentre vector of the master event;
- $t_{\mathrm{M}j}$ the arrival time at the $j$th station from the master event;
- $T_{\mathrm{M}j}$ the travel time to the $j$th station from the master event.

Since $\mathbf{x}_M$ and $\mathbf{x}$ are very close to each other one can make a Taylor expansion of the arrival time of the target event around the value $\mathbf{x}_M$. Subtracting the equations in (5.11) from each other gives

$$LOC_j(\delta\mathbf{x}) = (t_j - t_{\mathrm{M}j}) - \delta t_0 - \frac{\partial T_j}{\partial x_0}\Big|_{\mathbf{x}_\mathrm{M}} \delta x_0 - \frac{\partial T_j}{\partial y_0}\Big|_{\mathbf{x}_\mathrm{M}} \delta y_0 - \frac{\partial T_j}{\partial z_0}\Big|_{\mathbf{x}_\mathrm{M}} \delta z_0 \tag{5.12}$$

where $\delta\mathbf{x} = (\delta x_0, \delta y_0, \delta z_0, \delta t_0)^T$ is the vector of unknown corrections from the master event to the target event.

Note that the theoretical travel times do not occur in equation (5.12), but only their partial derivatives. If the ray path from event to station is approximated by a straight line, and noting that the velocity model needed to calculate travel times can be written in terms of $d_{Mj}$, $t_{Mj}$ and $t_{M0}$, where $d_{Mj}$ is the straight line from the master event to the $j$th station, the following misfit function can be written:

$$LOC(\boldsymbol{\delta}\mathbf{x}) =$$

$$\sum_{j=1}^{na} \left| \delta t_{Mj} - \delta t_0 - \frac{t_{Mj} - t_{M0}}{d_{Mj}^2} \left[ (x_{M0} - x_j)\delta_{x_0} + (y_{M0} - y_j)\delta_{y_0} + (z_{M0} - z_j)\delta_{z_0} \right] \right| \tag{5.13}$$

The misfit function (5.13) can be solved using any of the techniques mentioned in the previous section. The origin time difference can also be eliminated from equation (5.13) by means of centring.

In the case of the availability of master events, one can suggest the following procedure which combines objective and relative location:

- Locate the event by means of an objective location method, e.g. minimizing equation (5.4) and finding the global minimum.
- Select relevant master events which are close enough to the target event.
- Relocate the target event with respect to each relevant master event using the ATD procedure.
- Use the smallest residual, as defined by equation (5.13), to select the best location.

Since the relative location technique implicitly uses a wave velocity derived from the master event travel times and distances to the stations, it should produce a location at least as good as one found with a location method employing an *a priori* fixed velocity. However, this cannot be guaranteed, since it is largely dependent on the distance between master and target event as well as the difference in average wave velocity along the respective ray paths. The relative location technique can be augmented by taking into account similarity of waveforms (Poupinet *et al.* 1984; Thorbjarnardottir and Pechmann, 1987; Milev *et al.* 1995). This is based on the fact that events that have similar source mechanisms and that lie within a fairly small volume should generate similar waveforms. If one has two events which have similar waveforms, they are called doublets. If, in addition, the one event is a master event the arrival times for the target event can be found relative to those of the master event, since the waveforms are similar. To quantify the degree of similarity between two waveforms, the following cross-correlation function can be used:

$$C_{mt}(k) = \frac{\sum_{i=1}^{nsa} wm_i\, wt_{i+k}}{\sqrt{\sum_{i=1}^{nsa} wm_i^2 \sum_{i=1}^{nsa} wt_i^2}} \tag{5.14}$$

where $wm$ is the waveform of the master event and $wt$ is the waveform of the target event, $nsa$ is the number of samples and $k$ is the lag in samples.

Consider the following two events (magnitude $m_M$=1.4, $m_M$=1.1 respectively) which occurred 19 days apart on a gold mine in the Freegold Mines, South Africa. These two events were located a distance of 25 m from each other using equation (5.4). The average hypocentral distance to the stations that located the first event was 1874 m and that of the second event 1849 m. The two events are situated within 2% of these average hypocentral distances from each other. For three of the first five stations to record the event (those which have the same sampling rate), the cross-correlation was calculated for the three components of the signal (Fig. 5.3). For each of these three stations, the waveforms (P and S together) correlate more than 70% and the events can be considered as doublets.

The fact that two seismic events are spatially close does not necessarily imply the same source mechanisms or the production of similar waveforms. To illustrate this, two events (magnitude $m_M$=1.5, $m_M$=1.2), from the same mine as previously, were located within 40 m of each other (less than 5% of the average hypocentral distance to the stations). The cross-correlation was again calculated for three of the first five stations, and in Fig. 5.4 one can see that the correlation is low. These two events cannot therefore qualify as doublets. Milev *et al.* (1995) quantified the degree of similarity between two

**Fig. 5.3** Cross-correlation of the three component waveforms for each of three stations which recorded the potential doublets. The maximum values indicated correspond to the value of $k$ that maximizes $C_{mt}(k)$.

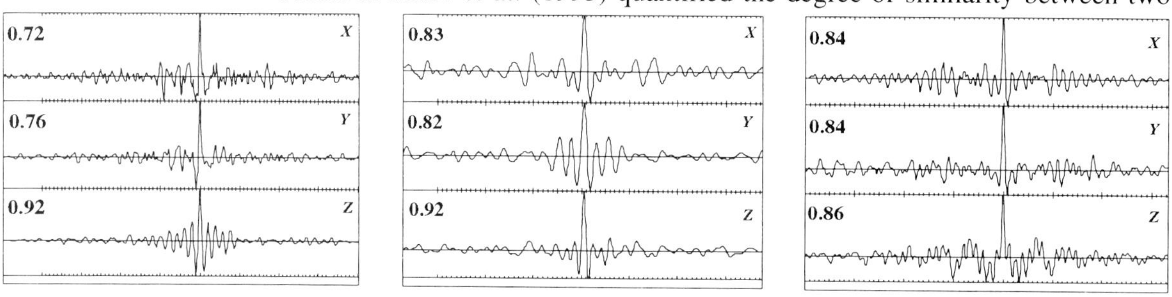

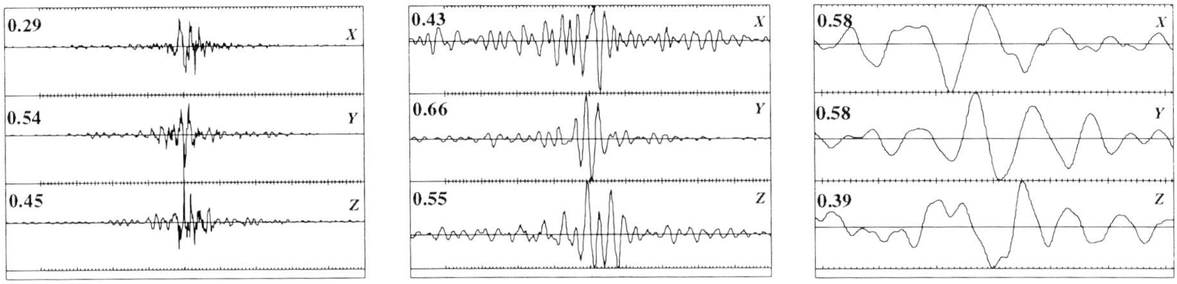


**Fig. 5.4** Cross-correlation of three component waveforms for each of three stations which recorded non-doublet events which are spatially close to each other. The maximum values indicated correspond to the value of $k$ that maximizes $C_{m\,i}(k)$. In this case, correlation is poor.

events by constructing a matching factor based on cross-correlation between waveforms for each phase at the stations that recorded both events and have the same sampling rate. If doublets have been identified, the relative location technique is applied and the time lags between the P (and S) phases of the two events are used as the arrival time differences.

## 5.3 Joint hypocentre and velocity determination for clusters of events

JHVD_C (Joint Hypocentre and Velocity Determination for Clusters) is an extension of the location problem, since not only are hypocentre parameters retrieved, but also the average velocities to the stations are treated as unknowns. In mining, it is often the case that seismic events are spatially clustered.

If an event generates *na* arrivals then four hypocentre unknowns have to be solved as well as *na* velocity unknowns, corresponding to the average velocities along the wave paths. This is an underdetermined system, since there are fewer arrival time equations (*na*) than unknowns (*na*+4), so it is necessary to use more than one event.

It is assumed that the average velocity of a wave from one event in the cluster to the *j*th station is the same as for all the other events in the cluster, thus enabling the system to have more equations than unknowns. It is necessary to bear in mind that there are now some additional restrictions on the problem. If the cluster of events is spatially too large there exists a physical contradiction: if two events in the cluster are far apart the assumption that the wave travels through the same rock is no longer valid; conversely, if the cluster is too small increased numerical instability may be found, since the arrival time equations are very similar. To indicate the size of the system of equations to be solved, consider the following example: if there are 50 events in a cluster and each event was recorded by the same ten stations, the problem consists of solving 500 equations and 210 (50×4+10) unknowns, if only P arrivals are used. One can minimize the following:

$$LOC(\mathbf{h}_1,\cdots,\mathbf{h}_{ne},\mathbf{V}) = \sum_{i=1}^{ne}\sum_{j=1}^{ns(i)} \left| t_{ij} - \bar{t}_i - [T_{ij}(\mathbf{h}_i, V_j) - \bar{T}_i(\mathbf{h}_i, V_j)] \right| \tag{5.15}$$

where:

| | |
|---|---|
| $\mathbf{h}_i$ | the unknown hypocentre of the $i$th event; |
| $\mathbf{V}$ | the velocity model with $V_j$ being the unknown average P or S wave velocity associated with the $j$th station; |
| $t_{ij}$ | the observed arrival time of the P or S wave from the $i$th event to the $j$th station; |
| $\bar{t}_i$ | the average arrival time for the $i$th event; |
| $T_{ij}$ | the unknown travel time from the $i$th event to the $j$th station; |
| $\bar{T}_i$ | the average travel time associated with the $i$th event; |
| $ne$ | the number of events; |
| $ns(i)$ | the number of arrivals of the $i$th event. |

Due to the large number of unknowns and local minima, the misfit function is difficult to solve using a nonlinear minimization technique. Therefore, the system of equations is usually linearized with a Taylor expansion. The size of the system can be reduced by incorporating the Spencer–Gubbins matrix partitioning procedure (Spencer and Gubbins, 1980) or the Pavlis–Booker approach (Pavlis and Booker, 1980), which separates the problem into discrete and continuous parts of the same inverse problem. These procedures result in the largest matrix to be inverted being dependent on the number of velocity unknowns, rather than the number of events. The JHVD_C procedure for mine data was applied by Mendecki (1981; 1987) and Drzezla and Mendecki (1982a,b). Jech (1989) used the time difference between P and S wave arrivals to relocate seismic events and to improve the velocity model in the Ostrava-Karvina mining district in Czechoslovakia. The case where the seismicity is not spatially clustered is considered by Maxwell and Young in Chapter 6 of this book.

As an example of the application of JHVD_C, data from the Elandsrand Gold Mine in South Africa were used. The objective was to obtain the P wave velocity model for different clusters of events, in order to improve location accuracy (Essrich, 1996). Nine clusters of events of approximately 300×200×400 m were chosen and the event locations and velocities were computed. To evaluate the results, a simplified geological model of the mine was assumed and P velocities were assigned. The model is shown in Table 5.1. Using this model, sections were drawn which intersected clusters and stations and average velocities were calculated along straight ray paths. The P wave velocities derived from the geological model were compared to those inverted and, in all cases, the difference was less than 5%.

**Table 5.1** Elandsrand Gold Mine simplified geological model

| Layer | Position | P velocity m/s |
|---|---|---|
| Pretoria group (shale & quartzite) | Surface to dolomite (0 – 750) | 5000 |
| Dolomite | dip 12° Sth 750 – 2000 | 6850 |
| Ventersdorp lavas | dip 25° Sth 2000 – 2400 | 5950 |
| Quartzite | dip 25° Sth 2400 – 3000 | 6050 |
| Gritty quartzite | dip 15° West 2400 – 2800 | 5750 |
| Booysen shale | dip 25° Sth 3000 – 3500 | 5500 |

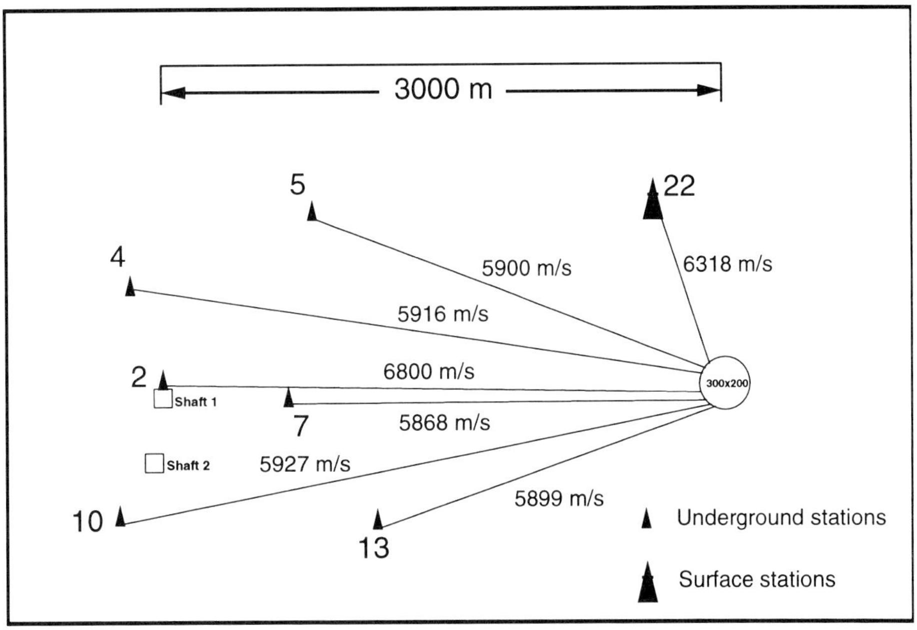


**Fig. 5.5** Inverted P wave velocities for a cluster of seismic events on Elandsrand Gold Mine. All stations seem to have an associated P wave velocity of about 5900 m/s, except for stations 22 and 2. Station 22 is a surface site, which explains the higher velocity. A thick dolomite layer (1200 m) lies between the surface and the quartzites and lavas being mined. Station 2 is situated 600 m above reef in this dolomite layer, explaining the higher P wave velocity in this case.

In Fig. 5.5, the results of the velocity inversion for one of the clusters are shown. Except for stations 22 and 2, all P wave velocities are of the order of 5900 m/s. Station 22 is a surface site and station 2 is situated in dolomite, 600 m above the reef, so in both cases the wavefront predominantly moves through dolomite. The other stations are all on reef, in quartzite, which explains their lower P wave velocities. JHVD_C is specifically useful for identifing sites which need significant velocity alteration in order to improve location accuracy. Results, corroborated by the geological model, suggest an increase in P wave velocity of about 500 m/s for station 2; current P wave velocity used is about 6300 m/s.

The following example uses data from a Zambian copper mine. In Fig. 5.6 the larger symbols indicate events located using fixed velocities; the filled symbols indicate locations obtained by JHVD_C. The relocated events are nearer to the structure of interest where the damage occurred (the diagonal line that represents the ore body).

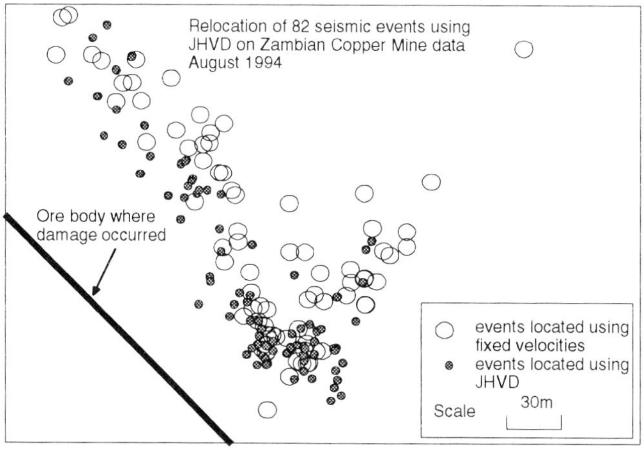


**Fig. 5.6** Events located by JHVD_C are on average 50 m closer to the ore body being mined than the same events located using fixed velocities.

## 5.4 Optimal spatial distribution of seismic stations

The location error depends on the following factors:

- The errors in arrival time determination: this source of errors can never be fully eliminated, since not only are both automatic and human pickers subject to errors but also the effective sampling rate of the instruments puts a bound on the accuracy of arrival times.
- Inadequate knowledge of the velocity model.

- Inaccuracy in the station coordinates.
- The method of solution: if the system is linearized a different location error will be found than if a nonlinear minimization technique is used. The applied norm used will also play a role in the location error.
- The spatial distribution of stations with respect to the event to be located.

A number of papers are devoted to the subject of the best or optimal distribution of seismic stations in order to minimize the location errors; see Gibowicz and Kijko (1994) for a comprehensive review and references. The following are a few examples particularly dedicated to engineering applications: Mendecki and van Aswegen (1986), Drzezla (1991), Kijko and Sciocatti (1993; 1995), Jones and Rayne (1994); and Ge and Hardy (1995). It is advisable to use one of the quantitative methods to design a new, or to reorganize an existing, network of seismic stations, rather than to rely on one's experience.

Two methods frequently used to design the configuration of mine seismic networks are described below.

### 5.4.1 Optimality with respect to location error – a statistical approach

The choice of one or another station configuration should depend upon the expected location error that would be found if the given configuration was used. Here the criterion quantifying the optimality of a given configuration is defined so that it depends upon a covariance matrix $\mathbf{C}_\mathbf{x}$ of the unknowns $\mathbf{x}=(t_0,x_0,y_0,z_0)^\mathrm{T}$ (Kijko, 1977; Kijko and Sciocatti, 1993; 1995).

The covariance matrix can be graphically interpreted in terms of a confidence ellipsoid, where the eigenvalues of $\mathbf{C}_\mathbf{x}$ form the lengths of the principal axes of the ellipsoid. Finding a configuration for which the volume of this ellipsoid is a minimum is referred to as *D*-optimal planning (John and Draper, 1975). The volume is proportional to the product of the eigenvalues, or equivalently, to the determinant of $\mathbf{C}_\mathbf{x}$. Assuming that $\mathbf{x}$ is estimated by means of any norm between 1 and $+\infty$, the covariance matrix of $\mathbf{x}$ is equal to $k(\mathbf{A}^\mathrm{T}\mathbf{A})^{-1}$, where $\mathbf{A}$ is the partial derivative matrix of computed arrival times with respect to $\mathbf{x}$, and $k$ is a constant (Gibowicz and Kijko, 1994). Since $\det[\mathbf{C}_\mathbf{x}]=(\det[\mathbf{C}_\mathbf{x}^{-1}])^{-1}$ minimizing $\det[\mathbf{C}_\mathbf{x}]$ will maximize $\det[\mathbf{A}^\mathrm{T}\mathbf{A}]$, satisfying the *D*-optimality criterion.

If one has two possible configurations and one event to locate, the optimal configuration will be that which results in the smallest $\det[\mathbf{C}_\mathbf{x}]$. However, one is not interested in only one event, but in all events which the network has to locate. Therefore, the optimal configuration will be that which minimizes the following:

$$\sum_{i=1}^{ne} p_h(\mathbf{h}_i)\, \lambda_{x_0}(\mathbf{h}_i)\, \lambda_{y_0}(\mathbf{h}_i)\, \lambda_{z_0}(\mathbf{h}_i)\, \lambda_{t_0}(\mathbf{h}_i) \tag{5.16}$$

where:

- $ne$ number of events of interest; these can be positioned within certain seismically active areas which are to be specifically monitored;
- $p_h(\mathbf{h}_i)$ relative importance of the event at $\mathbf{h}_i$; it would be a function of a probability of an event occurring in that vicinity, but it may also rely on factors such as the life-span of the specific mining area;
- $\lambda_{\mathbf{x}}(\mathbf{h}_i)$ eigenvalues of $\mathbf{C}_{\mathbf{x}}$.

Since for all events the partial derivative matrix **A** has the same number of rows, it is implicitly assumed that all $n$ stations in the network are triggered by each event. This problem can be addressed integrating equation (5.16) over the seismic energy range of interest

$$\sum_{i=1}^{ne} \sum_{E=E_{min}}^{E_{max}} p_h(\mathbf{h}_i)\, p_E(E)\, \lambda_{x_0}(\mathbf{h}_i)\, \lambda_{y_0}(\mathbf{h}_i)\, \lambda_{z_0}(\mathbf{h}_i)\, \lambda_{t_0}(\mathbf{h}_i) \tag{5.17}$$

where $<E_{min}, E_{max}>$ is the seismic energy range of interest, and $p_E(E)$ is the probability density of energy. It is possible to relate the energy of the seismic event to some detectability distance $r$, and such a relation is approximated by $E = r^q$ where $q$ is close to 2.

Now $\mathbf{C}_{\mathbf{x}}$ is a function of the detectability distance $r$, and thus of the seismic event energy, since the number of rows in **A** corresponds only to those stations that are within a distance $r$ from the event. Furthermore, following a derivation of the seismic energy distribution by Rikitake (1976), the probability density of detectability distance $r$ can be written as

$$p_r(r) = \frac{\alpha\, r^{-(1+\alpha)}}{\left(r_{min}^{-\alpha} - r_{max}^{-\alpha}\right)} \tag{5.18}$$

where $\alpha = -4b/3$ and $b$ can be found by plotting magnitude $m_M$ against the log of the cumulative number of events larger than $m_M$ and obtaining the slope of the line.

The criterion (5.17) for the optimal distribution of seismic stations now takes the final form

$$\sum_{i=1}^{ne} \sum_{r=r_{min}}^{r_{max}} p_h(\mathbf{h}_i)\, p_r(r)\, \lambda_{x_0}(r)\, \lambda_{y_0}(r)\, \lambda_{z_0}(r)\, \lambda_{t_0}(r) \tag{5.19}$$

In relation (5.19) $r_{min}$ must be at least the distance from a seismic event to the fifth nearest station, since it is assumed that five stations are needed to properly locate the event if only P wave arrivals are used. $r_{max}$ is a detection distance corresponding to the maximum energy release of interest $E_{max}$.

In some cases the requirement of the best estimation applies to some subsets of unknown parameters only, as with epicentre coordinates. If we only have a certain subset of parameters of interest, only the associated eigenvalues are used in the forming of the product.

The quality of seismic network configurations can also be evaluated from the maps of expected standard errors of location determination. This is necessary since thus far the procedure generates configurations that are optimal on average for a set of events or seismically active areas. Such a configuration is not necessarily optimal for a given area or event, so evaluation for given scenarios is needed.

Kijko (1977) incorporated expected error in velocity model and arrival time determination by defining $\mathbf{C_x} = (\mathbf{A}^T\mathbf{W}\mathbf{A})^{-1}$ where $\mathbf{W}$ is the diagonal matrix of $1/w_i$, where

$$w_i = \left(\frac{\partial T_i}{\partial V_i}\right)^2 \sigma^2_{V_i} + \sigma^2_{t_i}$$

in which $\sigma^2_{t_i}$ is the variance of the error in the $i$th arrival time determination and $\sigma^2_{V_i}$ is the variance of the error in the $i$th average velocity.

The diagonal elements of the matrix $\mathbf{C_x}$ are the variances of the errors associated with the unknowns $t_0, x_0, y_0$ and $z_0$ they are used for contour generation in the maps.

### 5.4.2 Optimality with respect to location error – a direct approach

The sensitivity of the solution of a system of nonlinear or linear equations to small changes in the data is measured by the condition number of the system. The optimal configuration of seismic stations is defined as that which has the smallest condition number of the system of nonlinear equations (Mendecki and van Aswegen, 1986). This procedure is also known as *C*-optimal planning. If the condition number is small, small changes in the errors of arrival time determination and velocity model should not severely influence the location. If they did, the location would be unstable and probably unreliable and the condition number would be large. The system of nonlinear equations to solve is

$$F(\mathbf{x},\mathbf{d}) = t_j - t_0 - T_j(\mathbf{h}, V_j, \mathbf{x}_{sj}) = 0 \qquad j = 1, \cdots, \text{ number of arrivals} \tag{5.20}$$

where:

| | |
|---|---|
| $\mathbf{h}$ | the coordinates of the hypocentre $(x_0, y_0, z_0, t_0)^T$; |
| $t_0$ | the origin time; |
| $t_j$ | the exact $j$th arrival time; |
| $V_j$ | the exact velocity associated with the $j$th arrival; |
| $\mathbf{x}_{sj}$ | the exact coordinates $(x_{sxj}, x_{syj}, x_{szj})^T$ of the station which recorded the $j$th arrival. |

In equation (5.20), $F$ is the system of nonlinear equations and $\mathbf{d} = [\mathbf{t}, \mathbf{V}, \mathbf{x}_s]^T$ is the data vector consisting of arrival time data $\mathbf{t}$, velocity model $\mathbf{V}$ and station coordinates $\mathbf{x}_s$ and $\mathbf{x} = (\mathbf{h}, t_0) = \mathbf{x}_0$, namely the true hypocentre and origin time.

The system of equations $F(\mathbf{x},\mathbf{d})$ can only be equal to zero if $F$, $\mathbf{x}_0$ and $\mathbf{d}$ are known exactly. In planning a configuration, one assumes $F$ and $\mathbf{x}_0$ and only $\mathbf{d}$ is unknown. Since the data vector $\mathbf{d}$ is contaminated with noise, we are trying to solve $F(\mathbf{x},\tilde{\mathbf{d}})=0$ where $\tilde{\mathbf{d}}$ is the data vector with the associated errors, and this is incorrect. However, if $F$ has a Lipschitz first derivative in the neighbourhood of $(\mathbf{x}_0, \mathbf{d})$, then $F(\mathbf{x},\tilde{\mathbf{d}})=0$ has a unique solution $\tilde{\mathbf{x}}$, where $\mathbf{x}_0$ is the true solution. Forming a Taylor expansion of $F(\tilde{\mathbf{x}},\tilde{\mathbf{d}})$ at $(\mathbf{x}_0,\mathbf{d})$, we have

$$F(\tilde{\mathbf{x}},\tilde{\mathbf{d}}) - F(\mathbf{x}_0,\mathbf{d}) = F_d'(\mathbf{x}_0,\mathbf{d})\,(\tilde{\mathbf{d}} - \mathbf{d}) + F_x'(\mathbf{x}_0,\mathbf{d})\,(\tilde{\mathbf{x}} - \mathbf{x}_0) + C \tag{5.21}$$

where $F_d'$ and $F_x'$ are matrices of partial derivatives and $C$ represents the higher order terms. At point $(\tilde{\mathbf{x}},\tilde{\mathbf{d}})$ the left hand side of equation (5.21) is zero and after multiplication with the pseudo-inverse of $F_x'(\mathbf{x}_0,\mathbf{d})$ and taking the norm, it readily follows (Woźniakowski, 1977) that

$$\frac{\|\tilde{\mathbf{x}} - \mathbf{x}_0\|}{\|\mathbf{x}_0\|} \leq cond(F,\mathbf{d})\,\|\tilde{\mathbf{d}} - \mathbf{d}\| + R \tag{5.22}$$

where $\|.\|$ is one of Holder's norms, $R$ is a term that is usually much smaller than $cond(F,\mathbf{d})\,\|\tilde{\mathbf{d}} - \mathbf{d}\|$ and

$$cond(F,\mathbf{d}) = \frac{\|[F_x^{/}(\mathbf{x}_0,\mathbf{d})]^{-1}\,[F_d^{/}(\mathbf{x}_0,\mathbf{d})]\|}{\|\mathbf{x}_0\|} \tag{5.23}$$

where $[\,]^{-1}$ refers to the pseudo-inverse of the matrix.

Equation (5.22) can be used to generate contours of expected location errors and states that the solution error, expressed as a fraction of the true solution, is constrained from above by $cond(F,\mathbf{d})\,\|\tilde{\mathbf{d}} - \mathbf{d}\|$.

The partial derivatives of $F(\mathbf{x}, \mathbf{d})$ with respect to $\mathbf{x}$ and the data vector $\mathbf{d}$ can be expressed as:

$$F_{\mathbf{x}}^{/}(\mathbf{x},\mathbf{d}) = \begin{bmatrix} \frac{x_0 - x_{sx1}}{V_1 d_1} & \frac{y_0 - x_{sy1}}{V_1 d_1} & \frac{z_0 - x_{sz1}}{V_1 d_1} & 1 \\ \vdots & \vdots & \vdots & \vdots \\ \frac{x_0 - x_{sxn}}{V_n d_n} & \frac{y_0 - x_{syn}}{V_n d_n} & \frac{z_0 - x_{szn}}{V_n d_n} & 1 \end{bmatrix} \tag{5.24}$$

$$F_{\mathbf{t}}^{/}(\mathbf{x},\mathbf{d}) = \begin{bmatrix} -1 & 0 & \cdot & 0 \\ 0 & \cdot & \cdot & \cdot \\ \cdot & \cdot & \cdot & 0 \\ 0 & \cdot & 0 & -1 \end{bmatrix} \qquad F_V^{/}(\mathbf{x},\mathbf{d}) = \begin{bmatrix} -\frac{d_1}{V_1^2} & 0 & \cdot & 0 \\ 0 & \cdot & \cdot & \cdot \\ \cdot & \cdot & \cdot & 0 \\ 0 & \cdot & 0 & -\frac{d_n}{V_n^2} \end{bmatrix} \tag{5.25}$$

$$F_{\mathbf{x}_s}^{/}(\mathbf{x},\mathbf{d}) =$$

$$\begin{bmatrix} \frac{x_{sx1} - x_0}{V_1 d_1} & \frac{x_{sy1} - y_0}{V_1 d_1} & \frac{x_{sz1} - z_0}{V_1 d_1} & 0 & \cdot & \cdot & \cdot & \cdot & 0 \\ \cdot & \cdot & \cdot & \cdot\;\cdot\;\cdot & \cdot & \cdot & \cdot \\ 0 & \cdot & \cdot & \cdot\;\cdot & 0 & \frac{x_{sxn} - x_0}{V_n d_n} & \frac{x_{syn} - y_0}{V_n d_n} & \frac{x_{szn} - z_0}{V_n d_n} \end{bmatrix} \tag{5.26}$$

$F_{\mathbf{x}}^{/}(\mathbf{x},\mathbf{d})$ can be inverted using singular value decomposition. The matrix norm used depends on the problem, but it was found that the spectral norm is adequate.

In practice, one needs an approximation $\tilde{\mathbf{x}}$ such that

$$\frac{\|\tilde{\mathbf{x}} - \mathbf{x}_0\|}{\|\tilde{\mathbf{x}}\|} < k \tag{5.27}$$

for $k$ such that, say, $10^{-5} < k < 10^{-2}$. This is only possible if $cond(F,\mathbf{d})\,\|\tilde{\mathbf{d}} - \mathbf{d}\|$ is of the order of $k$. It is impossible to compute a good approximation of $\mathbf{x}_0$ if $cond(F,\mathbf{d})$ is very large regardless of the method used. If the problem is extremely ill conditioned, $cond(F,\mathbf{d}) > 1/\|\tilde{\mathbf{d}} - \mathbf{d}\|$, then even a reasonable approximation cannot be found.

Minimization of the condition number can be taken as a deterministic criterion to design any experiment which leads to solving a system of nonlinear or linear equations.

### 5.4.3 Example of planning the spatial configuration of seismic stations

A shaft pillar is being monitored on one of the Freegold gold mines in South Africa. The area of interest has a plan view of 650 m × 650 m. Figure 5.7a indicates contours of expected errors for events with magnitude $m_M$=0.0 that occur at a depth of 1921 m in the area on and around the shaft pillar. These events have detectability distance such that all seven stations are triggered and used in the location procedure. The reef blocks still to be mined out in the shaft pillar are indicated in Fig. 5.7. In this example an arrival time determination error of 5 ms for all arrival times is assumed and a P velocity error of 150 m/s. The configuration of seismic stations that has, as a result, the contours in Fig. 5.7a and Fig. 5.7c is very close to the actual configuration currently used in this mine. As can be seen in Fig. 5.7a and Fig. 5.7c, the expected errors are acceptable, especially in the area of interest, namely the pillar area.

If one station is moved a distance of 65 m (64 m in the horizontal plane and 11 m in depth), then the situation depicted in Fig. 5.7b and Fig. 5.7d occurs. Clearly there is a strip right through the shaft pillar that has unacceptably high expected error in location. The reason for this bad behaviour is that now the seven stations in the network are close to being on the same plane and the location procedure struggles. The solution is extremely sensitive to errors in the arrival time and velocity model, due to the network geometry, and this is independent of the specific algorithm to be used. The point that this example is trying to make is that a small change in the spatial configuration may significantly change the location accuracy. Figure 5.7a and Fig. 5.7b were generated using the statistical approach. To verify these results, the direct approach was used to generate contours from the same configuration (see Fig. 5.7c and Fig. 5.7d). Qualitatively the two approaches have the same result, although quantitatively the statistical approach is about twice as conservative.

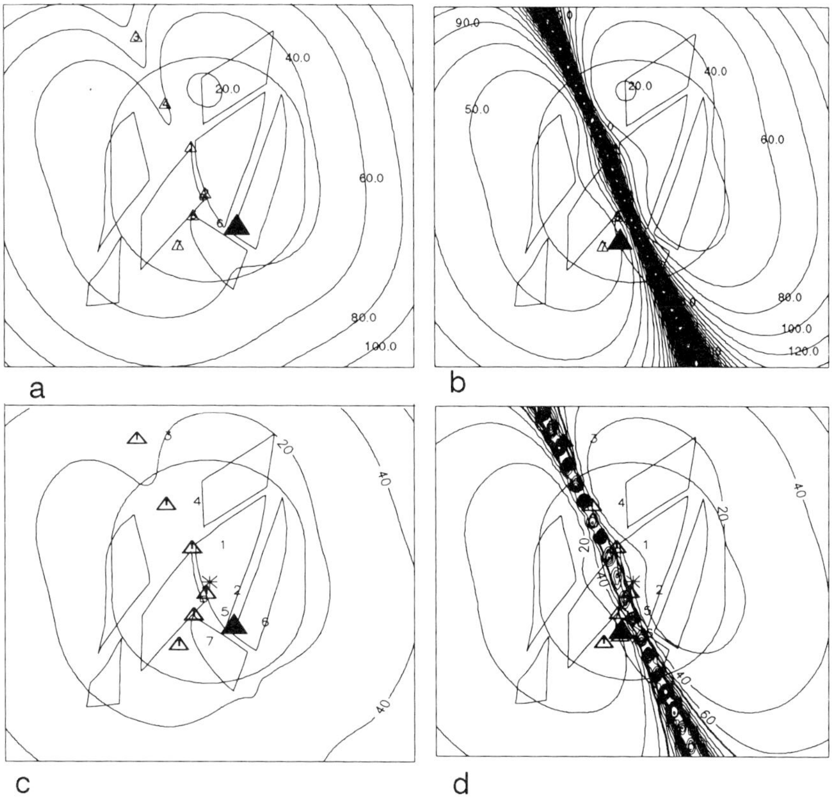


**Fig. 5.7** In **a.** the expected $x,y,z$ location error using a seven station seismic network at a shaft pillar in a South African gold mine. The area has dimension 650 m × 650 m.
If one station (filled triangle) is moved only 65 m the situation in **b** results. The quality of location goes down dramatically.
**a** and **b** are generated using the statistical approach.
**c** and **d** use the direct approach.

# 6 Seismic Velocity Inversion from Microseismic Data

## 6.1 Seismic tomography

Seismic tomographic imaging is the generation of maps of the spatial variation in a seismic parameter through a rockmass. Several seismic parameters may be imaged, such as attenuation or velocity, although P wave velocity images are most common because of the relative simplicity and accuracy of the measurement. P wave velocity images may be constructed in either one, two or three dimensions, using either actively detonated seismic sources (i.e. blasts) or passively monitored induced seismicity. Imaging has been used in a number of mining applications:

- lithological mapping;
- fracture mapping;
- blast damage mapping;
- stress mapping.

Velocity imaging has also been shown to be particularly useful in rockburst investigations (see Maxwell and Young, 1996, for details). Clearly, detailed velocity structures will improve the accuracy of seismic parameter calculation, in particular hypocentral location, as shown in Fig. 6.1. In rockburst investigations, comparisons between velocity images and seismicity have shown that the largest mining induced tremors are associated with regions of high velocity, while low velocity regions are generally aseismic. Interestingly, similar observations have been made by several researchers examining natural seismicity. The velocity–seismicity relationship is attributed to velocity varying as stress preferentially closed pre-existing cracks in the rockmass. Laboratory studies of intact and stress-relief cracked samples show a large velocity change at low confining stress, and almost constant velocity at stress levels above a few megapascals. However, for a real fractured rockmass significant velocity changes can occur at higher stress levels (Zimmerman and King, 1985; Xu *et al.*, 1990), depending on the relative roughness of the contacting fracture surfaces (Hopkins *et al.*, 1990). Controlled studies around a simpleunderground opening in homogeneous, unfractured granite have been able to show a strong correlation between velocity and stress. The relationship between velocity and

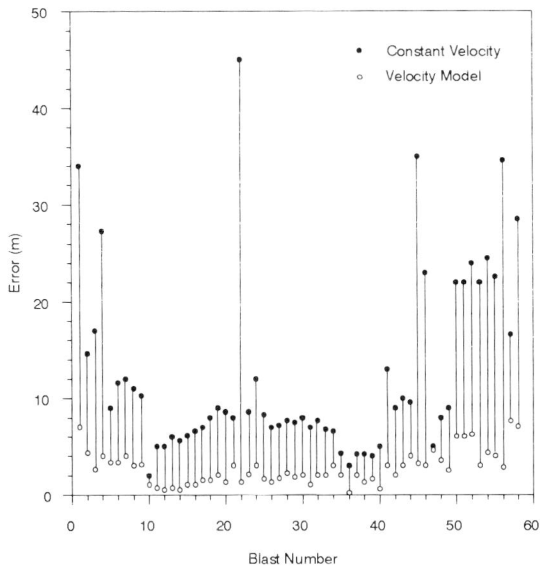


**Fig. 6.1** Measured hypocentral errors of controlled explosions, computed using a homogenous velocity model (closed circles) and a 3D velocity model (open circles).

seismicity can therefore be interpreted by frictional stability, whereby the largest tremors occur in regions of high velocity, and therefore high normal stress, where the rockmass is able to store significant strain energy. These studies have shown how a holistic interpretation of velocity images and seismicity can therefore be used to understand the seismic response of the rockmass to mining induced stress changes. A velocity image is therefore a powerful tool with which to monitor a seismically active rockmass, from which seismic potential can be assessed.

In a heterogeneous mine, velocity variations can also result from lithological changes. However, since the lithology is generally well understood in an active mine, the lithological variations can be incorporated into the *a priori* velocity structure, and relative variations from the known lithology based velocity interpreted in terms of stress state. Similarly, the ray path diffracting around underground openings larger than the wavelength can also be incorporated. Sequential images of temporal velocity variations can also be used to isolate stress changes. For example, repeated explosions at the same location can be used to compute images of the velocity changes. Sequential images can also be computed from induced seismicity data using a doublet (repeated events from the same location) technique. Arrival time differences of a set of doublets covering two time periods can be inverted for relative locations and velocity changes. Temporal investigations of velocity structure in mines have shown that the induced seismicity may be related to temporal velocity reductions, where decreased normal stress may unclamp critically stressed fractures resulting in induced tremors (Maxwell and Young, 1993a).

## 6.2 Arrival time inversion

Velocity inversion of seismic event data is an extension of standard controlled source inversion. A brief overview of the inversion is given here (further details can be found in Nolet, 1987, and Maxwell and Young, 1993b). From an estimated hypocentral location of an earthquake, $i$, the travel time, $T_{ij}$, to a sensor, $j$, can be estimated over a ray path (see Chapter 4 of this book) as:

$$T_{ij} = \int u \, ds \tag{6.1}$$

where $ds$ is an incremental ray path segment and $u$ is the slowness (reciprocal of velocity). Arrival time residuals, $r_{ij}$, can then be computed from the estimated earthquake origin time $\tau'_i$, and observed arrival times $t_{ij}$:

$$r_{ij} = t_{ij} - T_{ij} \, \tau_i' \tag{6.2}$$

By linearizing about the initial ray path, the residuals can be related to perturbations in the initial slowness and hypocentre parameters by the simultaneous equation:

$$r = AU + Bx \tag{6.3}$$

where $A$ and $B$ are matrices of the partial derivatives of arrival time with respect to slowness and hypocentre parameters. The slowness model can be constructed in a number of ways, the most common of which is with rectangular cells with either constant velocities in each or an interpolation scheme, typically trilinear, between the centre of the neighbouring cells.

There are several ways of solving this simultaneous linear equation. For example, Aki and Lee (1976) invert a single large partitioned matrix, and Lees and Crosson (1989) ignore the coupling and solve each component separately. Spencer and Gubbins (1980) proposed an efficient way of solving this large, sparse matrix. An analogous, more elegant method was proposed by Pavlis and Booker (1980), using a singular value decomposition of the matrix $B$ for a single earthquake:

$$B_i = U \Lambda V^T \tag{6.4}$$

such that the matrix $U$ can be partitioned as

$$U = [U_1 \, U_0] \tag{6.5}$$

where the null space matrix $U_0$ is orthogonal to $U$ and $B$, such that it will act as an 'annihilation' operator,

$$U_0 B_i = 0 \tag{6.6}$$

The simultaneous equation then reduces to

$$U_0 r_i = U_0 A_i u \tag{6.7}$$

or,

$$r' = A' u \tag{6.8}$$

which can be concatenated with other annulled simultaneous equations from different earthquakes.

The annulled simultaneous equation can be solved by a number of methods. For example, row action techniques such as simultaneous iterative reconstruction technique, SIRT, and algebraic reconstruction technique, ART, are commonly used in seismic tomography, but can be unstable in some cases. The conjugate gradient technique proposed by Paige and Saunders (1982) has also been widely used in recent years. However, neither technique provides a simple inversion quality and confidence estimates, although some studies have attempted to address this shortcoming. Damped least squares inversions of the form,

$$(A^T A + \lambda^2 I) = A^T y \tag{6.9}$$

are commonly used. The damped system of equations can then be solved using standard matrix inversion algorithms.

The actual damping factor used in the inversion can be chosen in a number of ways. For a 'stochastic inverse', the damping factor, $\lambda^2$, can be selected as the ratio of the expected variances of the data and model. In practice it is difficult to predict the variances *a priori*, although an 'optimal' damping factor can be chosen from a graph of the inverted slowness perturbations versus the residual. This graphical technique allows a tradeoff between goodness of fit to the data and smoothness of the inverted slowness model. The Levenberg–Marquardt method of nonlinear damped least squares inversion can be used with progressively smaller damping factors at successive iterations, and recomputed ray paths and hypocentre parameters after each iteration. It should be noted that nonlinearity from the hypocentre component of the inversion increases the total nonlinearity of the system compared to standard travel time inversion.

After the inversion, it is necessary to assess the quality of the results. Confidence in the estimated slowness model can be quantified in two ways. The first is by the resolution matrix,

$$R = (A^TA + \lambda^2 I)^{-1} A^TA \tag{6.10}$$

Rows of the resolution matrix are windows or filters relating the 'true' model to the estimated model. Proportionally large off-diagonal terms quantify the resolution 'smearing' between the corresponding model parameters. The second indicator of the image confidence is the covariance estimate of model errors. A standard covariance estimate is

$$C = \sigma^2 (A^TA + \lambda^2 I)^{-1} A^TA \tag{6.11}$$

where $\sigma^2$ is the estimated data variance. Alternatively, a total error estimate (Jackson, 1979) can be computed as

$$C_T = \sigma^2 (A^TA + \lambda^2 I)^{-1} \tag{6.12}$$

which has the property of including resolving errors. Resolution and errors can also be estimated by a comprehensive synthetic study of forward modelling arrival times from an assumed velocity model and inverting the data. The inversion results can then be directly compared with the initial synthetic model to check the ability of the data to resolve certain features, and assess the uniqueness of the inversion results. The importance of this step in any inversion study cannot be understated. The image resolution will also be related to the image grid geometry and accuracy of hypocentre parameters. Furthermore, the seismic wave properties will obviously sample non-ubiquitous features much smaller than the wavelength. The physical sampling will also be governed by the Fresnel zone of the waves.

Finally, seismic anisotropy effects, either stress induced or from petrographic or discontinuity fabrics, should be considered. If anisotropy is suspected, directional variations in velocity should be investigated. However, since apparent directional variations are possible from isotropic, heterogenous velocity structures depending on the ray path geometry, results should be validated with the geomechanical properties and shear wave splitting, if possible. Anisotropy can then be inverted by assuming some symmetry class and inverting for slowness in the unique directions of the symmetry system as well as the orientation (see for example Červený and Firbas, 1984). In practice, inverting for the anisotropy is difficult and resolution is limited by the azimuthal ray path coverage. A pragmatic strategy is to assume a uniform orientation and degree of anisotropy and invert for slowness in one direction, which is analogous to correcting for anisotropy so that the data 'appear' isotropic. Significant anisotropy, which is not accounted for with an isotropic inversion, may result in artefacts in the final image.

## 6.3 Application

Examples of the application of arrival time inversion to mining-induced seismicity investigations appear to be limited in the literature to Jech (1989), Mendecki (1987) and Maxwell and Young (1993a,b). Jech (1989) inverted a 3D structure of the relative velocity between P and S waves in the Ostrava-Karvina coal district from induced seismicity. The resulting velocity structure was used for source location calculations. Mendecki (1987) inverted for the average velocity from clusters of tremors to individual sensors. The results agreed well with velocity measurements from blasts, and can be used to improve hypocentre accuracy. Maxwell and Young (1993a) used a series of blasts as simulated tremors and compared travel time and arrival time inversion results. Good agreement was found between the inversions, independent of the calculation technique of the simultaneous equation for the travel time inversion. The resulting velocity structure was found to improve the hypocentral location accuracy of the blasts. Maxwell and Young (1993b) used a 3D velocity structure from Strathcona Mine in Canada to infer rockburst hazard. Temporal velocity changes associated with a rockburst were also investigated using a doublet method as previously described.

To demonstrate the applications of arrival time inversion, the results of a velocity imaging study at the Western Holdings Mine, in Welkom, South Africa, will be examined. In this case, seismic data recorded during 1993 were used to image the velocity structure. Arrival time measurements were used to compute initial hypocentral locations and error estimates, assuming a homogeneous velocity structure. From an initial total of 1820 events with at least five arrival times, the data were sorted to produce a data set minimized with respect to the following parameters:

- stochastic location error estimates;
- ratio of maximum to minimum coordinate error estimates;
- difference in locations between just P and P and S wave arrival times;
- rms arrival time residual.

The event sorting based on the above parameters indicated that the

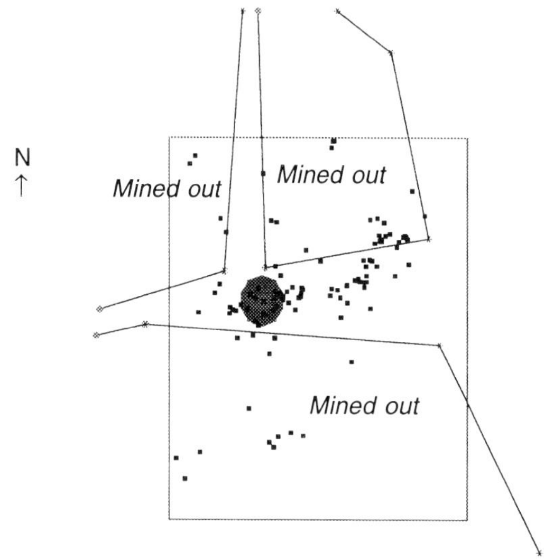


**Fig. 6.2** Plan view of the geometry of the area at the Western Holdings Mine, showing the mining faces (annotated areas), image grid (rectangle 800 × 1000 m for scale), selected hypocentres and location of the 1994 tremor.

parameters are not independent, so that an event which was dropped due to one of the parameters would also be rejected on the other parameters as well. The events were also sorted based on the spatial location, for events around a region of interest. A total of 118 events corresponding to 636 arrival times were selected. Short ray paths were ignored in the analysis due to relatively large residuals resulting from mislocations and timing errors. Alternatively, a weighting technique could have been used to reduce their influence. Figure 6.2 shows the location of the sorted earthquakes, local geologic structures, mining geometry and outline of the inversion grid (7×10 cells with five depth layers). Sensors are distributed around the grid to distances of 4 km. Station corrections were used to account for velocity variations outside the imaging grid.

The resulting set of arrival time residuals was found to have a large number of 'outliers', such that the residuals do not correspond to a normal distribution. Velocity structures were produced using damped least squares techniques, with an optimal damping factor (Maxwell and Young, 1994). However, no significant velocity anomalies were detected relative to stochastic error estimates computed using the variance of the residuals. Inversions were repeated with residuals being rejected if they exceeded a preset maximum, which was systematically decreased. A relatively small cutoff was required before statistically significant anomalies could be produced. The cause of these outliers is uncertain, but may have resulted from picking uncertainty, system timing uncertainty, or velocity variations. However, since the residuals were

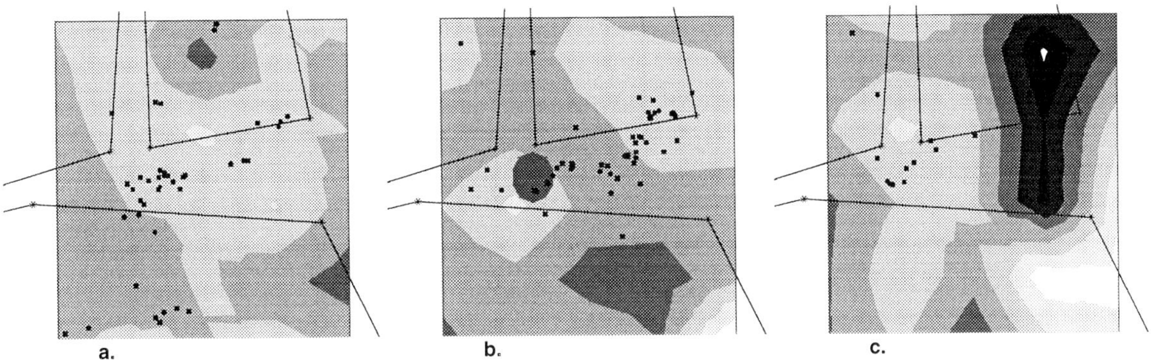


**Fig. 6.3** Plan view of the velocity image. 4840 (dark) to 5680 m/s (light), at depths of 1600 m (a), 1800 m (b) and 2000 m (c), superimposed with sorted hypocentres.

not significantly reduced through velocity inversions, it seems unlikely that the latter make a major contribution to the extreme residuals. The residuals were then rescaled to yield an $L_1$ norm solution, which resulted in a more Gaussian-type distribution without a residual cut off. Subsequent inversions resulted in statistically significant velocity anomalies.

Figure 6.3 shows the resulting velocity image at different depths, superimposed with well constrained hypocentres. The associated error estimates are approximately 200 m/s in the central seismically active regions, and the resolution matrix indicates well resolved features. There are three significant, steeply dipping, structural features in the region (see Fig. 6.2): a roughly NS striking fault which cuts off the orebody on the east; a second NS fault to the west; and a EW 'dyke-like' fault. Most of the seismicity occurs near the intersection of the EW fault and the eastern NS fault. The velocity image shows a high velocity anomaly near the intersection of the western NS fault and the EW fault, in the region of the concentrated seismicity. The western NS fault also has a high velocity region further north. Another high velocity anomaly can be seen along the eastern NS fault, south of the intersection with the EW fault (lower right hand side of Fig.6.3b andFig.6.3c). The intersection of the eastern NS and EW faults corresponds to a significant low velocity anomaly, which is also extended to the north. It is interesting to note that these general features are also apparent in the original, unscaled, damped least squares images.

In order to compare the velocity structure with the induced seismicity, the velocity was interpolated to each event hypocentre. Variations in source parameters could then be compared with variations in velocities at the source of the events, by plotting the parameters versus velocity. Figure 6.4 shows comparisons of moment magnitude, and static stress drop with velocity. The range of each of the parameters can be seen to be at a maximum at an intermediate velocity. At progressively higher and lower velocities, the

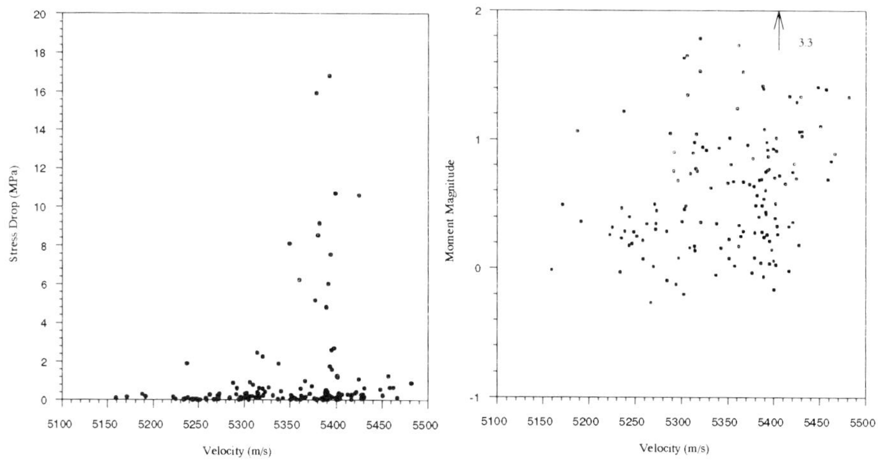


**Fig. 6.4** Cross-plot of hypocentral velocity versus static stress drop (left) and moment magnitude (right).

parameters become limited to relatively low values. Other studies of both natural and induced seismicity have shown similar results, although the largest tremors have been found to occur in the highest velocity zones (Maxwell and Young, 1994). A large moment magnitude 3.3 tremor which occurred in June 1994 in the high velocity region of the EW fault is also shown on the plots.

These observations can be interpreted based on fracture mechanics. The low velocity regions map relatively weak zones, incapable of storing sufficient strain energy to be the site of a strong event. In other words, the region is deforming through aseismic, stable sliding. High velocities correspond to strong, stiff regions, capable of storing significant strain energy. Other studies (Maxwell and Young, 1994) have shown that these stiff regions correspond to asperities with high seismic potential; however, the energy tends to be released by large tremors and not smaller background seismicity. The background seismicity, the source parameters of which peak at an intermediate velocity, may then correspond to a transition zone or possibly the nucleation point for the large tremors.

Also investigated was seismicity recorded in 1994 after the experiment, which included the magnitude 3.3 tremor previously mentioned. A visual comparison showed very good agreement between the main high and low velocity anomalies along the EW fault and both seismic index and apparent stress, as defined in Chapter 10 of this book. As previously stated, this is the region where the majority of the seismicity is concentrated and hence where the highest confidence in rockmass characterization from source parameters occurs. However, no induced seismicity is found along the eastern NS fault, near the high velocity anomaly. This region may also be an asperity with high seismic potential. The large tremor was found to be located in a region of very high velocity, supporting the earlier interpretation that the region corresponded with an asperity with high seismic potential.

## 6.4 Velocity inversion in a combined seismological and geomechanical investigation

The incorporation of velocity imaging in a seismic analysis helps to enhance the interpretation of the seismic response of the rockmass. After the identification of potential problem areas by this holistic interpretation of the seismic source and velocity properties, the next stage is to modify the mining sequences to minimize the seismic risk. One possible solution would be to mine in the low velocity, aseismic, destressed zones, as far away from the problem areas as possible. Clearly, the seismic risk will be related to safety and economic considerations, the perceived level of which must be offset by mining operational considerations.

Numerical stress models can be used to simulate mining induced stress changes. Recently, impressive advances have been made in developing sophisticated numerical codes to simulate complex rockmass responses (e.g. plasticity and discontinuities). However, greater simulation complexity increases the degrees of freedom of the model (increased range of responses and input parameters), and hence results in more uncertainty in the simulation. Therefore, some form of validation and calibration of the model is necessary to quantify and improve the model uncertainty. Traditionally, geomechanical instrumentation is used to measure the rockmass response, usually only at a few points, and compare it with the numerical model for validation. However, a holistic seismic interpretation of the rockmass response to stress redistributions using seismicity and velocity imaging may also be compared to modelling results to validate models. Seismic data offer the possibility of monitoring the rockmass in 3D, supplementing traditional geomechanical measurement. Once validated, numerical models can be run to compare the influence of different future mining scenarios on the identified problem areas.

For example, the simulation of the rockmass conditions at the Western Holdings Mine would require a plasticity model, capable of simulating rockmass failure and stress shedding. Such a model could be calibrated/validated with the velocity–seismicity data, in addition to standard geomechanical observations, to match the location and extent of the failed and stressed regions. The majority of the small magnitude earthquakes at the site were located in a region of velocity gradient, between a low velocity failed zone and a high velocity, highly stressed zone. The seismic potential for subsequent large tremors is relatively higher in the high velocity region, as corroborated by the occurrence of the magnitude 3.3 event in the centre of the high velocity region. Validated numerical models of these conditions could simulate stress changes in the high seismic potential/high velocity region, which would result from different future mining scenarios. The potential seismic risks of these scenarios could then be examined and offset by mining operational considerations. Velocity and seismicity information from future mining could then be used to further validate the models. However, the calibrated quantitative model would have to be used qualitatively to examine changes in the identified problem areas. Alternatively, new modelling approaches are being developed to numerically simulate the actual earthquakes. For example, the Particle Flow Code (Cundall, 1996) has been used to model complex macroscopic constitutive behaviour from simple shear and tensile behaviour of bonds in an assemblage of bonded grains. PFC is a dynamic model capable of directly simulating seismic events resulting from the breakage of these bonds. Potentially this model could be used to directly simulate the observed velocity structure and seismicity, in terms of both location and size. Once validated, such a model could be used quantitatively to assess the possibility of mining scenarios inducing large tremors in the problem regions.

Alternatively, after the identification of problem areas, some remedial measure might be taken to alleviate the problem. For instance, destress blasting maybe used to alleviate the stresses in the problem area. In this case, a sequential imaging technique could be used to assess the extent and degree of blast induced damage (e.g. Maxwell and Young, 1993a). Velocity measurements can also be made along specific ray paths, without reconstructing a velocity image. However, the sensitivity of such a velocity measurement will be dictated by the velocity anomalies encountered along the ray path. After an initial imaging experiment is performed and problem areas are identified, the velocity could be monitored along select ray paths. These measurements could be automated at the mine site and used in routine geomechanical measurements. Further sequential imaging could also be performed at critical periods, such as mining in close proximity to the problem areas or after rockbursts. Automated velocity measurements could be incorporated into a mines geomechanics program, and provide supplementary information to standard stress and strain measurements.

# 7 Seismic Source Radiation and Moment Tensor in the Time Domain

## 7.1 Radiation from the seismic source – far, intermediate and near fields

Seismic waves in an elastic medium are produced by an inelastic deformation at the seismic source. Let us assume that the seismic source occupies the volume $V$, in which the body forces $f$ are present, and in which the jump of displacement takes place on the surface $\Sigma$, composed of a positive side, $\Sigma+$, and a negative side, $\Sigma-$, with the unit vector, $\nu$, normal to the source surface and pointing outwards. If the jump of displacement on the surface $\Sigma$, when crossing it from $\Sigma+$ to $\Sigma-$, is $[u(\zeta,\tau)]$ at point $\zeta$ on the source and at time $\tau$, then the $n$ components of the displacement outside the source at a point $\mathbf{x} = \mathbf{x}(x_1, x_2, x_3)$, can be expressed in the general form by the representation theorem (Aki and Richards, 1980), assuming that sources of traction are not allowed:

$$\mathbf{u}_n(\mathbf{x},t) = \int_{-\infty}^{\infty} d\tau \iiint_V f_p(\eta,\tau)\, G_{np}(\mathbf{x}, t-\tau; \eta, 0)\, dV(\eta)$$

$$+ \int_{-\infty}^{\infty} d\tau \iint_\Sigma [\mathbf{u}_i(\zeta,\tau)]\, \mathbf{c}_{ijpq}\, \nu_j \partial\, G_{np}(\mathbf{x}, t-\tau; \zeta, 0)/\partial\zeta_q\, d\Sigma(\zeta) \qquad (7.1)$$

In the above formula $G$ is a Green function, and the tensor $\mathbf{c}_{ijkl}$ describes material constants. In the case of an isotropic body these values reduce to only two Lamé constants. In elastodynamics the Green function can be described as the displacement at the observation point $\mathbf{x}$ caused by the unit force applied at the source point $\zeta$. The Green function is a tensor because forces and displacements are three dimensional vectors. The '$np$' component of the Green function, $G(\mathbf{x},t; \zeta,\tau)$, is a $k$-component of the displacement at the observation point $\mathbf{x}$ at moment $t$, caused by the unit force component $n$, applied at the source point $\zeta$ at moment $\tau$.

Let us concentrate on the second term in equation (7.1), which describes the seismic source as the surface along which the jump of displacement takes place. In the absence of body forces the representation theorem can be written in a shorter form by using the convolution (*) convention:

$$\mathbf{u}_n(\mathbf{x}, t) = \iint_{\Sigma} [\mathbf{u}_i] \nu_j \mathbf{c}_{ijpq} * \partial \mathrm{G}_{np} / \partial \zeta_q d\Sigma \tag{7.2}$$

The discontinuity of displacement at the seismic source can be represented by the distribution of forces acting on the surface Σ (see '**f**' in Fig. 7.1).

The source is considered to be the result of internal forces only, so the acting forces should consist of the pairs (couples) of forces of opposite sign acting on the opposite surfaces of the source at the points ζ. The displacement due to the forces along the *n* direction, and with the dipole arm on the *q*-axis, is represented by $\partial G_{np} / \partial \zeta_q$.

The strength of these force dipoles, which can be found on the unit area of the surface Σ, is described by the term

$$[\mathbf{u}_i] \nu_j \mathbf{c}_{ijpq}$$

where *i*, *j*, *p* and *q* = 1,2,3.

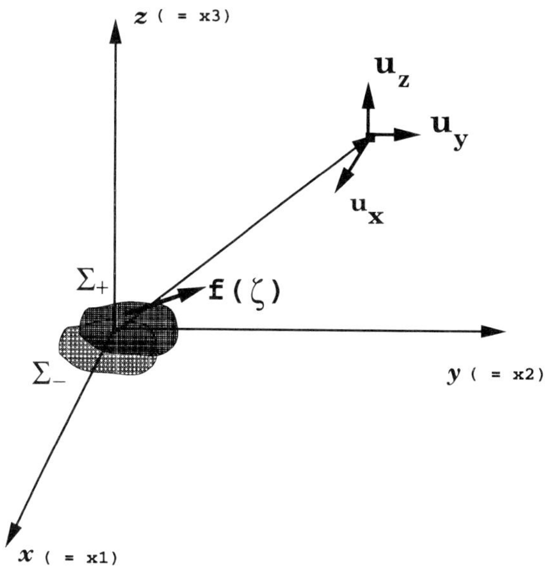


**Fig. 7.1** Internal forces on the surface Σ at the seismic source

For every pair (*p*,*q*) of forces (which are directed along the *p*-axis and are attached to the *q*-axis or vice versa) the above value has the dimension of force × length/unit area, and can be interpreted as force moment over unit of surface area, or the density of force moments. The values of $[\mathbf{u}]_i \, \nu_j \, \mathbf{c}_{ijpq}$, for $p,q = 1,2,3$, form a tensor, which has a special name: the moment tensor density, **m**. For the pair $(p,q)$ of forces the $pq$-component of the moment tensor density is $\mathbf{m}_{pq}$. In terms of the moment tensor density, the reciprocal theorem can be expressed as follows:

$$\mathbf{u}_n(\mathbf{x}, t) = \iint_{\Sigma} \mathbf{m}_{pq} * \partial G_{np} / \partial \zeta_q \, d\Sigma \tag{7.3}$$

For an isotropic body, the stiffness tensor, $\mathbf{c}_{ijkl}$, is expressed through Lamé constants, $\lambda$ and $\mu$ (in rock mechanics the symbol $G$ for rigidity is often used instead of $\mu$):

$$\mathbf{c}_{ijkl} = \lambda \delta_{ij} \delta_{kl} + \mu (\delta_{il} \delta_{jk} + \delta_{jl} \delta_{kl}) \tag{7.4}$$

In this case, the moment tensor density takes the form:

$$\mathbf{m}_{pq} = \lambda \nu_k [\mathbf{u}_k(\zeta, \tau)] \delta_{pq} + \mu \left( \nu_p [\mathbf{u}_q(\zeta, \tau)] + \nu_q [\mathbf{u}_p(\zeta, \tau)] \right) \tag{7.5}$$

where $p,q = 1,2,3$.

The nine possible components of the tensor **m** are shown in Fig. 7.2. Assuming that the displacement vector lies parallel to the source surface, $\Sigma$, $\nu_k[\mathbf{u}_k] = 0$, we get:

$$\mathbf{m}_{pq} = \mu \left( \nu_p [\mathbf{u}_q(\zeta, \tau)] + \nu_q [\mathbf{u}_p(\zeta, \tau)] \right) \tag{7.6}$$

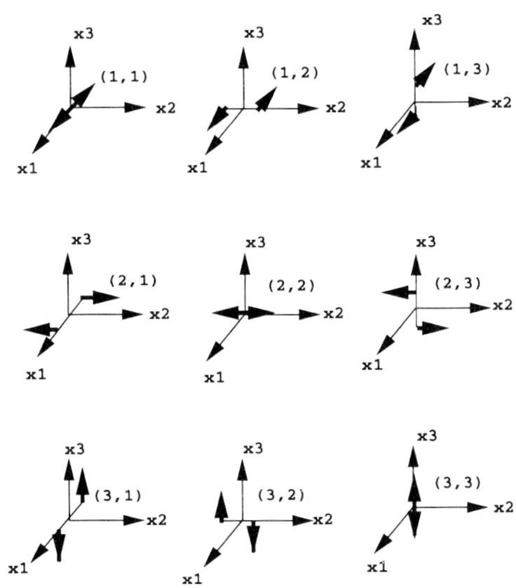


**Fig. 7.2** The nine possible components of the tensor **m**

As an example, let the source surface $\Sigma$ be a flat area with $x$ and $y$ non-zero components in a Cartesian $(x,y,z)$ coordinate system, and $\mathbf{u}$ a vector having all three components different from zero. Then the vector $\nu = (0,0,1)$, and the moment tensor density $\mathbf{m}$ has the form:

$$\mathbf{m} = \begin{bmatrix} \lambda[\mathbf{u}_z] & 0 & \mu[\mathbf{u}_x] \\ 0 & \lambda[\mathbf{u}_z] & \mu[\mathbf{u}_y] \\ \mu[\mathbf{u}_x] & \mu[\mathbf{u}_y] & (\lambda+2\mu)[\mathbf{u}_z] \end{bmatrix} \tag{7.7}$$

This matrix represents the nine vector dipoles (two of them equal to zero) which describe the seismic source in this example. In the case where the seismic source can be considered as occupying a small area $\Sigma$ it is convenient to introduce the notion of the moment tensor, $M$, defined as follows:

$$M_{pq} = \iint_{\Sigma} \mathbf{m}_{pq}\, d\Sigma \tag{7.8}$$

From the definition, $\mathbf{m}$ and $M$ are symmetric tensors (due to the symmetry of the stiffness tensor). The natural extension of the moment tensor definition to the sources, occupying a small (in comparison to the recorded wavelength) volume $V$, is, following from equation (7.8), given by:

$$M_{pq} = \iiint_{V} \mathbf{m}_{pq}\, dV \tag{7.9}$$

The density of the force dipoles $\mathbf{m}_{pq}$ in the source region of the volume $V$, where

$$\mathbf{m}_{pq} = \frac{d}{dV} M_{pq}$$

which, in this case, has the dimension of stress, can be considered as a change of stresses at the volume $V$ due to the action of the seismic source, and is sometimes called a 'stress glut' (Backus and Mulcahy, 1976a,b).

The displacement vector caused by the seismic source can be expressed through the moment tensor and Green function:

$$\mathbf{u}_n(\mathbf{x},t) = M_{pq} * \frac{\partial \mathrm{G}_{np}}{\partial \zeta_q} \tag{7.10}$$

or, in an integral representation, as:

$$\mathbf{u}_n(\mathbf{x},t) = \int_{-\infty}^{+\infty} M_{pq}(\tau)\, \partial G_{np}(\mathbf{x}, t-\tau; \zeta, 0) / \partial \zeta_q \, d\tau \tag{7.11}$$

In 1849 Stokes published the solution for the Green function, for an infinite homogeneous medium, as a displacement $\mathbf{u}(\mathbf{x},t)$, caused by the single force $F/(\zeta,\tau)$ applied at the point $\zeta = 0$, at moment $\tau = 0$, and acting in the $\mathbf{x}_m$ direction:

$$\begin{aligned}\mathbf{u}_n(\mathbf{x},t) &= \frac{1}{4\pi\rho}(3\gamma_n\gamma_m - \delta_{nm})\frac{1}{R^3}\int_{R/\alpha}^{R/\beta} \tau' F(t-\tau')\, d\tau \\ &+ \frac{1}{4\pi\rho\alpha^2}\gamma_n\gamma_m \frac{1}{R} F\left(t - \frac{R}{\alpha}\right) - \frac{1}{4\pi\rho\beta^2}(\gamma_n\gamma_m - \delta_{nm})\frac{1}{R}F\left(t-\frac{R}{\beta}\right)\end{aligned} \tag{7.12}$$

In equation (7.12) $R$ is the distance from the point of observation $\mathbf{x}$ to the point where the force is applied, $\gamma_n$ and $\gamma_m$ are the directional cosines of the vector $\mathbf{x} - \zeta$, $\delta$ is the Kronecker delta, $\alpha$ and $\beta$ are the velocities of P and S seismic waves in the medium with density $\rho$. Equation (7.12) allows the expression for the Green function to be written explicitly, the displacement caused by the unit force applied at the source at the moment $\tau$, where $\delta$ is the Dirac delta:

$$\begin{aligned}G_{nm}(\mathbf{x}, t; \zeta, \tau) &= \frac{1}{4\pi\rho}(3\gamma_n\gamma_m - \delta_{nm})\frac{1}{R^3}\int_{R/\alpha}^{R/\beta} \tau' \delta\,(t-\tau-\tau')\, d\tau' \; + \\ &\frac{1}{4\pi\rho\alpha^2}\gamma_n\gamma_m \frac{1}{R}\delta\left(t-\tau-\frac{R}{\alpha}\right) - \frac{1}{4\pi\rho\beta^2}(\gamma_n\gamma_m - \delta_{nm})\frac{1}{R}\delta\left(t-\tau-\frac{R}{\beta}\right)\end{aligned} \tag{7.13}$$

In equations (7.12) and (7.13), the terms in the first line describe the *near field*, and the terms in the second line the *far field*, separately for P waves (terms with $\alpha$ velocities) and for S waves (terms with $\beta$ velocities). In the far field, P waves have radiation factor $\gamma_n\gamma_m$, which specifies longitudinal displacement of the particles of the medium, and S wave $\lambda_n\lambda_m - \delta_{nm}$, which is the direction perpendicular to the vector $\mathbf{x} - \zeta$. As follows from equation (7.13) the Green function's dependence on time is in the form of the difference between the time at the receiver, $t$, and the at the source, $\tau$, so only $t-\tau$ terms are present.

To find the radiation from the double couples distributed over the source region with amplitude $M$, it is necessary to apply the double couples instead of single forces; see equation (7.12). To do that, calculations of spatial derivatives $(\partial/\partial\zeta)$ are needed, as follows from equations (7.10) and (7.12). Taking into account that:

$$\frac{\partial R}{\partial \zeta_q} = -\gamma_q$$

$$\frac{\partial \gamma_m}{\partial \zeta_q} = \frac{\gamma_m \gamma_q - \delta_{mq}}{R}$$

$$\frac{\partial}{\partial \zeta_q} \delta\left(t - \frac{R}{\alpha}\right) = \frac{1}{\alpha} \frac{\partial(-R)}{\partial \zeta_q} \frac{d}{dt} \delta\left(t - \frac{R}{\alpha}\right) \tag{7.14}$$

$$M_{pq}(t) * \frac{d}{dt} \delta\left(t - \frac{R}{\alpha}\right) = \frac{d}{dt} M_{pq}\left(t - \frac{r}{\alpha}\right)$$

the following equation determines the results for the $n$-component of the displacement vector caused by the action of the double couple of magnitude $M$ at the source (the moment tensor $M$ is assumed to be constant over the source area, which has small linear dimension in comparison with the radiated wave length):

$$\begin{aligned} \mathbf{u}_n(\mathbf{x}, t) = {} & RP^{\text{near}} \frac{1}{4\pi\rho} \frac{1}{R^4} \int_{R/\alpha}^{R/\beta} \tau M_{pq}(t-\tau)\, d\tau \\ & + RP^{\text{inter}(\alpha)} \frac{1}{4\pi\rho\alpha^2} \frac{1}{R^2} M_{pq}\left(t - \frac{R}{\alpha}\right) - RP^{\text{inter}(\beta)} \frac{1}{4\pi\rho\beta^2} \frac{1}{R^2} M_{pq}\left(t - \frac{R}{\beta}\right) \\ & + RP^{\text{far}(P)} \frac{1}{4\pi\rho\alpha^3} \frac{1}{R} \frac{d}{dt} M_{pq}\left(t - \frac{R}{\alpha}\right) - RP^{\text{far}(S)} \frac{1}{4\pi\rho\beta^3} \frac{1}{R} \frac{d}{dt} M_{pq}\left(t - \frac{R}{\beta}\right) \end{aligned} \tag{7.15}$$

where $RP^{\text{near}}$ is the radiation pattern in the near field, characterized by $1/R^4$ distance dependence; $RP^{\text{inter}(\alpha)}$ and $RP^{\text{inter}(\beta)}$ are the radiation patterns in the intermediate field, with $1/R^2$ distance dependence; and $RP^{\text{far(P)}}$ and $RP^{\text{far(S)}}$ are the radiation patterns in the far field for P and S waves respectively, characterized by $1/R$ distance dependence. These radiation patterns are combinations of directional cosines, $\gamma_j$, $j = n,p,q$:

$$
\begin{aligned}
RP^{\text{near}} &= 15\gamma_n\gamma_p\gamma_q - 3\gamma_n\delta_{pq} - 3\gamma_p\delta_{nq} - 3\gamma_q\delta_{np} \\
RP^{\text{inter}\,(\alpha)} &= 6\gamma_n\gamma_p\gamma_q - \gamma_n\delta_{pq} - \gamma_p\delta_{nq} - \gamma_q\delta_{np} \\
RP^{\text{inter}\,(\beta)} &= 6\gamma_n\gamma_p\gamma_q - \gamma_n\delta_{pq} - \gamma_p\delta_{nq} - 2\gamma_q\delta_{np} \\
RP^{\text{far}\,(\text{P})} &= \gamma_n\gamma_p\gamma_q \\
RP^{\text{far}\,(\text{S})} &= \gamma_n\gamma_p\gamma_q - \delta_{np}\gamma_q
\end{aligned}
\tag{7.16}
$$

By way of illustration, consider the simple case of the seismic source occupying the two dimensional surface $\Sigma$, covering the area $A$ of the $(x,y)$ plane in the Cartesian coordinate system $(x,y,z)$, with small linear dimension as compared to the distance from the source to the recording site and to the recorded wave length, and characterized by the time dependent discontinuity displacement at the source. Assume that the jump of displacement at the source has only two components different from zero, $[\mathbf{u}_y](t)$ and $[\mathbf{u}_x](t)$, and that the vector normal to the source is $\nu = (0,0,1)$ (the source lies in the $(x,y)$ plane). In that case the source can be represented by the moment tensor $M$:

$$
M = \begin{bmatrix} 0, & 0, & \mu[\mathbf{u}_x]A \\ 0, & 0, & \mu[\mathbf{u}_y]A \\ \mu[\mathbf{u}_x]A; & \mu[\mathbf{u}_y]A; & 0 \end{bmatrix}
\tag{7.17}
$$

Also assume that $[\mathbf{u}_x](t) = [\mathbf{u}_y](t) \equiv [\mathbf{u}](t)$, $A$ = 50 m × 50 m, $\mu = 3\times10^{10}$ Pa. The function $[\mathbf{u}](t)$ is shown in Fig. 7.3.

Equations (7.15) and (7.16) allow the calculation of the displacement components at point $\mathbf{x}$, which is $R$ length units distant from the source and for which the directional cosines are $\gamma_x$, $\gamma_y$ and $\gamma_z$.

After summation over repeated indices, for $x$ component of the near field displacement, the following is obtained:

$$
\mathbf{u}_x^{\text{near}}(\mathbf{x};t) = \frac{1}{4\pi\rho}\frac{1}{R^4} \times
\left\{ \left[30\gamma_x(\gamma_x\gamma_z) - 6\gamma_z\right] \int_{R/\alpha}^{R/\beta} \tau M_{xz}(t-\tau)\,d\tau + 30\gamma_x(\gamma_y\gamma_z) \int_{R/\alpha}^{R/\beta} \tau M_{yx}(t-\tau)\,d\tau \right\}
\tag{7.18}
$$

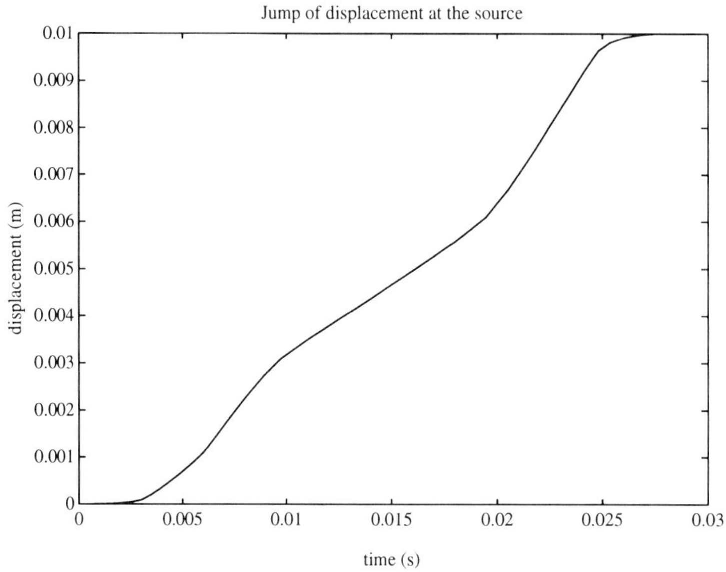


**Fig. 7.3** Jump of the displacement as a function of time at the source used in the example.

In equation (7.18), only non-zero terms are present. Taking into account that $M_{xz}$ (.,.) = $M_{yz}$ (.,.), equation (7.18) can be further simplified to:

$$\mathbf{u}_x^{\text{near}}(\mathbf{x};t) = \frac{1}{4\pi\rho}\frac{1}{R^4}\{30\gamma_x\gamma_z(\gamma_x+\gamma_y) - 6\gamma_z\}A\mu\int_{R/\alpha}^{R/\beta}\tau[\mathbf{u}](t-\tau)\,d\tau \tag{7.19}$$

The function :

$$\mathbf{u}_x(t) = \frac{\mathbf{u}_x^{\text{near}}(\mathbf{x};t)}{30\gamma_x\gamma_z(\gamma_x+\gamma_y) - 6\gamma_z}$$

is shown in Fig. 7.4, for three values of $R$, 750 m, 1000 m and 1500 m, and for some values of $\gamma_x$, $\gamma_y$ and $\gamma_z$. Particle velocities ($(d/dt)\mathbf{u}_x^{\text{near}}(t)$) recorded at the same three distances $R$, scaled as above, are shown in Fig. 7.5. Note that in the near field the radiated seismic field cannot be divided into P and S waves.

The segment of these values of $M(\tau)$ which are responsible for the radiation in the near field at the given moment $t$, $M_1$ ($t$-$R/\beta$) to $M_2$ ($t$-$R/\alpha$), is shown schematically in Fig. 7.6.

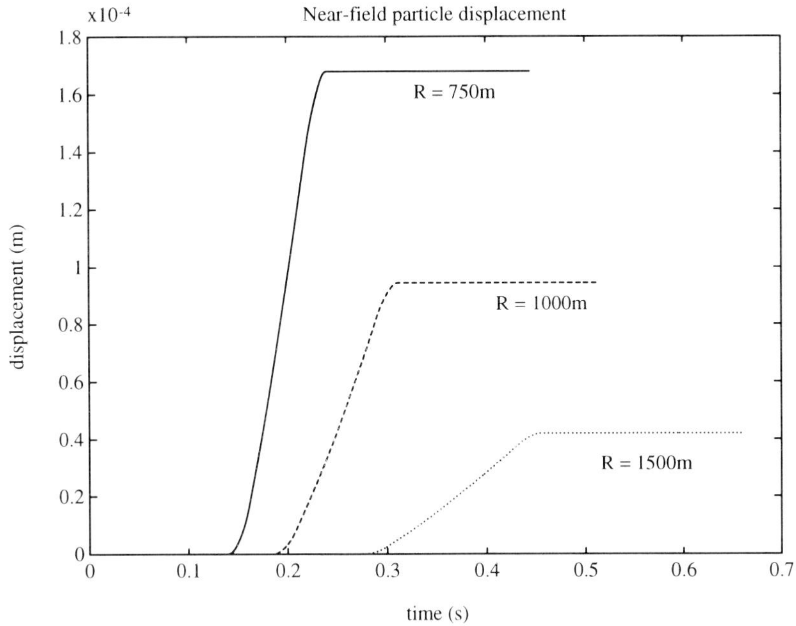


**Fig. 7.4** Near field particle displacement at different distances.

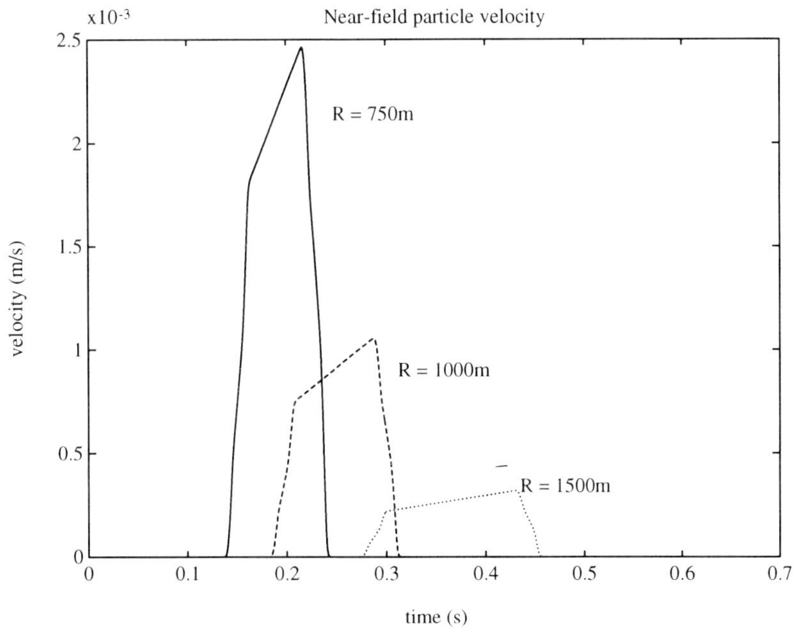


**Fig. 7.5** Near field particle velocities recorded at the three different distances.

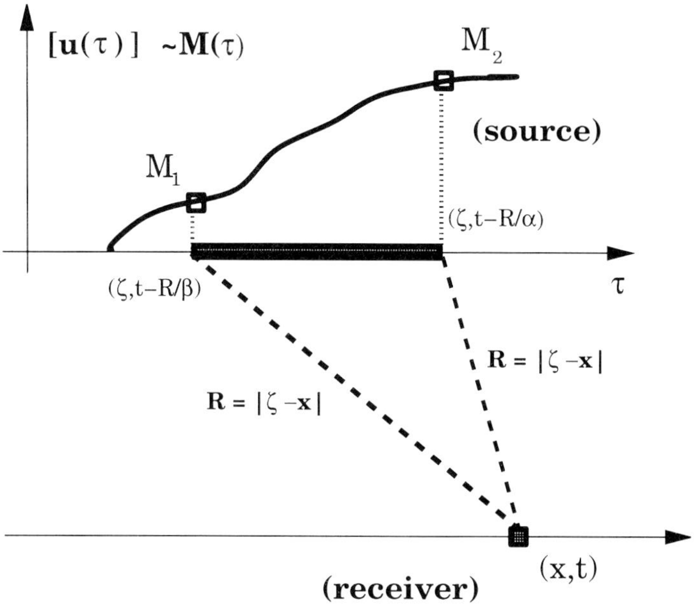


**Fig. 7.6** Time segment of the jump of displacement at the source used for calculation of the near field.

In the intermediate field the radiated seismic field depends on distance as $1/R^2$. In the example considered here, the $x$ component of the P wave particle displacement takes the form:

$$\mathbf{u}_x^{\mathrm{P(inter)}}(\mathbf{x};t) =$$

$$\frac{1}{4\pi\rho\alpha^2}\frac{1}{R^2}\left\{12\gamma_x\gamma_z(\gamma_x+\gamma_y)M_{yz}\left(t-\frac{R}{\alpha}\right)-2\gamma_z M_{yz}\left(t-\frac{R}{\alpha}\right)\right\} \tag{7.20}$$

$$=\frac{1}{4\pi\rho\alpha^2}\frac{1}{R^2}A\mu\{\,12\gamma_x\gamma_z(\gamma_x+\gamma_y)-2\gamma_z\}\left[[\mathbf{u}]\left(t-\frac{R}{\alpha}\right)\right]$$

It is interesting to note that in the intermediate field the radiated seismic field depends on the instant value of $M(\tau)$ at the proper moment of time, namely at $\tau = t-R/\alpha$ for P waves and $\tau = t-R/\beta$ for S waves, as shown in Fig. 7.7.

The displacement $\mathbf{u}_x^{\mathrm{inter(P)}}(t)$ and its time derivative, $((d/dt)\mathbf{u}_x^{\mathrm{inter}}(t))$, are shown in Fig. 7.8 and 7.9:

$$\mathbf{u}_x^{\mathrm{inter(P)}}(t) = \frac{\mathbf{u}_x^{\mathrm{P(inter)}}(\mathbf{x};t)}{\{12\gamma_x\gamma_z(\gamma_x+\gamma_y)-2\gamma_z\}}$$

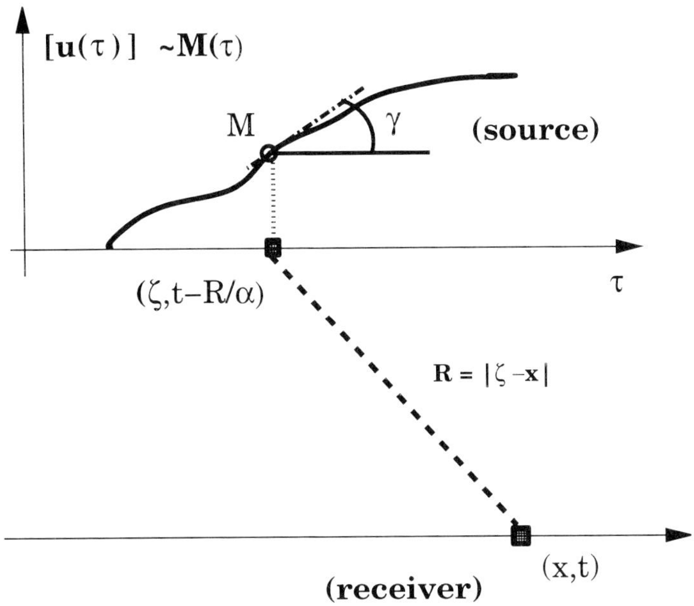


**Fig. 7.7** In the intermediate field, recorded at point $x$ at time $t$, the value of the jump of displacement responsible for the intermediate field at this point is the value of the jump of displacement at the source at point $\zeta$ at time $t-R/\alpha$. For the far field the derivative at the same point at the same time is proportional to the far field displacement recorded at point $x$ at time $t$.

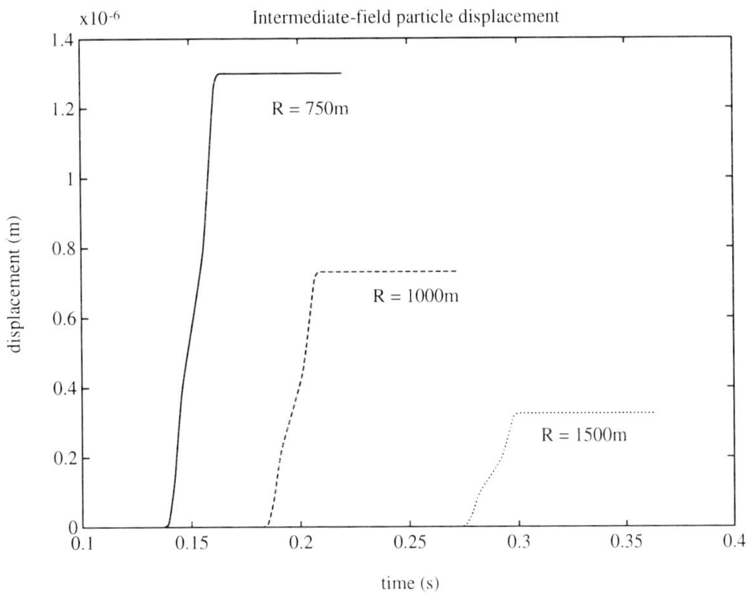


**Fig. 7.8** Intermediate field particle displacement at different distances from the source.

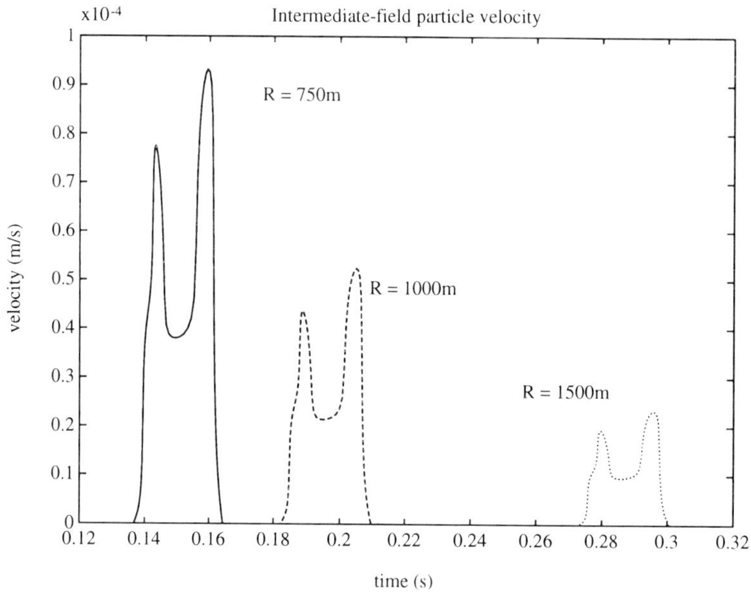


**Fig. 7.9** Intermediate field particle velocities at different distances from the source.

The far field radiation depends on the instantaneous time derivative at the time point $\tau = t - R/v$ (see Fig. 7.7), where $v = \alpha$ for P waves and $v = \beta$ for S waves. For P waves:

$$\mathbf{u}_x^{\mathrm{P(far)}}(\mathbf{x};t) =$$

$$\frac{1}{4\pi\rho\alpha^3}\frac{1}{R}\left\{\frac{d}{dt}[M_{xz}+M_{zx}]\gamma_x\gamma_z\gamma_x+\frac{d}{dt}[M_{yz}+M_{zy}]\gamma_y\gamma_z\gamma_x\right\}$$

$$=\frac{1}{4\pi\rho\alpha^3}\frac{1}{R}\left\{2\mu A\gamma_x\left[\gamma_x\gamma_z\frac{d}{dt}[\mathbf{u}_x]\left(t-\frac{R}{\alpha}\right)+\gamma_y\gamma_z\frac{d}{dt}[\mathbf{u}_y]\left(t-\frac{R}{\alpha}\right)\right]\right\} \tag{7.21}$$

$$=\mathbf{u}_x^{\mathrm{P}}(t)\gamma_x\{\gamma_x\gamma_z+\gamma_y\gamma_z\}$$

The $x$-component of the displacement field at the observation point $\mathbf{x}$ in this example is the result of the action of two double couples, (1,3) and (2,3) (see Fig. 7.2). The function $\mathbf{u}^{\mathrm{P}}_x(t)$ is shown in Fig. 7.10. From equation (7.21) it is easy to calculate the velocity of the particles at the given point $x$, which is simply the time derivative of equation (7.21). This is the physical value which is recorded by the velocity sensors:

$$\mathbf{v}_x^{\mathrm{P(far)}}(\mathbf{x};t) =$$

$$\frac{1}{4\pi\rho\alpha^3}\frac{1}{R}\left\{2\mu A\gamma_x\left[\gamma_x\gamma_z\frac{d^2}{dt^2}[\mathbf{u}_x]\left(t-\frac{R}{\alpha}\right)+\gamma_y\gamma_z\frac{d^2}{dt^2}[\mathbf{u}_y]\left(t-\frac{R}{\alpha}\right)\right]\right\} \qquad (7.22)$$

$$= v_x^{\mathrm{P}}(t)\gamma_x\{\gamma_x\gamma_z+\gamma_y\gamma_z\}$$

where:

$$v_x^{\mathrm{P}}(\mathbf{x};t) = \frac{1}{4\pi\rho\alpha^3}\frac{1}{R}2\mu A\frac{d^2}{dt^2}[\mathbf{u}]\left(t-\frac{R}{\alpha}\right)$$

Function $v_x^{\mathrm{P}}(t)$ is shown in Fig. 7.11, for the same point **x** as calculated for the particle displacement (see Fig. 7.10). In the above expressions $\gamma_i = x_i/R$, where $i = x,y,z$, or 1,2,3, if the source is located in the centre of the Cartesian coordinate system ($\zeta = 0$). To present the sign of the radial component of the first motion of the P wave, stereographic projection will be used. Assume the receiving stations are located around the source from the example, distributed on the lower hemisphere and on the upper hemisphere (Fig. 7.12). In Fig. 7.12, the signs of the first P motions (for stations located on the lower hemisphere) are shown by '+', when the first motion is directed from the source, and by '-', if the first motion is directed towards the source. The same convention is used for the first motion of P waves for stations on the upper hemisphere.

To calculate amplitudes of the far field displacement of the P wave and S wave at the given point of observation caused by the action of double couples of forces at the arbitrarily located seismic source, it is convenient to use a proper description of the location of the fault surface and the double couple in space. Conventional symbols to describe the fault parameters (Aki and Richards, 1980) are used in the following description. It is primarily necessary to define the Cartesian coordinate system which will be used for fault description. The *x*-axis is conventionally directed towards the north, the *y*-axis towards the east, and the *z*-axis down. The fault surface location in this coordinate system is described by means of two angles, $\Phi_s$, called the strike angle, and $\delta$, the dip angle (Fig. 7.13).

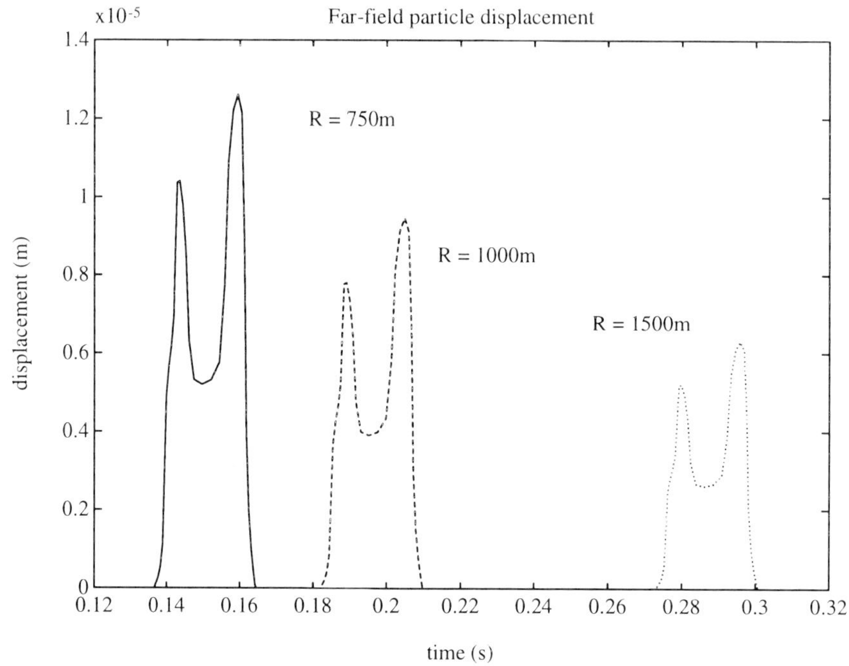


**Fig. 7.10** Far field particle displacement at different distances from the source.

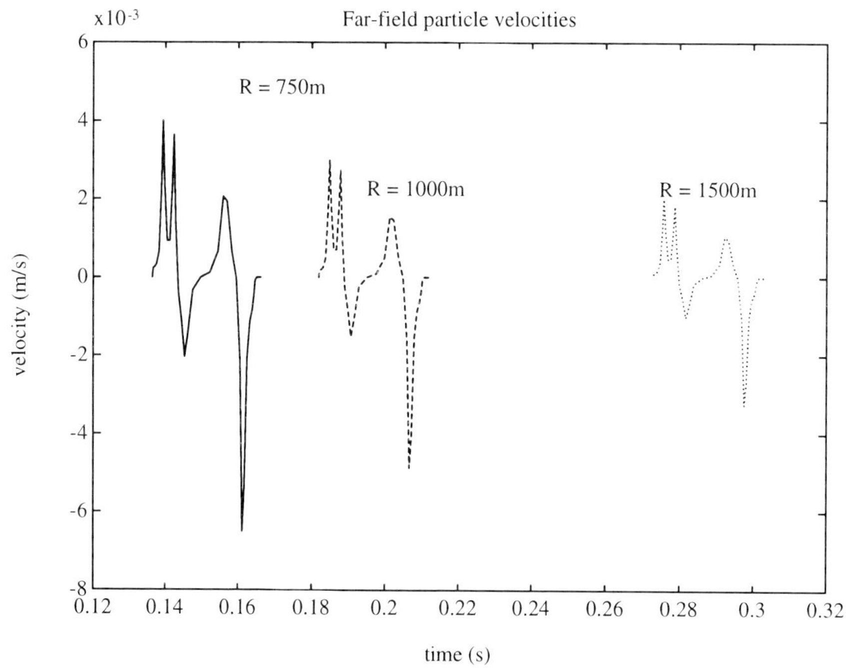


**Fig. 7.11** Far field particle velocities at different distances from the source.

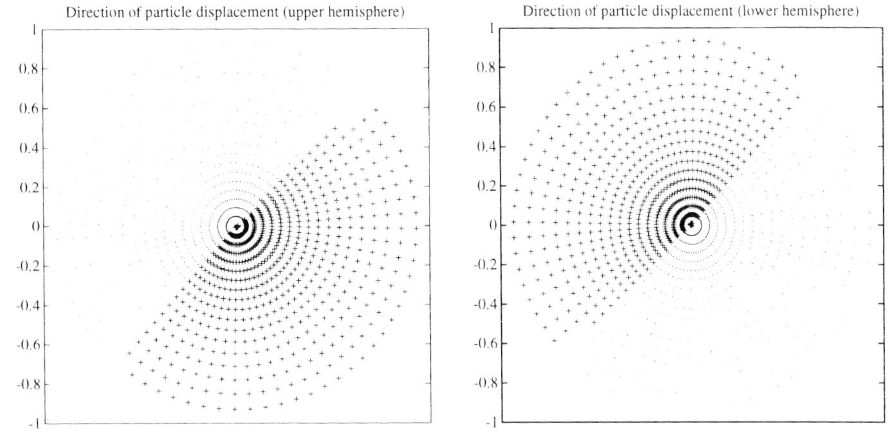


**Fig. 7.12** Direction of particle displacement (upper and lower hemisphere).

The strike angle is calculated from the north direction clockwise to the intersection of the fault with the *xy* surface, and $\delta$ is the angle between the *xy* surface and the fault surface, $0 \leq \Phi_s < 2\pi$, $0 \leq \delta \leq \pi/2$. The displacement on the fault of one surface in relation to another, or the direction of the double couple responsible for the jump of displacement, is given by the rake angle, $\lambda$, which is the angle between the horizontal line drawn on the fault and the direction of movement of the one fault surface against another (Fig. 7.14), $-\pi < \lambda \leq \pi$. The point of observation of the seismic wave is usually described by the angle $\Phi$, the azimuth of the seismic ray leaving the source in the direction of the receiver, and by $i_\zeta$ , the angle between the *z*-axis and the seismic ray (Fig. 7.15).

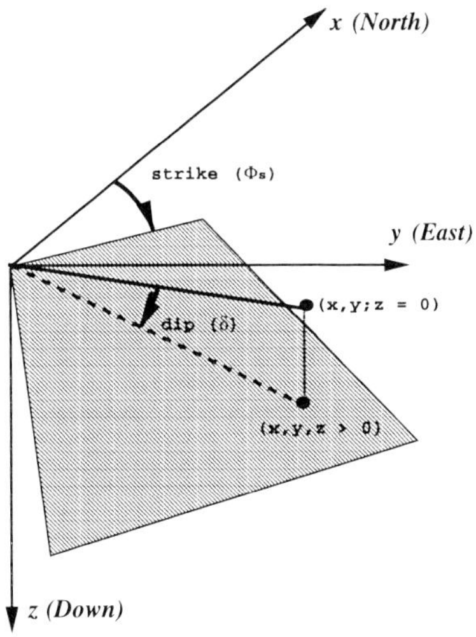


**Fig. 7.13** Fault surface location in the Cartesian coordinate system.

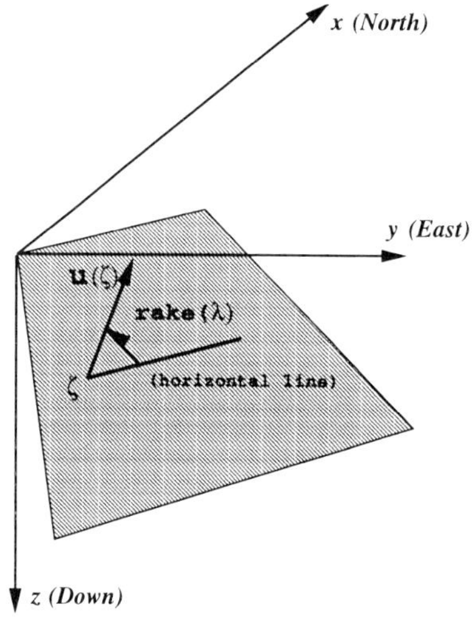


**Fig. 7.14** Direction of displacement of fault (rake λ).

Using coordinates describing the fault ($\Phi_s$, $\delta$), displacement on the fault ($\lambda$), and the coordinates pertinent to the seismic ray ($\Phi$, $i_\zeta$), it is possible to calculate radiation pattern coefficients and, for the far field P wave radiation pattern, $F^P$, this is (Aki and Richards, 1980):

$$\begin{aligned} F^P = \; & \cos(\lambda)\sin(\delta)\sin^2(i_\zeta)\sin[2(\Phi-\Phi_s)] \\ & -\cos(\lambda)\cos(\delta)\sin(2i_\zeta)\cos(\Phi-\Phi_s) \\ & +\sin(\lambda)\sin(2\delta)[\cos^2(i_\zeta)-\sin^2(i_\zeta)\sin^2(\Phi-\Phi_s)] \\ & +\sin(\lambda)\cos(2\delta)\sin(2i_\zeta)\sin(\Phi-\Phi_s) \end{aligned} \tag{7.23}$$

The radiation pattern for far field S waves can be divided into a radiation pattern for the S component parallel to the $(x,y)$ surface (see Fig. 7.15) and perpendicular to the seismic ray at the receiving point, $F^{SH}$, and a radiation pattern for the S component perpendicular to the seismic ray at the receiving point but lying in the vertical plane (parallel to the $z$-axis, Fig. 7.15), $F^{SV}$:

$$\begin{aligned} F^{SH} = \; & \cos(\lambda)\cos(\delta)\cos(i_\zeta)\sin(\Phi-\Phi_S) \\ & +\cos(\lambda)\sin(\delta)\sin(i_\zeta)\cos[2(\Phi-\Phi_S)] \\ & +\sin(\lambda)\cos(2\delta)\cos(i_\zeta)\cos(\Phi-\Phi_S) \\ & -0.5\sin(\lambda)\sin(2\delta)\sin(i_\zeta)\sin[2(\Phi-\Phi_S)]. \end{aligned} \tag{7.24}$$

$$\begin{aligned} F^{SV} = \; & \sin(\lambda)\cos(2\delta)\cos(2i_\zeta)\sin(\Phi-\Phi_S) \\ & -\cos(\lambda)\cos(\delta)\cos(2i_\zeta)\cos(\Phi-\Phi_S) \\ & +0.5\cos(\lambda)\sin(\delta)\sin(2i_\zeta)\sin[2(\Phi-\Phi_S)] \\ & -0.5\sin(\lambda)\sin(2\delta)\sin(2i_\zeta)\,[(1+\sin^2(\Phi-\Phi_S))] \end{aligned} \tag{7.25}$$

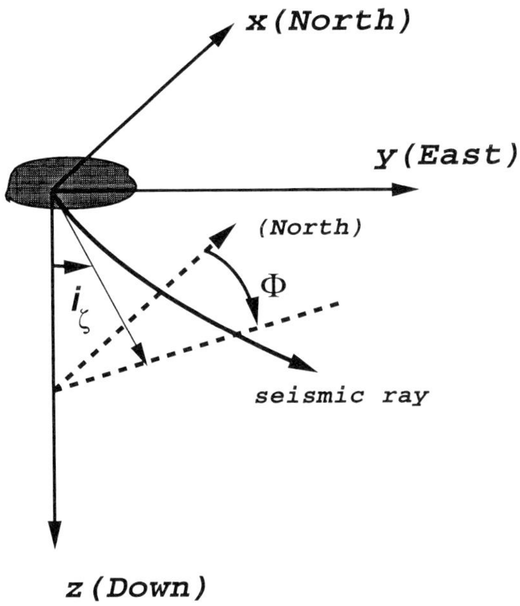


**Fig. 7.15** Ray parameters $i_\zeta$ and $\Phi$.

## 7.2 Moment tensor

The seismic displacement field, radiated from the seismic source, is described, according to equation (7.10), as a convolution in the time domain of the moment tensor $M$ and the spatial derivatives of the Green function. These derivatives represent nine double couples, with the intensity described by the components of the moment tensor. The representation of the general point seismic source is based on the eigenvalue decomposition of the moment tensor $M$ (Ben-Menahem and Singh, 1981; Jost and Herrman, 1989). Let the eigenvalues of the moment tensor $M$ be $m_1$, $m_2$ and $m_3$, and corresponding eigenvectors: $\mathbf{a}_1$, $\mathbf{a}_2$, $\mathbf{a}_3$. The eigenvalues $m_1$, $m_2$, $m_3$ represent the diagonalized moment tensor, $\mathbf{m}$ (this is the same tensor $M$, but in the new orthogonal system of coordinates, described by the unit vectors $\mathbf{a}_1$, $\mathbf{a}_2$, $\mathbf{a}_3$, in which $\mathbf{M}$ has only the diagonal components). The diagonalized moment tensor $\mathbf{m}$ can be decomposed into two major parts: a volumetric part, $\mathbf{m}_v$, and a deviatoric part, $\mathbf{m}_D$. The volumetric part $\mathbf{m}_v$, is represented by the trace of the tensor $M$ (or $\mathbf{m}$, the trace does not change with rotation of the system of coordinates):

$$\mathbf{m}_V = \tfrac{1}{3}(m_1 + m_2 + m_3)\,\mathbf{I} = \tfrac{1}{3}\,tr(M)\,\mathbf{I}$$

where $\mathbf{I}$ represents a unity matrix.

From the definition of the deviatoric part, it follows that

$$\mathbf{m}_D = \mathbf{m} - \mathbf{m}_V$$

The deviatoric part can be further decomposed in several ways. At least two such decompositions have a simple physical meaning: decomposition of $\mathbf{m}_V$ into three orthogonal vector dipoles, and into three double couples. The decomposition into three linear vector dipoles is as follows:

$$\mathbf{m}_D = \sum_{i=1}^{3} (m_i - \tfrac{1}{3}\, tr(M))\, \mathbf{a}_i \mathbf{a}_i$$

where $\mathbf{a}_i\mathbf{a}_i$ are dyadics. The dyadic $\mathbf{a}_i\mathbf{a}_i$ represents a dipole of unit strength, in which one arm points in the direction of $\mathbf{a}_i$, and the second arm in the direction of $-\mathbf{a}_i$. According to the above representation, the $i$th linear dipole is composed of the two equal forces of amplitude $m_i - \tfrac{1}{3}\, tr\,(M)$, when the first force points in the direction of $\mathbf{a}_i$ and the second one in the direction of $-\mathbf{a}_i$.

The second decomposition of the deviatoric part can be done by using the six independent pure double couples, every one of which can be represented as follows:

$$(m_i - \tfrac{1}{3}\, tr(M))\, (\mathbf{a}_i \mathbf{a}_i - \mathbf{a}_j \mathbf{a}_j)\,, \quad i,j = 1,2,3; \quad i \neq j$$

From this, it is possible to compose the three double couples, using only pairs of the eigenvectors:

$$\mathbf{m}_D = \tfrac{1}{3} \sum_{\substack{i,j=1 \\ i \neq j}}^{3} (m_i - m_j)\, (\mathbf{a}_i \mathbf{a}_i - \mathbf{a}_j \mathbf{a}_j).$$

In this case, every double couple is composed of two linear vector dipoles, each with the same absolute amplitude $\tfrac{1}{3}(m_i - m_j)$.

If one of these dipoles lies in the direction of the vector $\mathbf{a}_i$, and has amplitude, say, $\tfrac{1}{3}(m_i - m_j)$, then the second one has amplitude $-\tfrac{1}{3}(m_i - m_j)$ and lies on the direction of the vector $\mathbf{a}_j$. The couple $_{(ij)}$ can be understood as composed of one force from the dipole which lies in the direction of the $\mathbf{a}_i$ vector, and a second force from the dipole which lies in the direction of the $\mathbf{a}_j$ vector.

### 7.2.1 The case of a synchronous source and the delta source time function

Assuming a synchronous source means that we assume the same time dependency for every component of the moment tensor. For long waves, with periods much longer than the rise time of the moment tensor, we may assume that time derivative of the moment tensor can be treated as the product of the moment tensor and the delta function. If $H(t)$ is a Heaviside step function, and $\delta(t)$ a delta function, then:

$$M_{pq}(t) = M_{pq}H(t); \quad d/dt\, M_{pq}(t) = M_{pq}\delta(t)$$

In this case, formula (7.5) for the far field can be written in a simplified form:

$$\begin{aligned}\mathbf{u}_n(\mathbf{x},t) = \; & 1/(4\pi\rho\alpha^3 R)\, RP^{\text{far(P)}}\; M_{pq}\,\delta(t - R/\alpha) + \\ & 1/(4\pi\rho\beta^3 R)\, RP^{\text{far(S)}}\; M_{pq}\,\delta(t - R/\beta)\end{aligned} \tag{7.26}$$

where $M_{pq}$ represents the value of the $pq$ tensor component at the source, and the displacement component is recorded at time $t - R/\alpha$ for the P waves, and at time $t - R/\beta$ for the S waves; $RP^{\text{far(P)}}$ and $RP^{\text{far(S)}}$ are radiation patterns (see equation (7.15)).

In this case, the moment tensor elements can be calculated in the time domain or in the frequency domain (see Chapter 8) on the basis of equation (7.5).

### 7.2.2 The case of an asynchronous source and arbitrary source time function

The often made assumption that all moment tensor components have the same time dependency is a very restrictive one. In the case of seismic events in mines, where volumetric changes in the source region are significant, we need to take into account different time dependencies of different moment tensor components. Below we outline a procedure for the time dependent moment tensor inversion for P waves only (the extension of this procedure to the case of S waves is straightforward). We restrict ourselves to the far field and intermediate radiation field, neglecting the near field radiation. To calculate the six moment tensor components, we need data from at least six seismic stations. We will use only the radial P wave component of displacement. The far field displacement only has radial component in the far field, but, in the case of intermediate field, we need to remember that it has a transverse component as well.

In the far field, the radial component of the displacement, for which the moment tensor components from the main diagonal are responsible, can be expressed as follows:

$$U_r(t) = \frac{\gamma_i^2}{4\pi\rho\alpha^3 R}\frac{d}{dt}M_{ii}\left(t - \frac{R}{\alpha}\right) \tag{7.27}$$

where $\gamma_i$ is the directional cosine of the seismic ray, $\alpha$ is the P wave velocity, $\rho$ stands for the rock density, and $R$ is the distance from the source to the station along the seismic ray.

The radial component of the far field created by the off diagonal moment tensor components, $M_{ij}(.)$, $i,j = x,y,z$, depends on two directional cosines $\gamma_i$ and $\gamma_j$:

$$U_r(t) = \frac{2\gamma_i\gamma_j}{4\pi\rho\alpha^3 R}\frac{d}{dt}M_{ij}\left(t - \frac{R}{\alpha}\right) \tag{7.28}$$

For the intermediate field, from the representation theorem, one can get the following formulations:

$$U_r(t) = \frac{4\gamma_i^2 - 1}{4\pi\rho\alpha^2 R^2} M_{ii}\left(t - \frac{R}{\alpha}\right) \tag{7.29}$$

for the diagonal moment tensor components, $M_{ii}$ (.), and for the off diagonal moment tensor components, $M_{ij}$ (.) :

$$U_r(t) = \frac{8\gamma_i\gamma_j}{4\pi\rho\alpha^2 R^2} M_{ij}\left(t - \frac{R}{\alpha}\right) \tag{7.30}$$

In order to calculate the moment tensor components one needs to solve a system of differential equations, which follow from the last four expressions.

In Fig. 7.16 – 7.24 we present the results of the moment tensor calculation of two sources of seismic waves: one is a 'natural' seismic event, the second is the blasting of 250 kg of explosive material, and the figures are paired to provide comparison between the events.

Figure 7.16 and 7.17 show the radial components of displacement; Figs. 7.18 and 7.19 show six independent moment tensor components; Figs. 7.20 and 7.21 show three double couples for each event, while Figs. 7.22 and 7.23 show the stereographic projection of the smallest and largest eigenvectors.

Note that Fig. 7. 24 illustrates alternative representation of the deviatoric part of the moment tensor for the 'natural' rockburst shown in Fig. 7.20.

In Fig. 7.16 – 7.21, one unit on the vertical axis equals $10^{10}$ Nm.

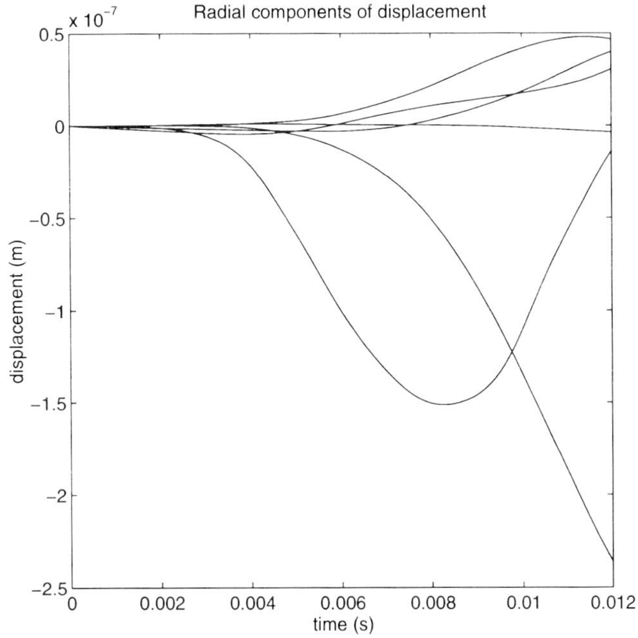


**Fig. 7.16** Radial components of displacement for a seismic event of magnitude $m_M$=1.1, recorded by six stations.

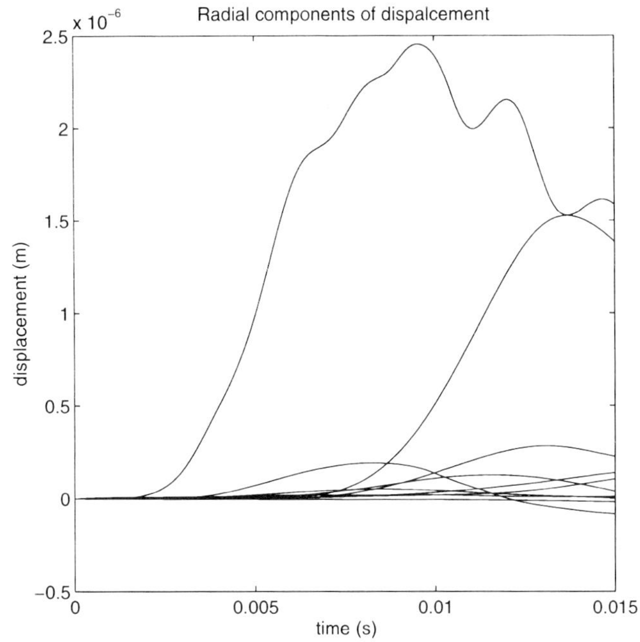


**Fig. 7.17** Radial component of displacement, caused by blasting of 250 kg of emulsion, recorded by nine stations. Note that all radial components are positive.

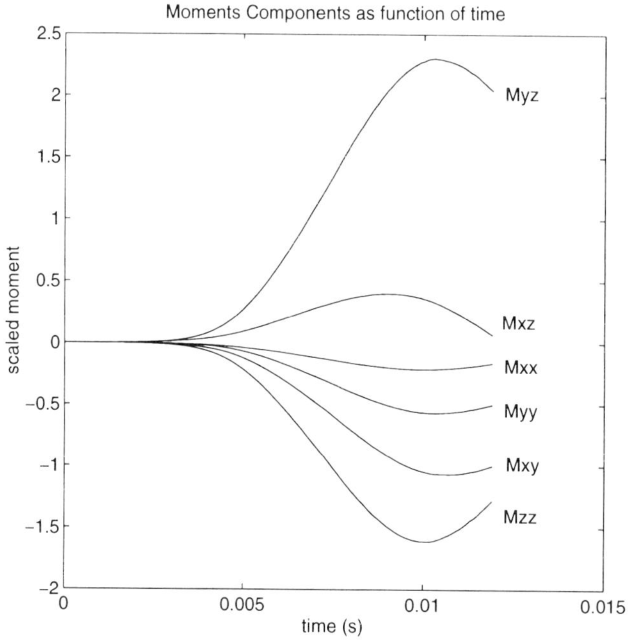


**Fig. 7.18** Six independent moment tensor components calculated in the mine coordinate system for a seismic event. Note the very short time when the moment tensor components were growing. One unit on the vertical axis is $10^{10}$ Nm.

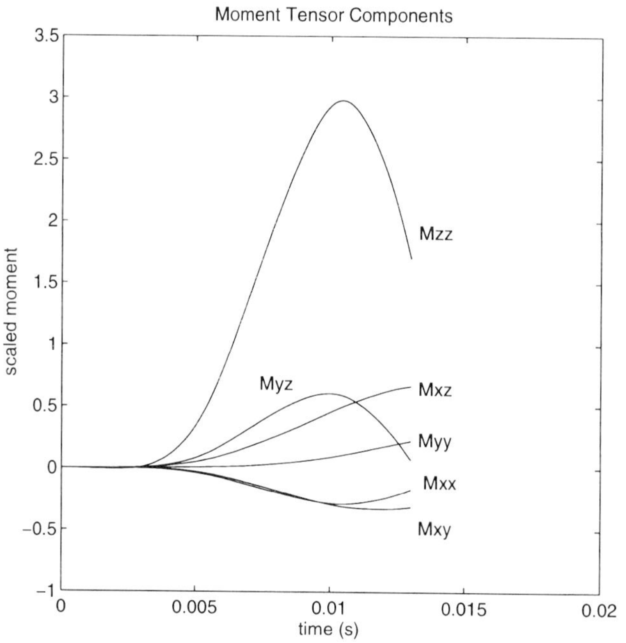


**Fig. 7.19** Six independent moment tensor components calculated in the mine coordinate scheme for the induced event (blast). One unit on the vertical axis is $10^{10}$ Nm.

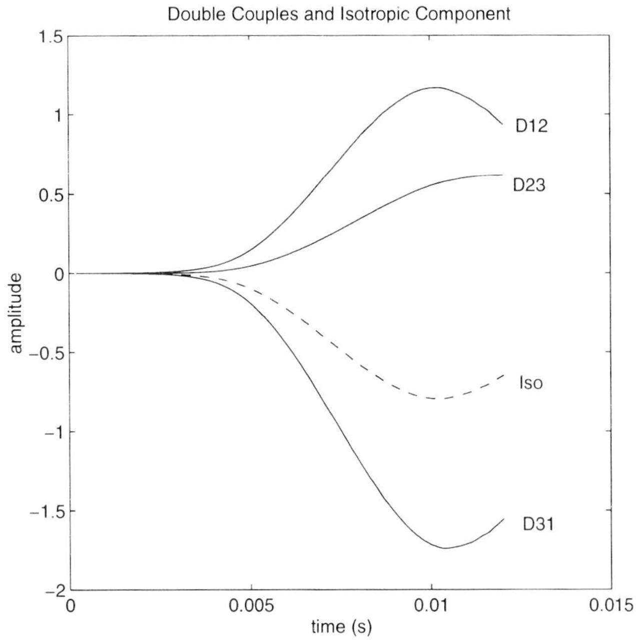


**Fig. 7.20** Three double couples which represent the deviatoric part of the moment tensor, for the seismic event. Together with the isotropic part (Iso), they have the same radiation as the six moment tensor components shown in Fig. 7.16.
One unit on the vertical axis is $10^{10}$ Nm.

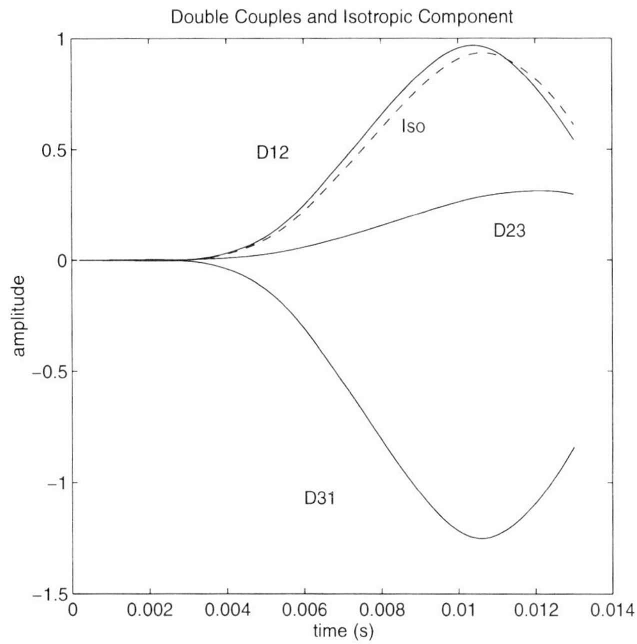


**Fig. 7.21** Three double couples which represent the deviatoric part of the moment tensor, $m_D$, of the induced event (blast). The volumetric component (Iso) is also shown. One unit on the vertical axis is $10^{10}$ Nm.

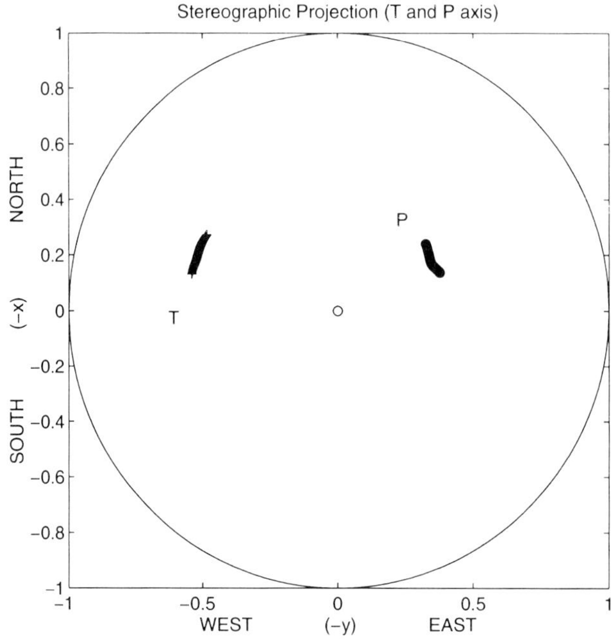


**Fig. 7.22** Stereographic projection of greatest (T) and smallest (P) eigenvectors of the seismic event. Wandering in space with time of the direction of the greatest linear vector dipole (V1 in Fig. 7.24) and the smallest (V3 in Fig. 7.24) is also shown.

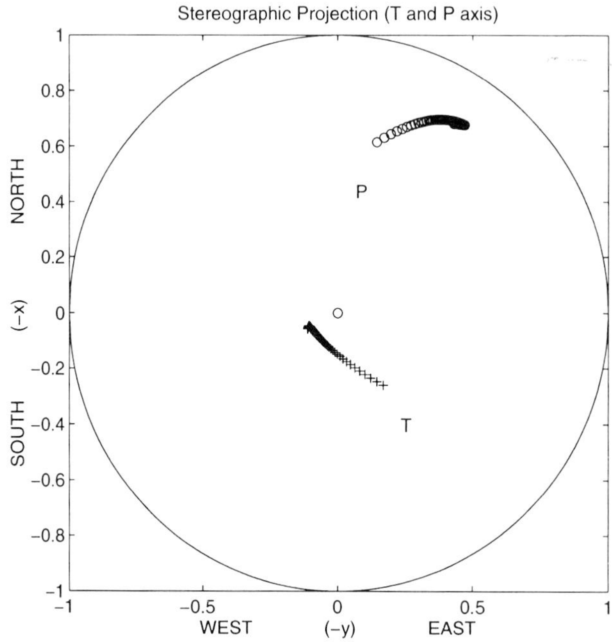


**Fig. 7.23** Stereographic projection of the greatest (T) and the smallest (P) eigenvectors of the induced event (blast).

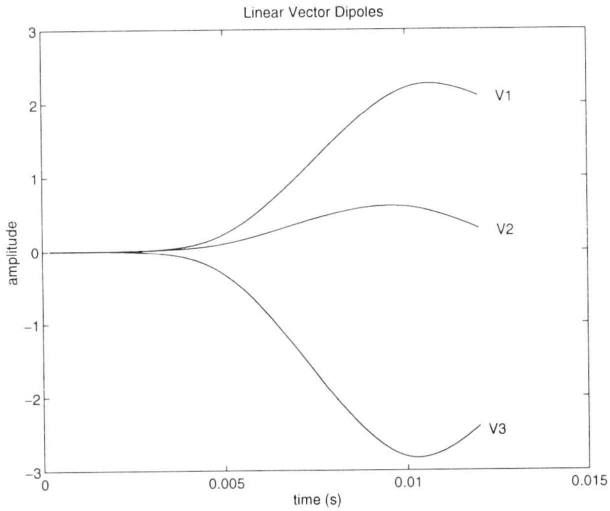


**Fig. 7.24** An alternative representation of the deviatoric part of the moment tensor: linear vector dipoles as a function of time for a seismic event.
One unit on the vertical axis is $10^{10}$ Nm.

This case is much more interesting from the seismological point of view than the previous one. It allows the evaluation of the duration of the seismic source and gives insight into the source complexity through the possibility of calculating the space wandering of the T and P axes. At the same time, the calculations are not limited to the far field only, but include the intermediate field as well. For distances of several hundred meters, this inclusion is important, because it allows more precise calculation of the seismic moment tensor components, avoiding errors of magnitude up to 20 or 30%.

# 8 Spectral Analysis and Seismic Source Parameters

## 8.1 Fast Fourier transform and multitaper

Spectral analysis, i.e. estimation of the frequency representation of time series, usually employs the fast Fourier transform (FFT) algorithm to perform a Fourier transform. This approach to spectral analysis is computationally efficient and produces reasonable results for a large class of signals. There are, however, several inherent performance limitations, specifically when dealing with a wide range of frequency and amplitude waveforms recorded in the intermediate or near fields of seismic sources. The first limitation is that of frequency resolution, i.e. the ability to distinguish the spectral responses of two or more time signals. The frequency resolution in hertz is roughly the reciprocal of the time in seconds over which sampled data is available. A second limitation is due to the implicit windowing of the data that occurs when processing with the FFT. Problems with windowing manifest as 'energy leakage' in the spectral features. Energy 'leaks' into the sidelobes obscuring other spectral features that are present and distorts, sometimes significantly, seismic source parameters like seismic moment and radiated seismic energy. Skilful selection of the windowing technique can reduce the sidelobe leakage but at the cost of reduced resolution.

The discussion below deals with the following problems:

- how to check that the applied Fourier transform gives proper results;
- how to eliminate the energy leakage through adaptive multitaper.

Assume that it is necessary to calculate a Fourier transform using the FFT of a continuous, real-valued signal $x(t)$, $t \in (0,T)$, where $t$ can be understood as time. If the signal $x(t) \in L^2$, which means that its 'energy' is limited, then its Fourier coefficients, $H(f)$, for the frequency $f$, are defined as follows:

$$H(f) = \int_0^T x(t)\, e^{2\pi i f t}\, dt \tag{8.1}$$

Parseval's theorem for the above transformation states that:

$$\int_0^T [x(t)]^2\, dt = \int_{-\infty}^{\infty} |H(f)|^2\, df \tag{8.2}$$

In practice we deal with discrete (sampled) values of the function $x(t)$. Assume that $t = k\cdot\Delta t$, where $\Delta t$ is the sampling period, $k = 0,1 \cdots, N-1$, and $N = 2^L - 1$, where $L$ is an integer. For time sampled input data one can calculate the discrete Fourier coefficients $h(f_n)$, where $f_n = n/(N\cdot\Delta t)$.

$$h(f_n) \cong \sum_{k=o}^{N-1} x(k\cdot\Delta t)\, e^{2\pi i f_n k\cdot\Delta t} \tag{8.3}$$

From (8.1) it follows that:

$$H(f_n) \cong \sum_{k=0}^{N-1} x(k\cdot\Delta t)\, e^{2\pi i f_n k\cdot\Delta t}\,\Delta t = \Delta t * h(f_n)$$

From Parseval's theorem it is easy to evaluate the energy of the signal by means of the Fourier coefficients. The energy of the signal (this value is proportional to the physical energy contained in the ground motion if $x(t)$ represents the velocity of the ground particles) in the time domain is expressed as an integral (for a continuous signal) or as a sum of the appropriate terms (in the discrete case).

By making use of Parseval's theorem, we obtain:

$$\int_0^T [x(t)]^2\, dt \cong \Delta t * \sum_{k=0}^{N-1} x^2(k\cdot\Delta t) = \frac{\Delta t}{N} \sum_{k=0}^{N-1} |h(f_n)|^2 \tag{8.4}$$

Note that for the Fourier coefficients the summation indices run from 0 to $N-1$. Due to the symmetry of these coefficients, it is convenient to write the right hand side of equation (8.4) in the form:

$$\frac{\Delta t}{N} \sum_{n=0}^{N-1} |h(f_n)|^2 = \frac{2\cdot\Delta t}{N} \sum_{n=1}^{(N-1)/2} |h(f_n)|^2 + \frac{\Delta t}{N} |h(f_0)|^2 \tag{8.5}$$

where

$$h(f_0) = \sum_{k=0}^{N-1} x(k\cdot\Delta t) \tag{8.6}$$

This equation allows the evaluation of the global performance of the FFT procedure used, when applied to the given input signal. The value of the 'error', $e$, can be defined as follows:

$$e = \frac{100\% \cdot \left\{ \sum_{k=0}^{N-1} [x(k \cdot \Delta t)]^2 - \frac{2}{N} \sum_{n=1}^{(N-1)/2} | h(f_n) |^2 - \frac{1}{N} \sum_{K=0}^{N-1} x(k \cdot \Delta t) \right\}}{\sum_{k=0}^{N-1} [x(k \cdot \Delta t)]^2} \tag{8.7}$$

The single, conventional windows which are used with FFT spectral analysis cause the central part of the input data to be emphasized, while weights are too small at the extremes of the time series. For example, the Hamming window discards over 0.6 of the statistical information in the time series. The spectral characteristics of conventional windows are also not as good as would be preferred: they have relatively wide central lobes and high side lobes, which cause spectral energy leakage in the calculated spectrum from a component with frequency $f$ to components with neighbouring frequencies $f'$. The best windows, which are characterized by the lowest possible spectral leakage from outside the frequency band $|f - f'| < W$ where $2W$ is some chosen bandwidth, are the so-called discrete prolate spheroidal sequences (Slepian, 1978), also called multitaper windows. These windows are the eigenvectors of the matrix eigenvalue problem:

$$C \cdot \mathbf{v} = \lambda (N, W) \cdot \mathbf{v} \tag{8.8}$$

where the matrix $C$ is as follows:

$$C_{ik} = \frac{\sin [2\pi W(i-k)]}{\pi (i-k)} ; \quad i,k = 0,1, \ldots, N-1, \tag{8.9}$$

$\mathbf{v}$ data vector.

The eigenvectors are mutually orthogonal and should be normalized to 1. The eigenvalues $\lambda_i$ are all positive due to the structure of the matrix $C$. The value $W$ is usually chosen to be equal to $l/(N \cdot \Delta t)$, where $l$ = 2,3,or 4, $N$ is the length of the input time series, and $\Delta t$ is the sampling period. An increase of the length of the resolution band in the frequency domain, $2W$, increases the number of multitaper windows with low leakage properties and decreases the variance of the calculated spectra, but at the same time the spectral resolution becomes worse.

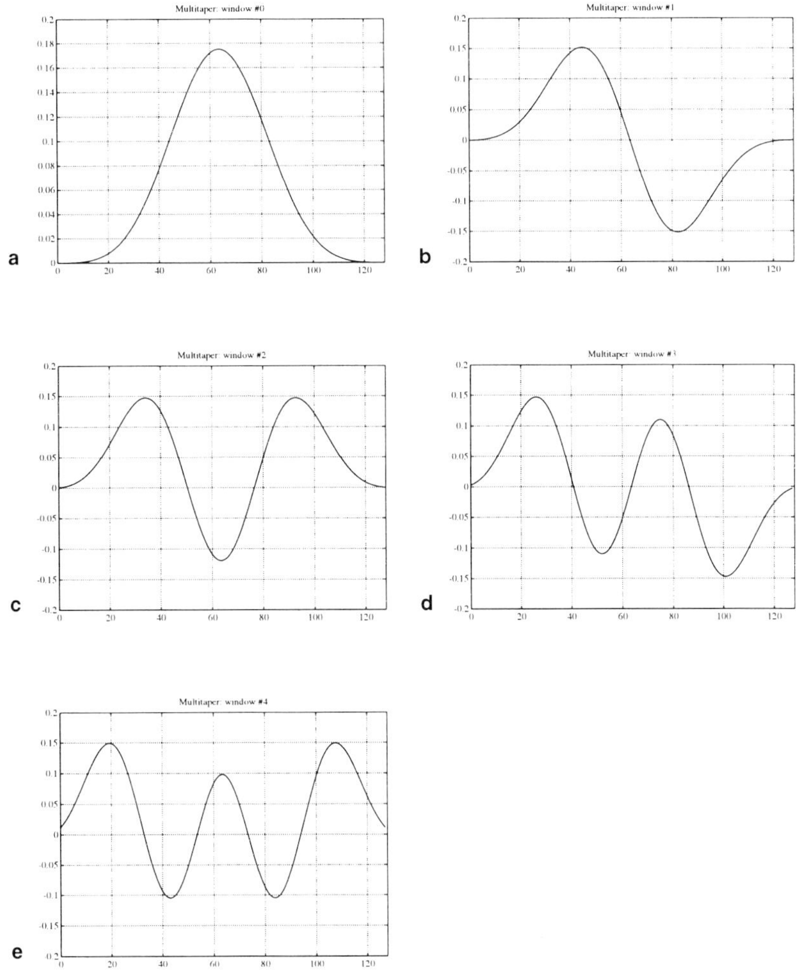


**Fig. 8.1** First five discrete prolate spheroidal sequence data tapers for $N$=128 points, $W = 4/(N \cdot t)$.

In the series of examples shown below, $W = 4/(N \cdot t)$ and the first five eigenvectors with the biggest eigenvalues were used as the multitaper windows. These multitaper windows are shown in Fig. 8.1.

To calculate the spectrum using multitapers, it is necessary to calculate the Fourier transform for every multitaper window separately and then to use some adaptive weighting procedure to obtain the final result (Thomson, 1982; Park *et al.*, 1987b). One such procedure is outlined below.

Let the consecutive weights for frequency $f$ be $w_m^k(f)$ and let $S_k(f)$ be the Fourier transforms for the $k$th multitaper window and frequency $f$. The adaptive multitaper spectral estimate, $S_m^a(f)$, for the $m$th step of the iteration, can be expressed in terms of $w_m^k(f)$ and $S_k(f)$ :

$$S_m^a(f) = \frac{\sum_{k=0}^{K-1} \lambda_k \, [w_m^k(f)]^2 \cdot S_k(f)}{\sum_{k=0}^{K-1} \lambda_k \cdot [w_m^k(f)]^2} \tag{8.10}$$

where

$$w_m^k(f) = \frac{S_{m-1}^a(f)}{\lambda_k S_{m-1}^a(f) + (1-\lambda_k)\,\sigma^2} \tag{8.11}$$

The starting value for $w_m^k(f)$ may be assumed to be a simple average of $S(f)$, i.e initial spectral estimates obtained through the use of the $k$th multitaper window and input data. In practice, the convergence of this procedure is quite fast, so the final number of iterations does not need to be high.

By computing the errors defined by equation (8.7), the behaviour of the given FFT algorithm can be checked when applied to the given data set, for example, as a function of the length of the input data. An example of such error calculations is shown in Fig. 8.4. The input data used for these examples were created by an AR($K$) (autoregressive, of order $K$) process. If $K$ numbers $a[i], i=1,\cdots,K$, are defined, and $\epsilon$ is a driving white noise with variance $\sigma^2$, then the time series of the AR($K$) process can be described by the equation:

$$X(i\cdot\Delta t) = \sum_{k=1}^{K} a[k] \cdot X(i-k, \Delta t) + \epsilon_i \tag{8.12}$$

Two such processes serve input data: the first, for $K = 2$, characterized by low spectral dynamic range, where $a[1]=0.75, a[2]=-0.5$ and the second, for $K = 4$, with high spectral dynamic range, when $a[1]=2.7607, a[2]=-3.8106, a[3]=2.6535, a[4]=-0.9238$. The resulting signals for 1024 points are shown in Fig. 8.2 (for the AR(2) process) and in Fig. 8.3 (for the AR(4) process).

Fortunately, the spectral density function for the AR process can be calculated analytically. For an AR($K$) process the spectral density function, $S(f)$, can be expressed by the means of the coefficients of the process, $a[i]$, its white noise variance, $\sigma^2$, and sampling rate, $\Delta t$, avoiding the direct use of any FFT algorithm:

$$S(f) = \frac{\sigma^2 \cdot \Delta t}{\left| 1 - \sum_{k=1}^{K} a[k] \, e^{2\pi i f k \cdot \Delta t} \right|^2} \tag{8.13}$$

Equation (8.13) is valid for all frequencies from 0 to the Nyquist frequency (it is a continuous function of frequency). Note that $S(f_n)$ for some discrete frequencies, can be approximated by the Fourier coefficients:

$$S(f_n) = \frac{\Delta t}{N} \mid h(f_n) \mid^2 \; ; \quad \text{for} \quad f_n = \frac{n}{N \cdot \Delta t}$$

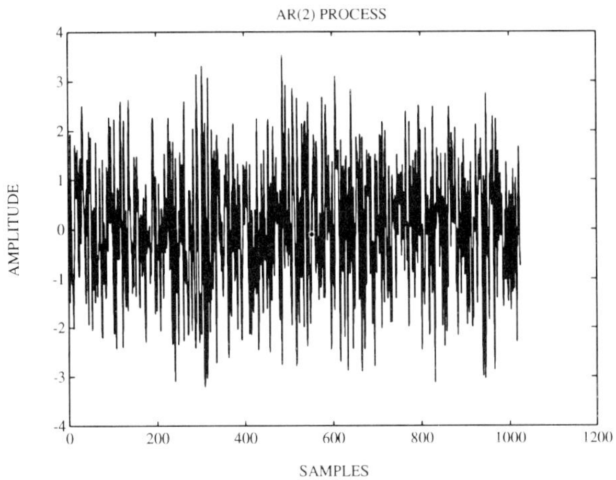


**Fig. 8.2** Auto-regressive process of the second order, generated using equation (8.12), calculated for 1024 points.

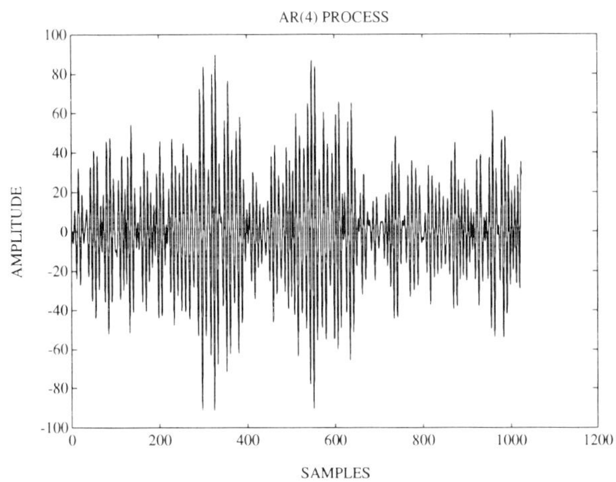


**Fig. 8.3** Auto-regressive process of the fourth order, generated using equation (8.12).

Using Parseval's theorem, one can properly interpret the physical meaning of the calculated spectral density, $S(f_n)$, as the energy of the signal in the frequency band

$$\left(f - \frac{0.5}{N \cdot \Delta t}\ ,\ f + \frac{0.5}{N \cdot \Delta t}\right) \tag{8.15}$$

and the proper name for the function $S(f_n)/\Delta t$ is the power spectral density.

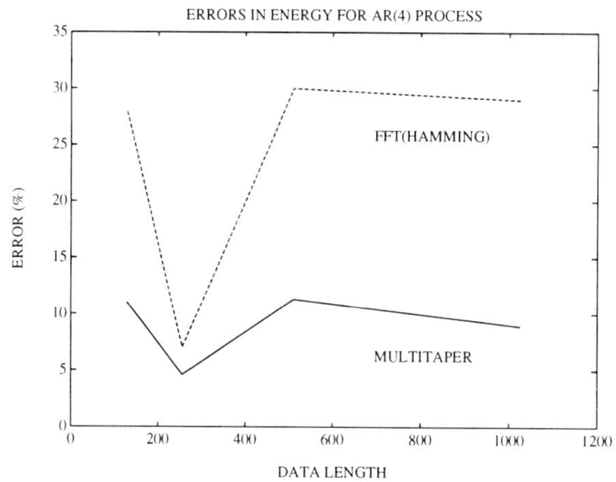


**Fig. 8.4** Dependence of relative error percentage on number of points in input data for FFT, using Hamming window and adaptive multitaper spectral estimation.

Figure 8.4 presents the results of above defined error calculations for a FFT with a single Hamming taper and the resulting errors when the adaptive multitaper spectral estimation was used. Data lengths of 128, 256, 512 and 1024 points are used, as shown on the $x$-axis. It is clearly seen that the calculated values of the error are higher for the FFT with Hamming window than when the adaptive multitaper estimation was used.

The above quantification gives a global evaluation of a performance of the spectral calculation procedure. By using the same input data one can also show that the local variance of the spectral densities, in the case of the FFT with Hamming window, is very high, giving an inaccurate estimation of the extremely important low frequency level of the amplitude spectrum (seismic moment), $h(f)$, which is a simple function of $S(f)$:

$$h(f) = \frac{N}{\Delta t}\sqrt{S(f)} \tag{8.16}$$

This is shown in Fig. 8.5 for the case of the AR(2) process, when the spectral density function is represented by a jagged curve for the FFT with

Hamming window, and by a smooth broken curve for real values (Fig. 8.5a), evaluated by the equation valid for the AR(2) process. The same, but for the adaptive multitaper FFT, is shown in Fig. 8.5b. Similar results are displayed in Fig. 8.6 for AR(4) and FFT with Hamming window, (Fig. 8.6a) and for adaptive multitaper FFT (Fig. 8.6b).

It has been already pointed out that the high level of the local variance can lead to a serious under- or overestimation of the low frequency level of a

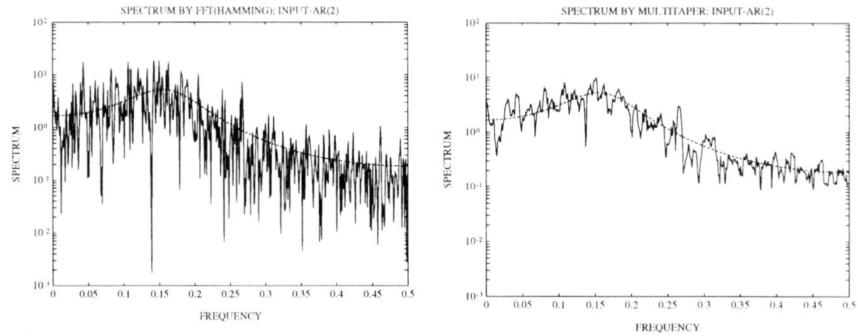


a b

**Fig. 8.5** Spectra of the AR(2) process calculated with a single Hamming Window (left hand side) and with multitaper (right hand side). The broken line shows the correct theoretical spectrum.

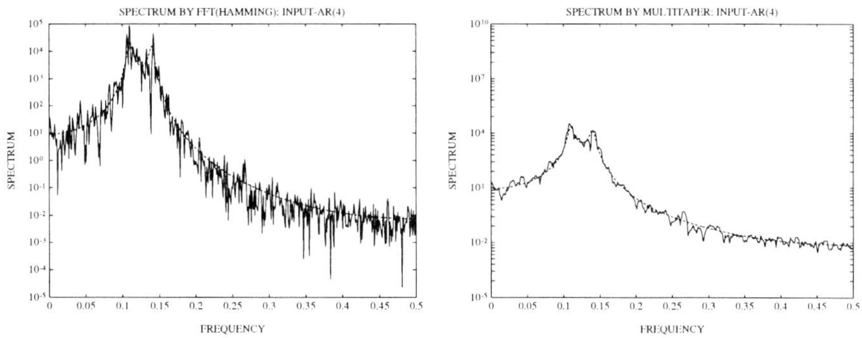


a b

**Fig. 8.6** The same as Fig. 8.5 but for the AR(4) process.

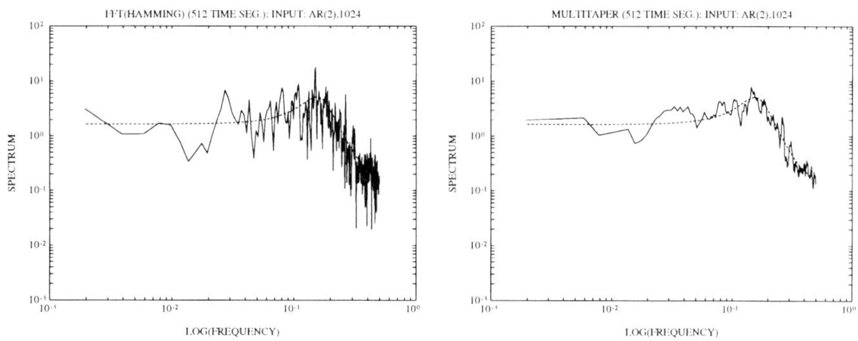


a b

**Fig. 8.7** Spectra of the AR(2) process calculated by segmentation in time using a single Hamming Window (left hand side) and multitaper (right hand side).

displacement spectrum for evaluation of seismic source parameters. As can be seen from Figs. 8.5 and 8.6, the FFT with a single window gives a spectrum which, on average, has smaller values than appropriate, leading to slight underestimation of the energy radiated by the seismic sources. This error can be partially diminished by dividing the input data into several time segments, computing a spectrum for every segment, and averaging for all time segments. The results of such a procedure for an AR(2) process with non-overlapping time segments for the FFT with Hamming window and for adaptive multitaper are shown in Fig. 8.7, once again revealing much better performance when the adaptive multitaper is used.

## 8.2 Source parameters from spectra

Seismic moment, corner frequency and seismic energy are traditionally inverted from the instrument, distance and $Q$-corrected spectrum of each waveform (component) and then averaged. In some cases, one may try to fit the attenuation parameter during the inversion to improve the match between the observed and theoretical spectra – on microseismic events in mines, for example, see Spottiswoode (1993) and Mendecki (1993). Many mine seismic networks record well over 1000 microseismic events daily. To get reliable source parameters every event should be recorded by 5 to 10 three component stations, preferably surrounding the source to ensure reasonable location and adequate sampling of the radiation pattern (Mendecki, 1993). Ten stations, each having recorded 5000 events, present 150 000 spectra to be inverted, which may strain the CPU resources of the seismic network.

A more efficient method is to correct the individual P, SH and SV spectra for instrument response, distance, attenuation and scattering ($Q$) and site effect, if known, then average (or stack) each spectral component over all the stations involved, and finally determine the source parameters from the stacked spectra. This method tends to average out incoherent site effects, source directivity and random fluctuations of the high frequency spectrum. In cases where there is no prior knowledge of $Q$, the best average $Q$ may be determined for all stations during the stacking procedure – the best $Q$ is that which results in the least difference between the Brune model spectrum fitted to the average spectrum and all contributing spectra.

If the site response is coherent, it could easily be enhanced rather than reduced by the conventional averaging of spectra over a number of stations. An effective way to correct spectra for such propagation effects is to employ homomorphic deconvolution in the cepstral domain (e.g. Dysart *et al.*, 1988; Mendecki, 1993). An advantage of this technique is that one needs no prior knowledge about the site effect one is trying to remove. This filtering procedure starts with the displacement amplitude spectrum. A complex function in the frequency domain is then constructed of which the real part is the logarithm of the amplitude spectrum and the imaginary part is zero. Taking the logarithm transforms the product to a sum so that the cepstrum contains the superposition of the source and the spike train. The inverse Fourier transform of this zero phase spectrum is taken and the resulting 'cepstrum' is filtered by excising peaks or troughs and/or by zeroing portions of the trace after the first or second zero crossings. A forward Fourier transform is then performed on the filtered cepstrum and the filtered amplitude spectrum is constructed from the antilog of the transformed function.

A general model to fit the displacement spectra as a function of frequency $f$ of both P and S waves can be given as

$$\Omega(f) = \frac{\Omega_0 e^{-[\pi f R/(V_c Q)]}}{\left[1 + (f/f_0)^{\gamma n}\right]^{1/\gamma}} \tag{8.17}$$

where $\Omega_0$ is the long period amplitude, $f_0$ the corner frequency, $R$ is the hypocentral distance, $V_c$ is the velocity of the respective body waves, $n$ the high frequency fall off rate on a log-log plot and $\gamma$ controls the sharpness of the corner. If $R$ = 0, $n$ =2 and $\gamma$ = 1, equation (8.17) reduces to the spectral shape proposed by Brune (1970):

$$\Omega(f) = \frac{\Omega_0}{1 + (f/f_0)^2} \tag{8.18}$$

Fig. 8.8 illustrates the Brune model of instrument and attenuation corrected far field S wave acceleration, velocity and displacement spectra for a seismic event of $m_M$=0.5.

Boatwright (1980) proposed a modified version of the spectral shape with $\gamma$ = 2 that produces a sharper corner than the Brune model. Abercrombie (1995) found that $\gamma$ = 2 in equation (8.17) produced a better model for the spectra of nearby earthquakes recorded by a three component station positioned in a 2.5 km borehole in southern California.

The lower curve in Fig. 8.9a shows the accumulated square error between the contributing corrected spectra and a Brune model spectrum fitted to the mean, as a function of the value of $Q$ used in the correction. The minimum error is found at $Q$=400. The upper curve shows how the value of the Brune model corner frequency $f_0$ varied at the same time. Fig. 8.9b shows the spectra, raw (+) and after correction for station hypocentral distances and

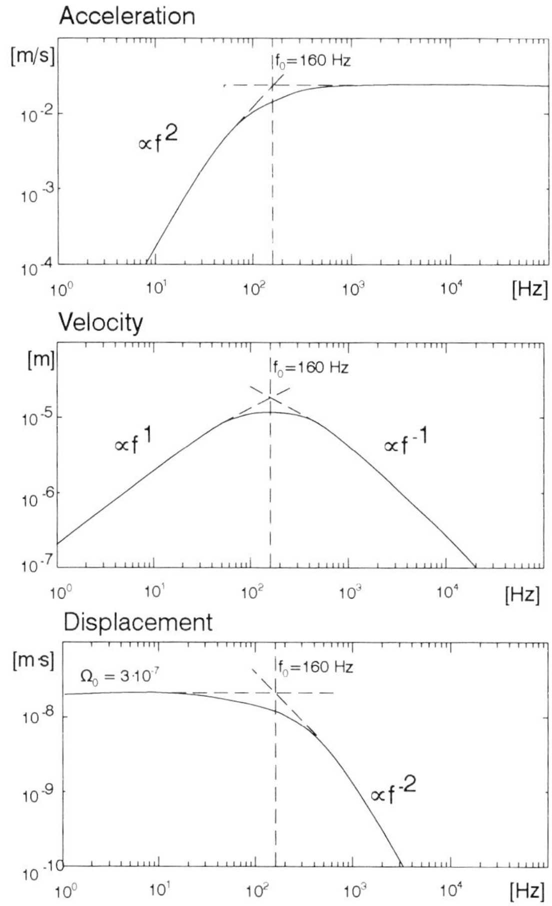


**Fig. 8.8** Models of instrument and attenuation corrected far field S wave acceleration, velocity and displacement spectra for a seismic event of moment magnitude $m_M = 0.5 = (2/3) \log M - 6.1$ and seismic moment $M$=7×10$^9$ Nm, where S wave velocity is 3500 m/s and rock density is 2700 kg/m$^3$.

$Q$=400 (·). The solid line is the best fit of Brune's model to the corrected spectra. The same functions are shown in Fig. 8.10 for a different event. As is usual when $Q$<300, the minimum error is much more clearly defined.

An objective automatic procedure for determining source parameters from spectra is a necessity in daily processing of large numbers of microseismic events in mines. A very efficient algorithm for calculating source parameters from the following velocity $S_{V2}$ and displacement $S_{D2}$ power spectra, has been proposed by Andrews (1986):

$$S_{V2} = 2\int_0^\infty V^2(f)\,df; \qquad S_{D2} = 2\int_0^\infty D^2(f)\,df \tag{8.19}$$

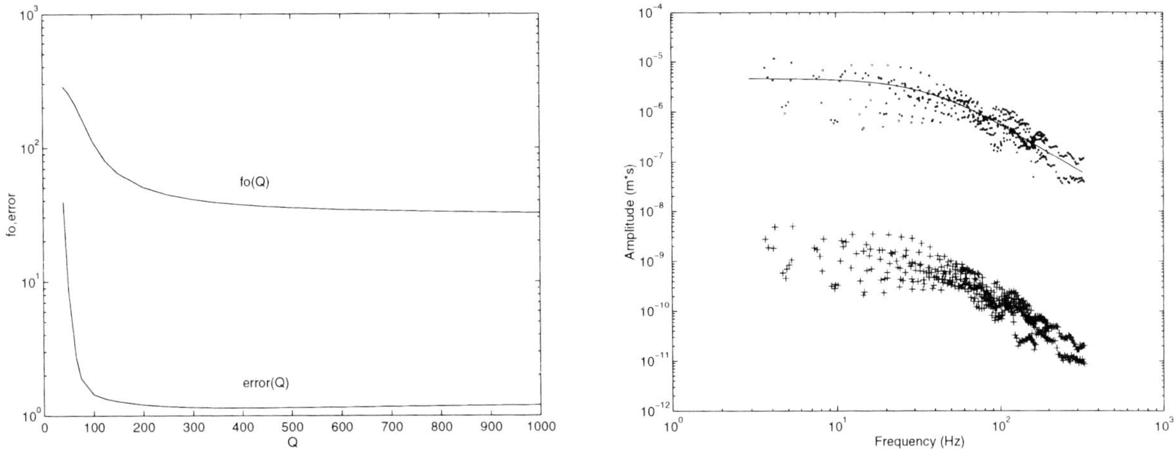


a. b.

**Fig. 8.9**
**a** The lower curve 'error($Q$)' shows the squared differences between the spectra corrected on the attenuation $Q$ and the Brune model based on the same correction. The best choice for $Q$ is $Q$ = 400. The second curve, '$f_0$(Q)' shows the values of the corner frequency calculated for different $Q$ (seismic event of magnitude $m_M$=1.1).
**b** Spectra before correction (+) and after correction for station hypocentral distances and for $Q$=400 (·). The solid line is the best fit of Brune's model to the corrected spectra.

Seismic moment $M$, radiated energy $E$ and Brune model corner frequency $f_0$ can then be calculated from the respective power spectra as follows:

$$M = 4\pi \rho V_c^3 \Omega_0 \cdot R_p \tag{8.20}$$

$$E = 4\pi \rho V_c S_{V2} \tag{8.21}$$

$$f_0 = \frac{1}{2\pi} \sqrt{S_{V2}/S_{D2}} \tag{8.22}$$

$$\Omega_0 = 2 \cdot S_{V2}^{-1/4} \cdot S_{D2}^{3/4} \tag{8.23}$$

In equations (8.20 – 8.23), $V_c$ stands for a given body wave velocity at the source region, $D$ for displacement, ρ for rock density and $R_p$ takes into account averaged radiation pattern, $R_p$=0.55 for P and $R_p$=0.63 for S wave respectively (Aki and Richards, 1980).

In practice, the integrations in equation (8.19) are performed between finite limits. The low frequency limit, $f_1$, may be set at $1/\Delta T$ by the duration $\Delta T$ of the data available for the Fourier transform, which does not produce statistically reliable results for lower frequencies, or it may be restricted by the sensor or system frequency response, or by low frequency noise. Similarly, the highest useable frequency, $f_2$ may be set by the Nyquist frequency ($1/(2\Delta t)$, where $\Delta t$ is the sampling interval) or by the sensor or system response, or noise characteristics.

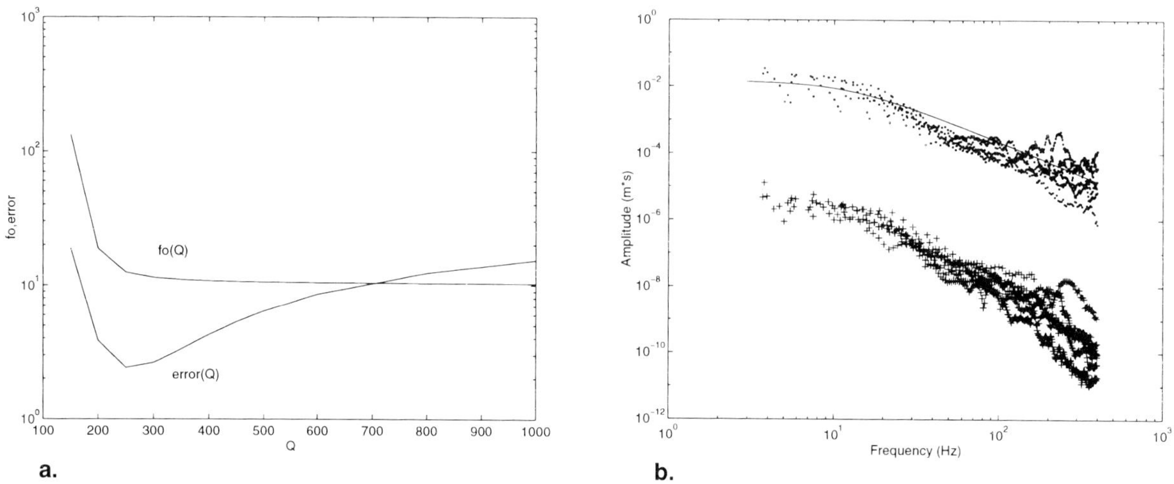


**Fig. 8.10** The same as Fig. 8.9 but for the case where the best choice for $Q$ is 250 (seismic event of magnitude $m_M$=3.8).

Since we have assumed the Brune model spectrum for $D(f)$, it is possible to calculate the effect of reduced bandwidth on $S_{V2}$ and $S_{D2}$ and thus on the source parameters (Di Bona and Rovelli, 1988). Specifically, if we let

$$u_1 = f_1 / f_0$$

$$u_2 = f_2 / f_0$$

and

$$A = \arctan(u_2) - \arctan(u_1)$$

$$b = \frac{u_2}{1 + u_2^2} - \frac{u_1}{1 + u_1^2}$$

then

$$\hat{S}_{V2} = \int_{f_1}^{f_2} V^2(f)\mathrm{d}f = S_{V2} \cdot \frac{A - B}{\pi/2}$$

$$\hat{S}_{D2} = \int_{f_1}^{f_2} D^2(f)\mathrm{d}f = S_{D2} \cdot \frac{A - B}{\pi/2} \tag{8.24}$$

where $\hat{S}$ indicates the underestimated value from the band limited signal.

While it is not possible to invert equation (8.24) for $S_{V2}$ and $S_{D2}$, because of the dependence on $f_0$, the results are monotonic, allowing a numerical solution.

An example of raw and corrected spectra is shown in Fig. 8.11a. The effect has been artificially enhanced in Fig. 8.11b by further restricting the bandwidth of the same data used in Fig. 8.11a, resulting in a correspondingly larger correction. Discarding high frequency data has also caused the estimate of $Q$ for the P wave to increase.

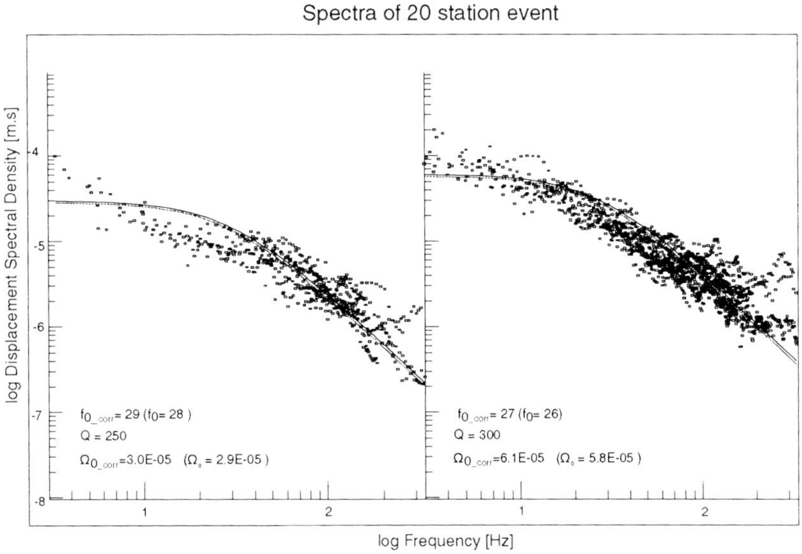


**a.**

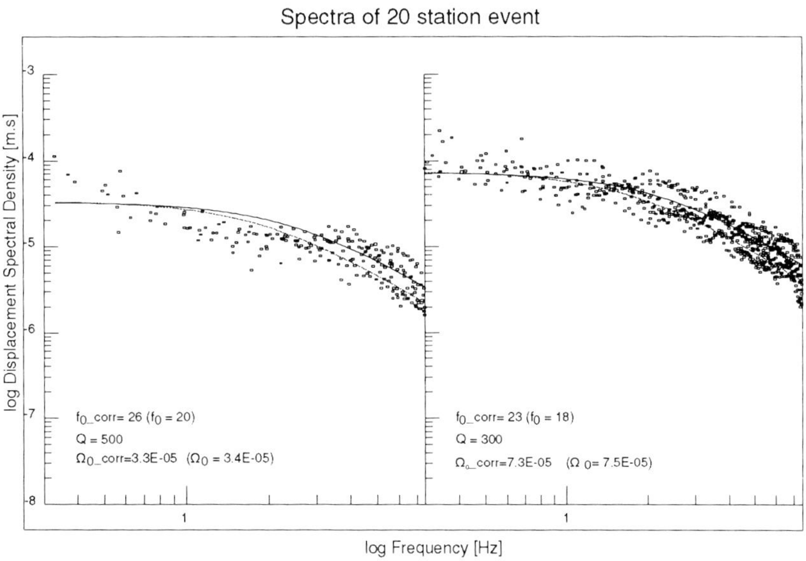


**b.**

**Fig. 8.11** The effects of correction for finite system bandwidth. The individual attenuation corrected spectra, (squares) for P wave (left panel) and S wave (right panel), are shown together with the Brune model spectrum fit, before (thin line) and after (thick line) correcting for the effects of finite system bandwidth. The same original data are shown, but the frequency range has been artificially restricted to 75 Hz in **b**, producing a larger correction than the full 350 Hz range used in **a**. Discarding high frequency data makes the effect of $Q$ less pronounced, and has caused the estimate of $Q$ for the P wave to change in **b**.

The seismic moment for the S wave is calculated as a geometrical average of moments for SH and SV waves:

$$M_s = \sqrt{M_{SH}^2 + M_{SV}^2} \tag{8.25}$$

and the total scalar moment is assumed to be the arithmetic average of moments for the P and S wave

$$M = (M_P + M_S)/2 \tag{8.26}$$

The magnitude $m_\mathrm{M}$ of a seismic event can then be estimated according to the Hanks and Kanamori (1979) moment-magnitude scale

$$m_M = (2/3)\ log M - 6.1 \tag{8.27}$$

The total radiated seismic energy is

$$E = E_P + E_{SH} + E_{SV} \tag{8.28}$$

and an average corner frequency

$$f_0 = (f_{0P} + f_{0SH} + f_{0SV})/3$$

The source radius $r$ can be estimated from the Brune model as

$$r = \frac{k\ V_c}{2\pi\ f_0} \tag{8.29}$$

where $k = 2.34$ for the S wave, assuming instantaneous stress release. For a quasi-dynamic circular source, Madariaga (1979) assumed $k$ for the P wave to be 2.01 and for the S wave to be 1.32. An average stress drop can then be calculated given the mean estimates of seismic moment and source radius (Brune, 1970),

$$\Delta\sigma = \frac{7M}{16 r^3} \tag{8.30}$$

Note that when using equations (8.22) and (8.23), the Brune stress drop and apparent stress $\sigma_A = \mu E/M$ are not independent parameters for a spectrum of any shape (Andrews, 1986).

# 9 Nonlinear Dynamics of Seismic Flow of Rock

Nonlinearity and chaos are common features of many systems in nature, (Gutzwiller, 1991; Radu, 1995) though this has been acknowledged largely only during the past two decades. The phenomenon which goes under the name 'chaos' is an intrinsic feature of dynamics and comprises the large sensitivity to initial conditions of the evolution of a nonlinear dynamical system. In practice, it may happen that the response of a physical system varies greatly when small changes occur in the environment containing the physical system. This type of behaviour is regarded as chaotic.

Special methods and techniques have emerged from the study of systems that exhibit the phenomenon of chaos (Heiss *et al.*, 1995). Their main merit is that they can sometimes offer alternative avenues to the traditional approach which consists of solving exactly the relevant equations of motion. Moreover, in the absence of a self-consistent model describing the dynamics of the process under study, it is not possible to gain any insight into the problem without the aid of more refined techniques. It is accepted that the occurrence of seismic events cannot currently be modelled with easy-to-use analytic equations. A study of the physics of seismicity is therefore expected to rely mostly on accurate data interpreted with the aid of nonlinear dynamics.

The seismic flow of rock is regarded as a highly complex process, intermittent in space and time, that provides for the occurrence of turbulence (Kagan, 1992; Chapter 10 of this book). In turn, turbulence is commonly associated with chaotic behaviour (Radu, 1995). It therefore appears that it is sensible enough to assume chaos may set in within the dynamics underlying the seismic flow of rock. In real systems, chaos and order coexist. By order we mean structures exhibiting a high degree of spatio-temporal regularity. Chaos and order both arise from the same type of nonlinear laws (Gutzwiller, 1991) and are often inseparable.

In order to make predictions about the occurrence of rockmass instabilities that may develop during the mining process, we need to manipulate the novel features of chaos theory. There is a connection between chaotic behaviour and instability. A system that is completely random but non-chaotic has a very stable configuration that may persist even in the presence of external perturbations. On the other hand, a system that is ordered may become unstable in the presence of external forces as its entropy has the opportunity to increase. The coupling between a regular system and external perturbations is in general nonlinear and leads to chaotic behaviour. Rockmass subjected to external conditions (e.g. mining) is regarded as a nonlinear dissipative system far from equilibrium. In such a system, instability is the uncertainty about the path to be followed by the system in its phase space, a concept to which we turn next.

## 9.1 Phase space

When a nonlinear-type analysis is performed, the main object of study is the system's 'phase space'. The phase space of a physical system is the set of all possible states that the system can have. That collection of states can be spanned by a number of canonically independent variables (Gutzwiller, 1991). To render the notion of phase space clearer, we illustrate below the simple case of the one dimensional harmonic oscillator. This physical system is described by the following energy conservation relation:

$$E = \frac{m\omega^2}{2}x^2 + \frac{p^2}{2m} \tag{9.1}$$

where $x$ is the position coordinate, $p=m\dot{x}$ is the momentum coordinate, $m$ is the mass of the oscillator, $\omega$ is the oscillation frequency and $E$ is the energy of the oscillations. The variables that will span the phase space are $x$ and $p$. If we represent the above equation in the system's phase space we get an ellipse. All the possible states of the system (harmonic oscillator) lie on that particular ellipse (see Fig. 9.1). This configuration corresponds to a perfectly stable kind of motion.

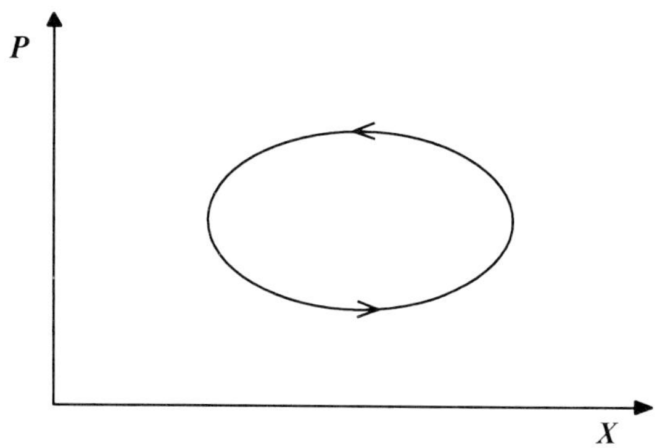


**Fig. 9.1** Phase space configuration of the harmonic oscillator.

We can construct phase space configurations for any physical system. Such configurations are, for instance, in use in the study of instability and failure modes in geomaterials (Mróz, 1985). If we denote by $u$ the displacement at a typical point of interest, then the plane $(u, \dot{u})$ (where $(\dot{u})$ is the rate of change in displacement) will constitute the phase space of that particular material point. If consecutive points in the phase space are joined, one arrives at a curve along which the dynamics of the process evolves with time. We have generated analytically three deformation modes which we present for illustration together with the corresponding stress–strain diagrams, respectively. The first case is a 'stable mode' deformation where the displacement varies according to the equation

$$u(t) = \frac{t^2}{t^2 + 1} \sin(3t) \tag{9.2}$$

where $t$ is time. The resulting phase space $(u, \dot{u})$ is a curve which tends towards a limiting cycle, as shown in Fig. 9.2a. The corresponding stress–strain diagram can be derived according to the following discussion: strain $(\epsilon)$ is proportional to $u$ while stress $(\sigma)$ is proportional to the applied force which in turn is proportional to $\ddot{u}$, $\ddot{u}$ being the acceleration of the point of interest. We show in Fig. 9.2A the stress–strain diagram that corresponds to this stable deformation mode.

In the case of a cyclic unstable mode of deformation that can be modelled with the formula

$$u(t) = t^2 \sin(9t), \tag{9.3}$$

the corresponding phase space consists of a spiral that diverges as time increases. This is illustrated in Fig. 9.2b. In Fig. 9.2B we display the stress–strain diagram that corresponds to this type of deformation.

Finally, we consider the burst which can be described by

$$u(t) = \frac{(t^7 + t)}{10} \tag{9.4}$$

which leads to an almost sudden increase with time. The phase space diagram is a monotonous curve that diverges as time increases. The stress–strain diagram, however, indicates a very sharp drop in the stress as soon as the softening regime is entered. Figure 9.2c and Fig. 9.2C show these features.

From the analysis of the above deformation modes we notice that the first situation is the most stable while the burst is least stable. The conclusion is that the more phase space a trajectory can explore within a certain time, the more unstable the corresponding deformation is.

The dimension of the phase space is determined by the number of degrees of freedom. If $x_1, x_2, \ldots, x_n$ and $p_1, p_2, \ldots, p_n$ are the canonical parameters that fully describe the dynamics of the system under study, then the dimension of the phase space is $2N$. As time evolves, the state of the system moves along a trajectory in the phase space. States that happen to be on a closed boundary at a given time remain on a closed curve at any time in the future.

If two nearby trajectories tend to diverge exponentially from each other, that behaviour is termed 'chaotic'. The onset of chaos is in general a consequence of the nonlinear coupling (Radu, 1995; Heiss *et al.*, 1995) that takes place between the canonical variables. Such a coupling is frequently seen in the differential equations that usually relate the canonical variables to each other. Rockmass response to mining is a complex process that qualifies as a good example. The coupling between a regular system and external forces is seen, for instance, when a rock cavity is subjected to the external stress of the surrounding rockmass. A number of quantitative measures have been introduced to quantify the degree of chaos in a given system. The most commonly used one is the Lyapunov exponent (Gutzwiller, 1991) which measures the average exponential divergence rate of nearby trajectories.

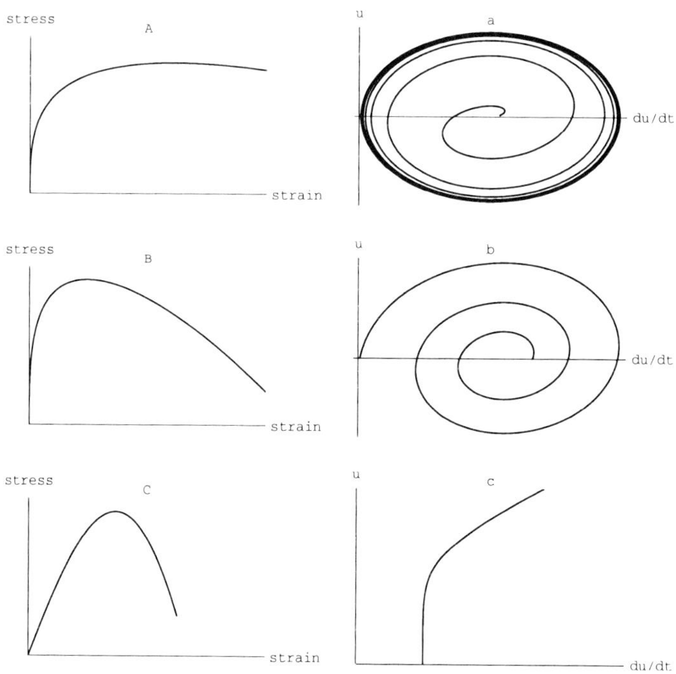


**Fig. 9.2** Trajectories in phase space for plastic deformations.

In conservative systems – where the total energy of the system remains constant – the volume of the phase space is constant. The flow of the trajectories winds around a toroidal shape which does not deform in time (see Fig. 9.3). With increasing chaotic behaviour the tori may break down and disintegrate, yet the volume of the phase space stays the same. A completely different situation is encountered in the case of dissipative systems which are kept far from thermodynamic equilibrium by mass and energy exchange with the environment. There, as the energy dissipates, the dynamics tends to vanish or reach a steady state.

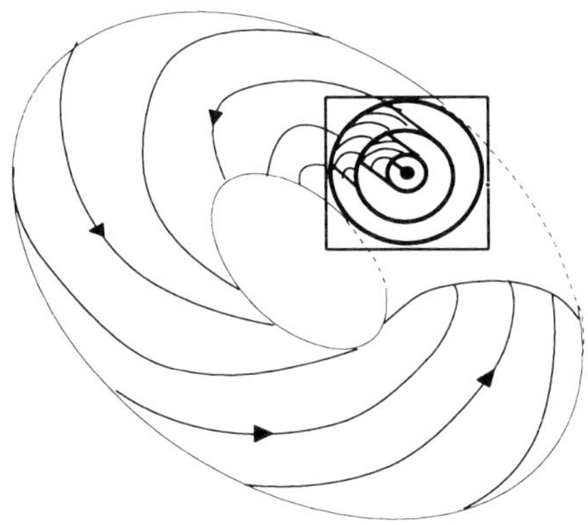

**Fig. 9.3** Invariant tori in the three dimensional surface of constant energy.

This is seen in the phase space as a continuous contraction of the volume. Moreover, the trajectories are gradually 'attracted' to and 'captured' by an attractor. An attractor is a submanifold of the phase space (e.g. a point, a closed curve, etc.) towards which all the trajectories tend to settle. If the dissipative system is also chaotic, the attractor usually has a fractal dimension and is called a 'strange attractor'. In such a case, to study the dynamics of the system, it suffices to only investigate the attractor. Figure 9.4 illustrates the Lorenz attractor which is used in weather and climate studies. There, we clearly distinguish two limit cycles (the attractor itself) surrounded by basins of attraction. A basin of attraction is the neighbourhood surrounding the attractor within which any trajectory will eventually 'sink'. In general, dissipative systems need not contract all volumes in the phase space, some sets may expand their volume with time, but such submanifolds do not contain attractors.

Making predictions about the physical parameters of the system under study amounts to estimating the evolution of the states in the corresponding phase space. As the seismic flow of rock is *par excellence* dissipative, we will need to identify and investigate the relevant attractors that may be encountered in the phase space. Recall that the complexity of a real system is measured by the dimension of the attractor describing its asymptotic motion. Using nonlinear techniques, the basin of each attractor can be divided into subsets of given levels of predictability (e.g. low, moderate, high). The ultimate aim is to map present states of the system into the future, based on knowledge about the time history of the system. Depending on the local level of predictability on the attractor – which in fact results from the computation of Lyapunov exponents – one can evaluate prediction functions that will interpolate into the future various seismic parameters such as apparent volume, energy index, seismic viscosity, etc. Although the technique does not lead to exact solutions, it can supply good estimates.

## 9.2 Reconstruction of the phase space from seismic data

In Chapter 10 of this book, a number of seismic parameters which quantify rockmass stability, such as seismic moment, seismic energy, energy index, apparent volume, apparent stress, seismic viscocity, seismic Deborah number and seismic Schmidt number are described. To completely reveal the dynamics behind rockmass instability, the phase space needs to be constructed. Although the seismic parameters should be considered as phase variables, it is not known whether they are all independent and canonical. Moreover, their number may also be smaller than the minimum number of independent physical variables that 'fully' span the phase space. A solution to this apparent inconvenience is to avoid the direct study of the physical phase space. Instead, an analogous space is introduced – the embedded phase space. The embedded phase space aims at reconstructing the original phase space – which, for some reason, cannot be constructed directly. If $a$, $b$, $c$, $d$, ... are the parameters that completely describe the dynamics of the system and their time series are known, one would be able to construct the 'real' phase space. However, since the parameters $a$, $b$, $c$, $d$, ... are related to each other by 'the same dynamics', a theorem by Takens (1984) ensures that it is sufficient to use only one physical variable to 'simulate' the actual phase space. The method that achieves this is called 'embedding'. An attractor that exists in the 'real' phase space will be seen in the 'embedded' phase space as well. The difference between the two attractors, 'real' and 'embedded', can be recovered by smooth transformations. This means that a thorough knowledge about the 'embedded' attractor is equivalent to understanding the behaviour of the trajectories on the 'real' attractor.

In the embedding of the phase space a given physical parameter is chosen, such as the seismic moment, $M$. The time series of the seismic moment is denoted by $M_i$. The 'embedded' phase space is then constructed by using as

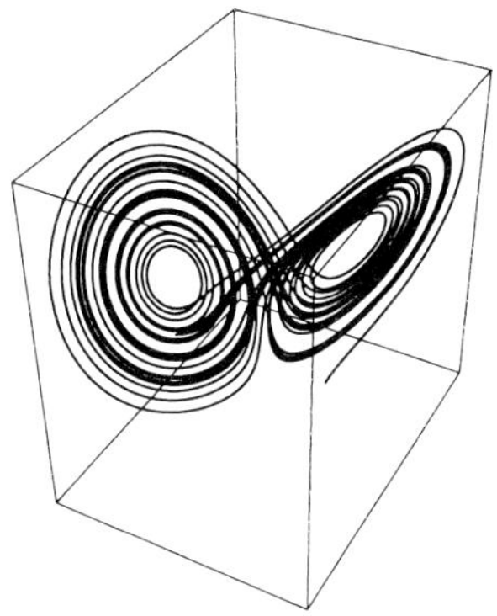

**Fig. 9.4** Illustration of the Lorenz attractor.

orthogonal axes the vectors ($M_i$, $M_{i+D}$, $M_{i+2D}$, ..., $M_{i+(d-1)D}$), where $d$ is the dimension of the 'embedded' phase space and $D$ is the so-called delay lag. Notice that since the attractor is a submanifold of the 'real' phase space, it is not necessary to use a value of $d$ equal to the dimension of the 'real' phase space. However, $d$ should be sufficiently large to embed the attractor that needs to be studied. The delay $D$ is chosen as the first zero of the linear autocorrelation function (Abarbanel, 1990) defined by:

$$C(m) = \frac{\sum_{i=1}^{N-M} (M_i - \bar{M})(M_{i+m} - \bar{M})}{\sum_{i=1}^{N} (M_i - \bar{M})} \qquad (9.5)$$

where:

$$\bar{M} = \frac{1}{N} \cdot \sum_{i=1}^{N} M_i$$

In Fig. 9.5 a cut (Poincaré map) through the embedded phase space of log$M$, using data from the area P1PILLAR (President Brand Mine, Freegold, South Africa), is shown. The appearance of an attractor, with its basin of attraction (lower density of points), can be clearly discerned.

The significance of the lag $D$ is that only $D$ consecutive events are linearly dependent on each other. Therefore, the axes $M_i$ , $M_{i+D}$, ...., $M_{i+(d-1)D}$ are linearly independent, hence orthogonal, and can fully span the phase space if $d$ is properly determined. The lag $D$ can also be used to define the spatial correlation range of seismic events.

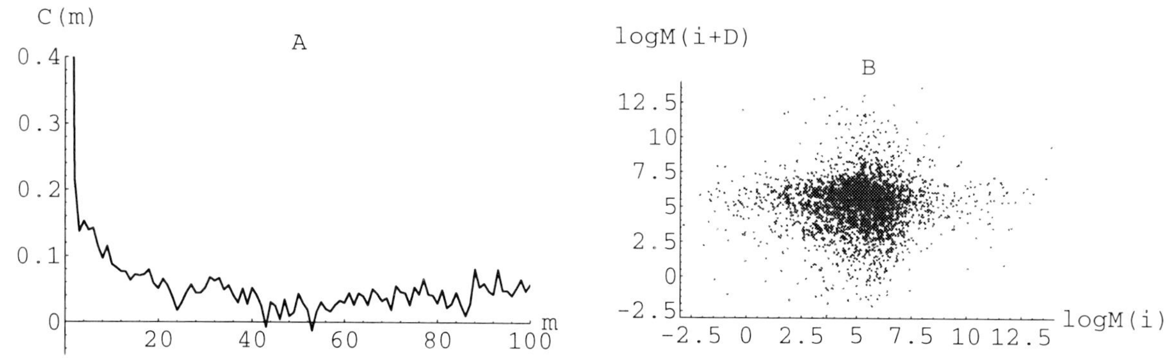


**Fig. 9.5**
(A) Autocorrelation function of the log$M$ and (B) Poincaré map of the log$M$ attractor as at P1PILLAR area.

We first consider the first $D$ events that occurred in chronological order. Since these events are by definition linearly correlated, we define the spatial correlation range of this group of events as the maximum distance between the first event and the remaining D−1 events of the group:

$$l_1 = \max_{i=2}^{D} \|\mathbf{r}_1 - \mathbf{r}_i\| \tag{9.6}$$

Next, we consider the events 2, 3, ..., $D$+1, again as they occurred chronologically. The spatial correlation range of this new group of events we define as:

$$l_2 = \max_{i=3}^{D+1} \|\mathbf{r}_2 - \mathbf{r}_i\| \tag{9.7}$$

We can repeat this procedure up to the event $N$–$D$+1:

$$l_{N-D+1} = \max_{i=N-D+2}^{N} \|\mathbf{r}_{N-D+1} - \mathbf{r}_i\| \tag{9.8}$$

Finally, we define as the spatial correlation range of the entire cluster of events the following average:

$$l_c = \frac{1}{N-D+1} \sum_{i=1}^{N-D+1} l_i \tag{9.9}$$

The spatial correlation range of a cluster of events depends, however, on the parameter used to calculate the first zero of the autocorrelation function, which is the lag $D$. One can therefore average over the spatial correlation ranges calculated in the case of log$E$, log$M$, (x,y,z)-location and time between consecutive events ($\Delta t$) and derive the 'average spatial correlation range' of a cluster of seismic events:

$$<l_c> = \frac{1}{4}\left(l_c^{(E)} + l_c^{(M)} + l_c^{(x,y,z)} + l_c^{(\Delta t)}\right) \tag{9.10}$$

## 9.3 Fractal correlation dimension

As already mentioned, the dynamics of chaotic systems that are dissipative exhibits strange attractors, i.e. attractors whose dimensions are fractal. There are various methods that are used to calculate the dimension of an attractor. In seismological studies, the most popular one calculates the correlation integral (Eneva, 1996) which is defined as:

$$C(r) = \frac{2}{(N-D+1)(N-D)} \cdot \sum_{i,j} \Theta\left(r - \|\mathbf{r}_i - \mathbf{r}_j\|\right) \tag{9.11}$$

where $\Theta(x)$ is the Heaviside function, i.e. $\Theta(x) = 1$ if $x>0$ and $\Theta(x) = 0$ otherwise. $\mathbf{r}_i$ and $\mathbf{r}_j$ are the physical coordinates of any two points of a discrete version of the attractor and $\|\mathbf{r}_i - \mathbf{r}_j\|$ is the distance between the points $i$ and $j$, given by the equation:

$$\|\mathbf{r}_i - \mathbf{r}_j\| = \sqrt{\sum_{k=1}^{d} \left(r_{i+(k-1)D} - r_{j+(k-1)D}\right)^2} \tag{9.12}$$

where $D$ is the lag (delay) corresponding to the parameter $r$ and $d$ is the embedding dimension. The parameter $r$ can be, for instance, location $(x,y,z)$, log$E$, log$M$, time between consecutive events, etc.

In the case of a fractal, the correlation integral scales as:

$$C(r) \propto r^{d_a} \tag{9.13}$$

where $d_a$ is the dimension of the fractal. For a given embedding dimension $d$, the fractal correlation dimension can be estimated as the slope of the linear portion of the plot of log$C(r)$ versus log$r$, which generically looks like Fig. 9.6.

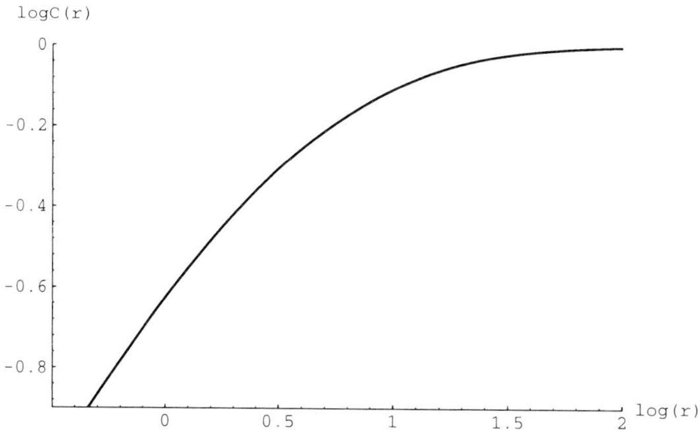


**Fig. 9.6** Generic plot of $\log C(r)$ versus $\log r$. The slope of the linear portion represents the correlation dimension.

As $r$ increases, $C(r)$ reaches its maximum value (i.e. 1) as the distances $\|\mathbf{r}_i - \mathbf{r}_j\|$ can never be larger than some $r_{max}$.

Finally, plotting $d_a$ versus $d$, one may establish the minimum embedding dimension which fully embeds the attractor. This is very important as an improper embedding of the attractor may lead to severe errors when estimates about the future behaviour of a physical parameter of interest are made. Generically, the plot of $d_a$ versus $d$ looks like Fig. 9.7. When $d$ reaches a certain value $d_{min}$, the fractal correlation dimension $d_a$ no longer increases. At that point ($d_{min}$) a number of results are established: the fractal dimension of the attractor is the value of $d_a$ corresponding to $d_{min}$. Moreover, if $d_a$ is fractional, there is chaos in the system, as strange attractors – attractors of fractal dimension – can be exhibited by chaotic systems only. Finally, the value $d_{min}$ denotes the 'proper' embedding dimension necessary to reconstruct the phase space and to make accurate estimates of physical parameters of interest.

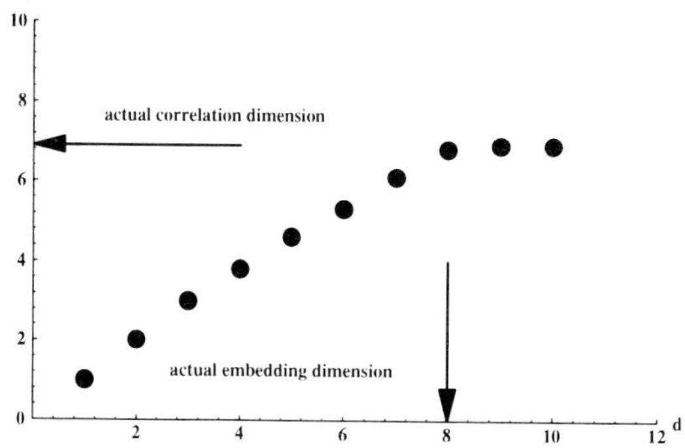


**Fig. 9.7** Generic plot of correlation dimension versus embedding dimension. The flattening point of the plot determines the dimension ($d_{min}$) of the reconstructed phase space and the dimension $d_a$ of the attractor.

## 9.4 Numerical results

Clusters of seismic events from the areas WHPOST, P1PILLAR and PB2EAST in the Welkom goldfield have been investigated in connection with possible chaotic behaviour. In Table 9.1 we summarize the main features of these areas.

**Table 9.1** Features of interest in the investigated areas

| | Area: WHPOST | | Area: P1PILLAR | | Area: PB2EAST | |
|---|---|---|---|---|---|---|
| Parameter | $(D)$ | SCR (m) | $(D)$ | SCR (m) | $(D)$ | SCR (m) |
| log(Moment) | 171 | 335.1 | 43 | 309.7 | 9 | 162.6 |
| log(Energy) | 171 | 335.1 | 43 | 309.7 | 9 | 162.6 |
| $x$ location | 26 | 263.8 | 23 | 251.8 | 22 | 213.7 |
| $y$ location | 47 | 288.5 | 12 | 200.0 | 25 | 221.5 |
| $z$ location | 72 | 307.1 | 8 | 149.1 | 32 | 236.9 |
| $s_e$ | 25 | 261.1 | 20 | 239.6 | 22 | 213.7 |
| $s_{ph}$ | 25 | 312.1 | 20 | 269.4 | 22 | 256.2 |
| $\Delta t$ | 136 | 329.1 | 31 | 278.6 | 11 | 174.0 |
| Average correlation length (m) | | 303.9 | | 250.98 | | 205.1 |

$(D)$ – Delay lag; SRC – Spatial Correlation Range

A few comments are in order at this stage. First of all, we distinguish between two kinds of distance between seismic events: the distance between the spatial location of the events ($s_e$) and the distance between the events taking into account the source size ($s_{ph}$). The relation between the two distances is:

$$s_{ph} = s_e + r_i + r_j \tag{9.14}$$

where $r_i$ and $r_j$ are the radii of the apparent volumes of the events $i$ and $j$ respectively. From the results tabulated, it appears that the number of linearly dependent events is the same whether $s_{ph}$ or $s_e$ is used as input for the

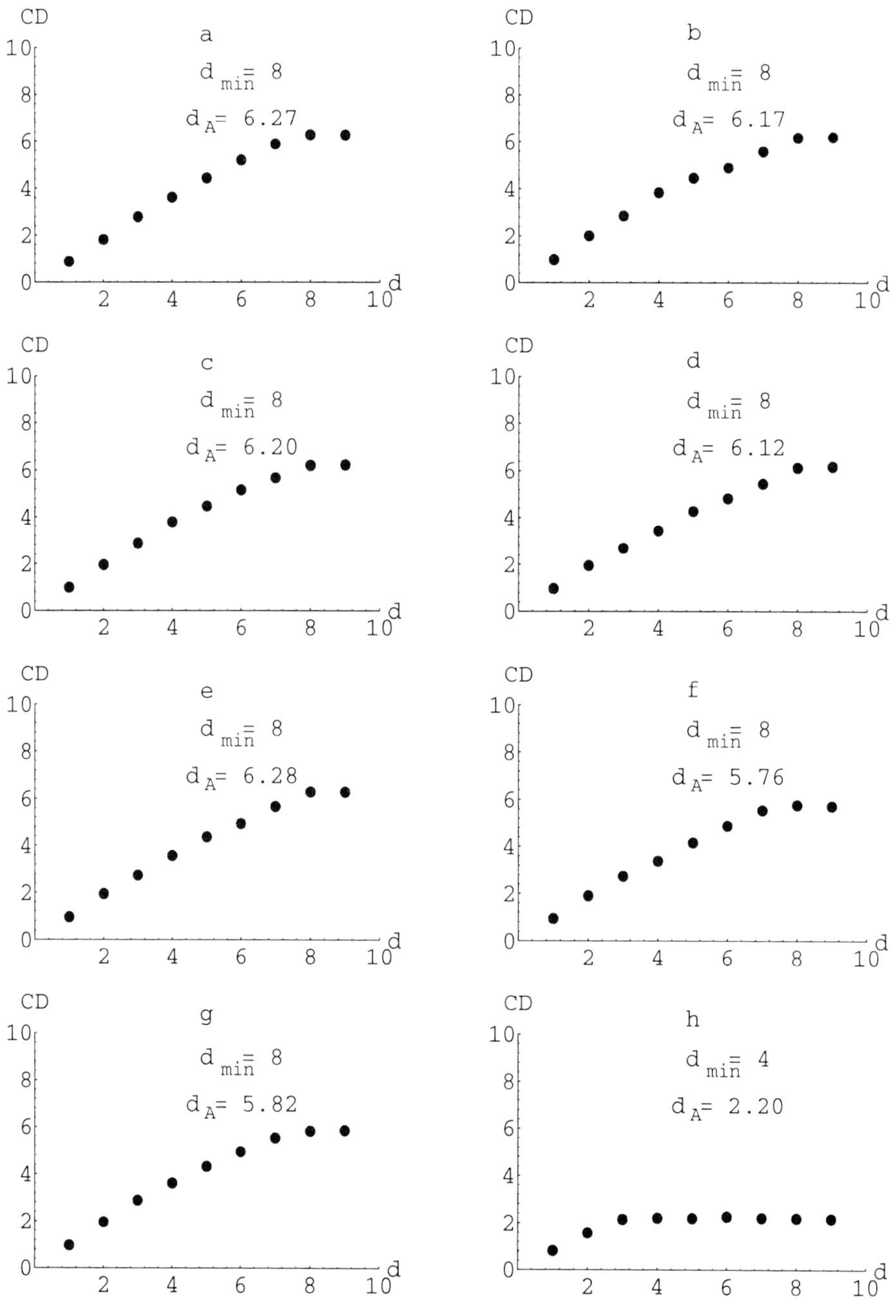


**Fig. 9.8** Correlation dimension analysis for the WHPOST area, where the parameter under study in each figure is respectively:
a. log$E$
b. log$M$
c. x location
d. y location
e. z location
f. distance between locations of consecutive events
g. distance between locations of consecutive events including the source size
h. time between consecutive events

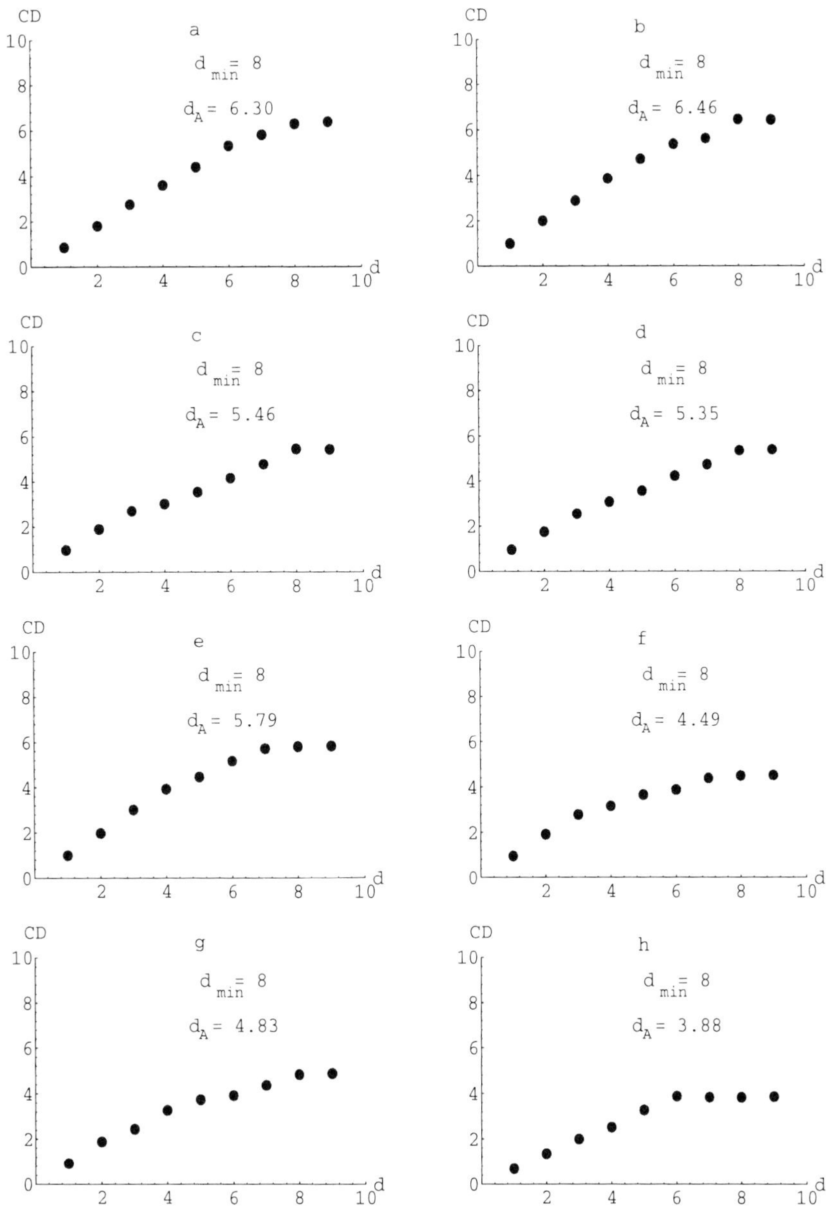


**Fig. 9.9** Correlation dimension analysis for the P1PILLAR area, where the parameter under study in each figure is respectively:
a. log$E$
b. log$M$
c. x location
d. y location
e. z location
f. distance between locations of consecutive events
g. distance between locations of consecutive events including the source size
h. time between consecutive events

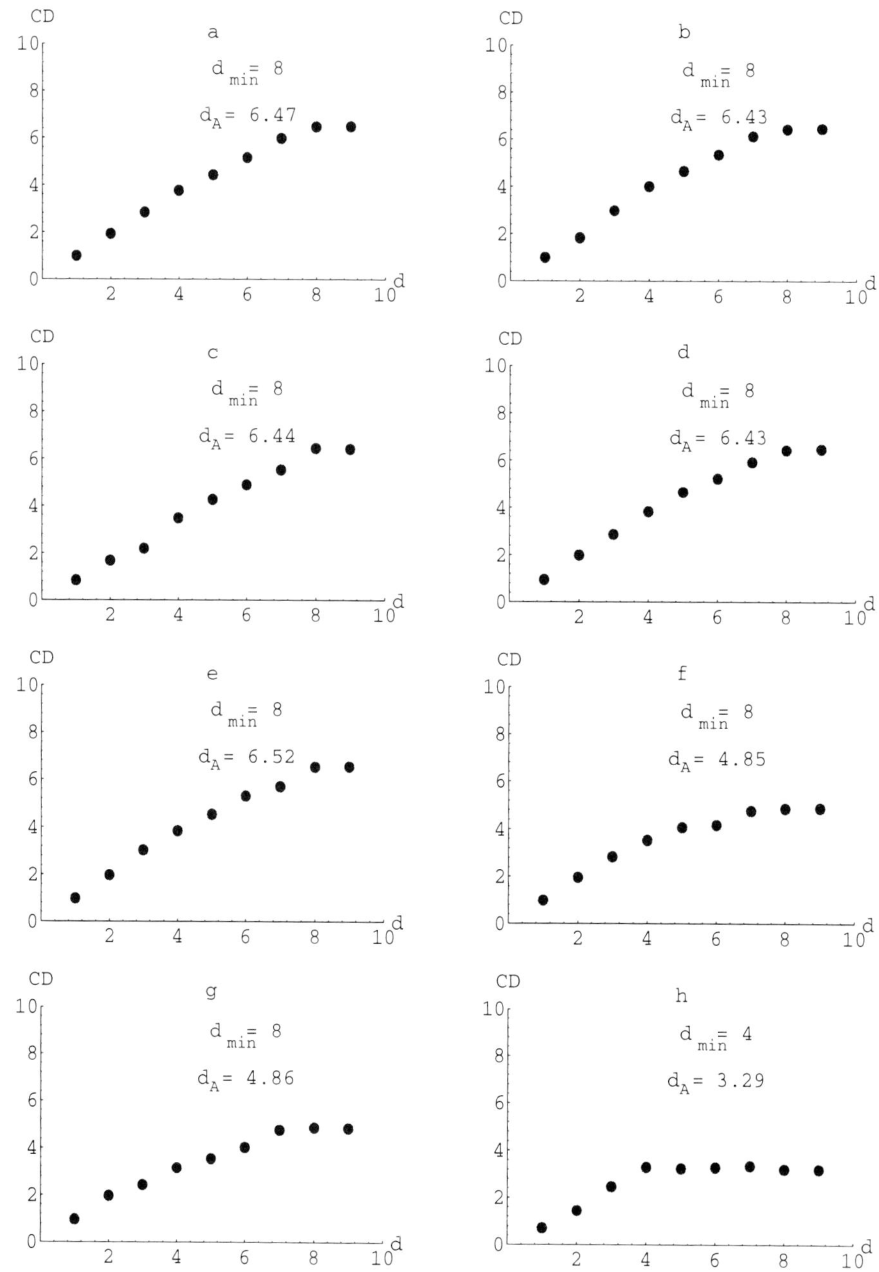


**Fig. 9.10** Correlation dimension analysis for the PB2EAST area, where the parameter under study in each figure is respectively:
a. log$E$
b. log$M$
c. x location
d. y location
e. z location
f. distance between locations of consecutive events
g. distance between locations of consecutive events including the source size
h. time between consecutive events

autocorrelation function. However, the spatial correlation range is always longer in the case when $s_{ph}$ is used. Finally, the fact that log$E$ and log$M$ give rise to the same number of linearly dependent events enhances the assumption that there is a simple relation connecting the two physical parameters.

The results of the analysis of fractal correlation dimension are presented in terms of plots of fractal correlation dimension $d_a$ versus embedding dimension $d$. The value of $d$ where such a plot flattens is denoted as the embedding dimension $d_{\min}$ corresponding to the parameter under study. The fractal dimension of the respective parameter is related to $d_{\min}$. Such plots for the three areas under study are shown in Figs. 9.8–9.10.

We observe that the most frequent embedding dimension is 8. The flattening of the curves indicates the existence of coherent structures which, in fact, are strange attractors, demonstrating the presence of chaotic behaviour.

## 9.5 Lyapunov exponent and limits of predictability

The phase space of a chaotic system contains regions of stochasticity where a description of the motion in terms of analytic formulae is not possible. The alternative to such a situation is to treat the motion in a more statistical sense. Instead of calculating individual orbits, one may seek to determine the evolution of certain average quantities that describe the degree of disorder. There are various ways to describe the motion in a stochastic region using terms like ergodicity, mixing, K-systems, Lyapunov exponents, etc.

We seek to evaluate the chaotic behaviour of time series of seismic parameters by calculating Lyapunov exponents. Lyapunov exponents are a computable measure of the chaotic behaviour of a trajectory and express the mean exponential rate of divergence of trajectories surrounding it.

Let us assume that we have managed to reconstruct the phase space of a seismic parameter by embedding its times series. We denote by $P$ the dimension of such a phase space. Next, we consider a trajectory in this $P$ dimensional phase space and a nearby trajectory with initial conditions $\mathbf{x}_0$ and $\mathbf{x}_0 + \Delta\mathbf{x}_0$ respectively. These trajectories evolve with time, yielding the tangent vector $\Delta\mathbf{x}\,(\mathbf{x}_0, t)$ having the Euclidian norm $d\,(\mathbf{x}_0, t) = \|\Delta\mathbf{x}\,(\mathbf{x}_0, t)\|$.

The time evolution for $\Delta\mathbf{x}$ in the case of time series can be found using the Wolf algorithm (Wolf *et al.*, 1985). Finally, we introduce the mean exponential rate of divergence of two initially close trajectories:

$$Ly(\mathbf{x}_0, \Delta\mathbf{x}) = \lim_{\substack{t\to\infty \\ d(\mathbf{x}_0, 0)\to 0}} \frac{1}{t} \log_2 \frac{d(\mathbf{x}_0, t)}{d(\mathbf{x}_0, 0)} \tag{9.15}$$

It can be shown that $Ly$ exists and is finite (Lichtenberg and Liebermann, 1983). Furthermore, there is a $P$ dimensional basis for $\Delta\mathbf{x}$, such that for any $\Delta\mathbf{x}$, $Ly$ takes on one of the $P$ values $Ly_i$ which are the Lyapunov characteristic exponents. Moreover, the Lyapunov exponents are independent of the choice of metric for the phase space. These exponents are all zero for integrable systems (no divergence of trajectories) and at least one is positive finite for chaotic systems (divergence of trajectories) and indicates how unpredictable and incomputable a time series is. In the case of finite time series, as it is the case in practical applications, formula (9.15) can only be applied for the range of time over which one has taken the measurements. This leads us to define the so-called 'local Lyapunov exponent' (*LLE*):

$$LLE = Ly(\mathbf{x}_0, \Delta\mathbf{x}) = \lim_{d(\mathbf{x}_0, 0)\to 0} \frac{1}{t} \log_2 \frac{d(\mathbf{x}_0, t)}{d(\mathbf{x}_0, 0)} \tag{9.16}$$

The *LLE* can now be used to quantify the minimum predictability limit. With the aid of the Wolf algorithm we can determine for a time series the maximum local Lyapunov exponent ($LLE_{max}$) out of all *LLE*s that characterize the chaotic behaviour in the respective $P$ dimensional phase space. $LLE_{max}$ represents the maximum amount of information lost by the time series during two consecutive event occurrences. The amount of information is quantified by the lag $D$ corresponding to the respective parameter: the last event in the time series has a linear information link up to the next $D$ events in the future. However, this amount reduces to $D-LLE_{max}$ after the first event in the future occurs, then to $D-2LLE_{max}$ after the second event occurs, etc. If $n$ is the number of events in the future after which $D-nLLE_{max}$ becomes zero, we denote $n$ as the maximum number of events that we can predict based on the present knowledge of our time series. This number is given by:

$$n = \text{int}\left[\frac{D}{LLE_{max}}\right] \tag{9.17}$$

where int[$q$] represents the integer part of the quantity $q$.

The temporal limit of predictability can now be constructed by determining the correlation time associated with the time series under study. Let $t_1$ be the time elapsed from the occurrence of event 1 to event $n$ in the time series, $t_2$ from event 2 to event $n+1$, ..., and $t_{N-n+1}$ the time from event $N-n-1$ to event $N$. Averaging over $t_1, t_2, \ldots, t_n$ results in the correlation time of the time series ($t_c$) under study, according to the parameter that yielded the Lyapunov exponent $LLE_{max}$:

$$t_c(n) = \frac{\sum_{i=1}^{N-n+1} t_i}{N - n + 1} \tag{9.18}$$

This expression represents the minimum limit of predictability of the parameter analysed.

The Lyapunov exponents and corresponding limits of predictability calculated for various seismic parameters at WHPOST, P1PILLAR and PB2EAST are presented in Table 9.2.

**Table 9.2** Limits of predictability

| | Area: WHPOST | | Area: P1PILLAR | | Area: PB2EAST | |
|---|---|---|---|---|---|---|
| Parameter | $LLE_{max}$ | Min. limit of predictability (days) | $LLE_{max}$ | Min. limit of predictability (hours) | $LLE_{max}$ | Min. limit of predictability (hours) |
| log(Moment) | 1.03 | 24.9 | 1.13 | 22.2 | 1.02 | 8.3 |
| log(Energy) | 1.00 | 25.6 | 1.12 | 22.2 | 0.97 | 8.3 |
| $x$ location | 1.14 | 3.1 | 1.21 | 11.3 | 1.05 | 20.8 |
| $y$ location | 1.09 | 6.7 | 1.07 | 6.5 | 1.03 | 23.0 |
| $z$ location | 1.20 | 9.3 | 1.07 | 13.6 | 0.88 | 31.2 |
| Distance between locations of consecutive events | 1.21 | 2.8 | 1.45 | 8.3 | 1.22 | 17.2 |
| Distance between locations of consecutive events including source size | 1.21 | 2.8 | 1.43 | 8.3 | 1.30 | 16.3 |
| Time between consecutive events | 2.67 | 7.9 | 3.51 | 5.4 | 2.87 | 3.8 |
| Average minimum limit of predictability | | 10.3 | | 12.3 | | 16.5 |

The value of the Lyapunov exponent does not depend on the monitoring for a given area, as the Lyapunov exponent is an invariant of the dynamics underlying the time series under study. In this regard, we have performed tests in which we filtered the time series of different parameters at a given area in terms of magnitude. While the Lyapunov exponent stayed the same at all levels of magnitude where we could perform the embedding analysis, the predictability limit increased with lower magnitude, as shown in Table 9.3. This indicates that greater sensitivity leads to a higher limit of predictability for seismic parameters. Hence, we stress that nonlinear dynamics techniques are applicable only to areas with increased sensitivity.

**Table 9.3** P1PILLAR area limits of predictability for various sensitivities.

Period of Monitoring: 30 September 1995 to 29 December 1995.
Parameter: log(Moment)

| Minimum recorded magnitude | Minimum limit of predictability (hours) |
|---|---|
| −2.3 | 22.26 |
| −1.6 | 16.70 |
| −1.3 | 12.05 |
| −1.0 | 8.45 |

To summarize, we stress that the dynamics of the systems analysed appears to be eight dimensional. It may be speculated that the underlying physics of seismic rockmass response to mining is eight dimensional. The remaining problem is to further identify the eight physical parameters that give rise to this dynamics. The systems studied all exhibit the phenomenon of chaos for the following reasons:

a) The existence of attractors has been established.
b) The dimensionality of the attractors is always fractal.

The occurrence of chaos supports the view which assumes the turbulent nature of the seismic rockmass response to mining (Kagan, 1992). In terms of considering the apparent volume in calculating the distance between events, notice that, although the number of linearly dependent events does not depend on whether one uses $s_e$ or $s_{ph}$, it appears that the fractal dimension of the attractor of the distance between events is underestimated if the apparent volume is not taken into account. This is more clearly seen when the number of events is larger, as at P1PILLAR. Analysis indicates that the $x,y,z$ locations, log$E$, log$M$, $s_e$ and $s_{ph}$ are all true reflections of the same underlying physical process which appears to be eight dimensional. At this stage, we cannot determine the eight canonically independent variables which constitute the axes of the phase space. This did not, however, hinder the determination of the limits of predictability.

These limits have been determined on the basis of the Lyapunov exponent, an invariant of dynamics which quantifies statistically the amount of existent chaos. It turned out that the limits of predictability in preliminary tests were finite and manageable, ranging from a few hours to about three weeks. It has also been shown that an increase in the predictability limit can be achieved if the monitoring sensitivity at a given area is increased.

# 10 Quantitative Seismology and Rockmass Stability

## 10.1 Quantitative description of a seismic event

Mining excavations, whether underground or open cast, induce elastic (reversible) and then inelastic (irreversible) deformation within the surrounding rock. Elastic deformation is defined as a process during which no new microdefects are nucleated while all existing microdefects convect with the mass without growing in size (Krajcinovic and Mastilovic, 1995). The potential energy accumulated during elastic deformation may be unloaded or it may be released gradually or suddenly during the processes of inelastic deformation.

The inelastic deformation of brittle-like rock is mainly due to fracturing and frictional sliding called cataclastic flow. Fracture and slip create inelastic deformation within a certain volume of rock and radiate seismic waves. The amplitude and frequency of seismic waves radiated from such a source depend, in general, on the strength and state of stress of the rock, the size of the source of seismic radiation, and on the rate at which the rock is deformed during the fracturing process. The following relations apply here (all other parameters being the same):

- the amplitude and frequency increase with an increase in rock strength and stress;
- the predominant frequency (i.e. the frequency at which the most energy is radiated) decreases with an increase in source size;
- the amplitude and frequency increase with an increase in coseismic deformation rate.

A seismic event can then be considered as a sudden inelastic deformation within a given volume of rock that radiates detectable seismic waves. To give some idea of scale, the volume of rock involved in the coseismic inelastic deformation, called source volume, varies from a fraction of a cubic metre for cracks of about a metre in length, to a fraction of a cubic kilometre for large seismic events with source dimension of a few hundred metres. The average strain change within the source volume varies from $10^{-4}$ to $10^{-2}$. The average velocity of the deformation varies from a few centimetres per second to a few metres per second, generally being slower within softer and less homogeneous rock and at lower differential stress. The average stress drop varies from $10^4$ Pa for slow events occurring within softer rock or along existing weak geological features to $10^8$ Pa for fast and large events rupturing hard highly stressed and, to a great extent, intact rock.

In general, a seismic event is a result of an instability in the flow of rock and, like eddies in the flow of fluid, can be considered as a coherent dissipative structure containing newly created useful information. Creation of that information must be associated with increased dissipation rates, i.e. an increase in deformation rate at the source generates more seismic radiation.

Seismic sensors detect waves caused by inelastic deformation within a certain distance. Since the attenuation of seismic waves increases with their frequency, the higher the amplitude and the lower the frequency, the longer the distance over which one can receive waveforms at a reasonable signal to noise ratio.

Proper quantitative seismological processing of recorded waveforms can routinely provide information on the following parameters pertaining to the source of the seismic radiation:

- time of the event, $t$;
- location, $X = (x,y,z)$;
- seismic moment, $M$, which measures coseismic inelastic deformation at the source and its tensor which defines an overall direction of principal stresses acting at the source and the nature of the coseismic strain change in terms of its isotropic and deviatoric components;
- seismic energy radiated from the source of the seismic event, $E$;
- size and stress release estimates.

Having recorded and processed a number of seismic events within a volume of interest, one can then quantify the changes in the strain and stress regime and in the rheological properties of the rock mass associated with the seismic radiation. That, in turn, gives a unique insight in space and time into the development of processes which may lead to unstable behaviour of rock.

### 10.1.1 Seismic moment, source size and stress drop

The most general description of the processes at the seismic source $V$ is by the distribution of forces or moments equivalent to the inelastic deformation. One can describe the inelastic processes at the source as the stress-free change of size and shape of an elastic body without alteration of the elastic properties of the region (Backus and Mulcahy 1976a,b). If change in size and shape can be expressed as a change in strain $\Delta\epsilon_{kl}$, then the equivalent stress change, or change in moment per unit volume, can be given by

$$\Delta\sigma_{ij} = c_{ijkl}\Delta\epsilon_{kl} \tag{10.1}$$

where $c_{ijkl}$ are elastic constants and $\Delta\sigma_{ij}$ is called the seismic moment density tensor or inelastic stress or stress glut. Therefore, the stress glut represents the internal stress necessary to cancel the strain produced by the internal inelastic process, i.e. to restore the source to its original shape and size (see Fig. 10.1). The stress glut is not exactly the stress drop – the difference between the initial equilibrium stress and the final equilibrium stress. The stress drop is not limited to the source volume but stress glut $\Delta\sigma_{ij}$ vanishes outside $V$.

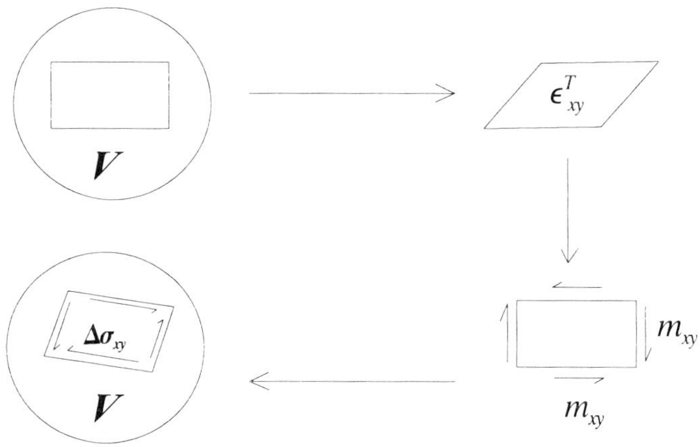


**Fig. 10.1** The concept of inelastic stress or stress glut. The rectangular element in the unstressed configuration (top left) undergoes a spontaneous change in strain. If it were not surrounded by the rest of the body, it would take the deformed shape shown on the top right. The seismic moment, inelastic stress, or stress glut is the stress that is needed to bring it back to its initial rectangular shape (bottom right). Finally, if the element is reinserted into the unstressed body and the stresses are allowed to readjust a final state of stress is obtained. This is the stress change (stress drop or increase) created by the inelastic strain (from Madariaga,

The total moment integrated over the source volume is the seismic moment tensor, $M_{ij}$. For long waves compared to the source size, the whole volume $V$ can be considered to be a system of couples located at, say, the centre of $V$, and the moment tensor components can be defined by (Aki and Richards, 1980):

$$M_{ij} = \int_V c_{ijkl}\Delta\epsilon_{kl}\,dV = \int_V \Delta\sigma_{ij}\,dV. \tag{10.2}$$

The moment tensor measures the inelastic deformation at the source during the seismic event and its value at the end of the source process measures the permanent inelastic strain produced by the event.

If one interprets the source volume $V$ as the region with the largest inelastic shear strain drop $\Delta\epsilon$ (of the order of stress drop $\Delta\sigma$ over rigidity $\mu$,

$\Delta\epsilon = \Delta\sigma/\mu$), the scalar seismic moment, $M$, may then be defined as the product of the estimated stress drop and source volume (Madariaga, 1979),

$$M = \mu \, \Delta \epsilon \; V = \Delta \sigma \; V \tag{10.3}$$

An average stress drop, $\Delta\sigma = M/V$, is then interpreted as a spatial density of the seismic moment and the source volume $V$ is expressed as

$$V = \frac{M}{\Delta \sigma} \tag{10.4}$$

The seismic moment of the idealized planar source is usually defined as

$$M = \mu \bar{u} A \tag{10.5}$$

where $\bar{u}$ is an average displacement (slip) over the source area $A$.

The seismic moment divided by rigidity is called the source strength (e.g. Abercrombie *et al.* 1995) or potency, $P$ (Ben-Zion and Rice, 1993),

$$P = \frac{M}{\mu} = \Delta \epsilon \; V \quad \text{or} \quad P = \bar{u} A \tag{10.6}$$

for volume or planar source respectively.

The seismic moment tensor can be decomposed into isotropic (or volume change) and deviatoric components providing an additional insight into the nature of the coseismic strain drop. However, a simple shear dislocation source can have non-zero isotropic components, e.g. if the source surface is irregular or embedded in an anisotropic material (Backus and Mulcahy, 1976b.)

For a homogeneous body, the coseismic volumetric change, $\Delta\theta$, can be calculated from (Aki and Richards, 1980)

$$\Delta\theta = tr\left(M_{ij}\right) / \left(3\lambda + 2\mu\right) \tag{10.7}$$

where $\lambda$ is the second Lamé elastic constant.

The eigenvalues and corresponding eigenvectors of the deviatoric component of the seismic moment tensor describe the magnitude and orientation, respectively, of the principal moment axes (neglecting gravity) acting at the source. These principal moment axes are uniquely determined by moment tensor inversion. Principal moment orientation data can provide sufficient information to find the best stress tensor (Michael, 1987).

Figure 10.2 shows the results of moment tensor inversions for two seismic

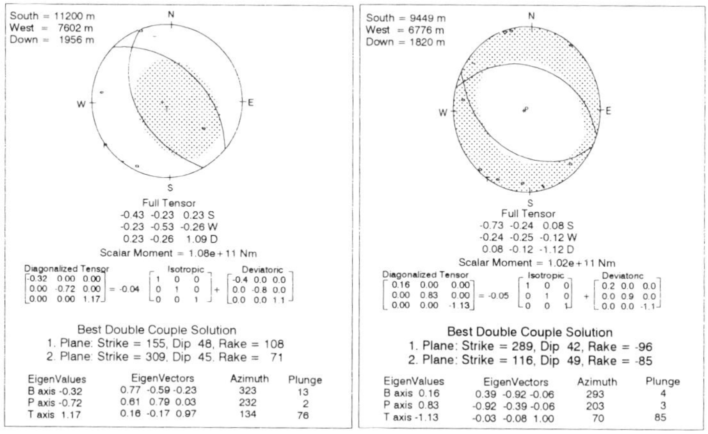


a b

**Fig. 10.2** Results of the moment tensor inversion and the decomposition of seismic events associated with two blasts.
a. Blast 1: six boreholes each 16 m long, 25 kg of emulsion per hole, delayed initiation.
b. Blast 2: two boreholes each 23 m long, 125 kg of emulsion per hole, simultaneous initiation.

events associated with two underground blasts designed to trigger a seismic event and/or prefracture the hard rockmass:

- blast 1: six boreholes approximately 16 m long, 25 kg of emulsion per hole, delayed initiation;
- blast 2: two boreholes each 23 m long, 125 kg of emulsion per hole, simultaneous initiation.

In both cases, seismic events of almost identical scalar seismic moment have been recorded by a number of three component digital seismic stations. Seismic moment tensor inversion and its decomposition, however, show a difference in the nature of the coseismic strain drop, i.e. blast 2 having a slightly larger double couple component. Source parameters, source volumes and coseismic volumetric changes of seismic events associated with blast 1 and blast 2 are shown in Table 10.1.

Table 10.1 Source parameters of seismic events associated with blasts

| | $M$ [Nm] | $\Delta\sigma$ [Pa] | $E$ [J] | $V=M/\Delta\sigma$ [m$^3$] | $\Delta\Theta$ [m$^3$] |
|---|---|---|---|---|---|
| Blast 1 | 1.08E11 | 1.39E5 | 1.99E4 | 7.8E5 | 0.1 |
| Blast 2 | 1.02E11 | 6.73E5 | 3.28E4 | 1.5E5 | 0.1 |

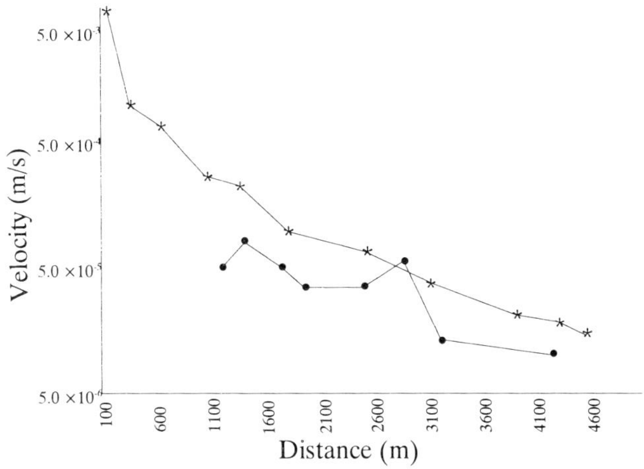


**Fig. 10.3** Amplitude of recorded maximum ground motions versus distance for two seismic events associated with underground blasts.
• Blast 1: six boreholes approximately 16 m long, 25 kg of emulsion per hole, delayed initiation
* Blast 2: two boreholes each 23 m long, 125 kg of emulsion per hole, simultaneous initiation.

Although 100kg less explosives were used, blast 1 should be considered more successful than blast 2 since it induced an inelastic deformation in a softer manner and over a larger volume. Amplitudes of recorded ground motions (Fig. 10.3) confirm that the larger the non-double couple component, the slower (or sweeter – Fröhlich, 1993) is the event and the weaker are the ground motions (McGarr, 1993).

Seismic moment $M$ is proportional to the integral of the far field displacement pulse of the seismic radiation. This property is frequently used to determine seismic moment from recorded body wave observations in the time or in the frequency domain.

Since radiation from a seismic source is of finite duration, the spectrum of the far field displacement has a finite width. Seismologists estimate the spectral width from the intersection of a constant low frequency trend and a high-frequency asymptote of the attenuation corrected displacement spectrum and call this intersection the corner frequency $f_0$. Brune (1970) proposed using corner frequency as a method for calculating source dimension, $r$, and stress drop, $\Delta\sigma$:

$$r = c_1 f_0^{-1}$$

$$\Delta\sigma = c_2 M f_0^3, \quad c_1, c_2 \text{ are model dependent constants.} \tag{10.8}$$

This is a strongly model dependent technique but it may provide reasonable estimates and a fair comparison between different sources from the same region recorded by the same seismic system.

Heterogeneities within the rockmass produce a complexity of radiated pulse shapes. It was found (e.g. Madariaga, 1979) that for multiple or complex sources the seismic moment is smaller than for simple planar sources of the same area and stress drop. A simple interpretation of this result is that the unbroken sections of the source reduce overall distortion at the source region reducing seismic moment. The stress drop of such a complex event, estimated by the usual circular source formula, is only a fraction of the actual stress drop on the broken section of the source. The actual and estimated stress drops are in the ratio of the total source size to the size of a typical subevent (Kostrov and Das, 1988).

### 10.1.2 Seismic energy

The energy release during fracture and frictional sliding is due to the transformation of elastic strain into inelastic strain. This transformation may occur at different rates ranging from slow creep-like events to very fast dynamic seismic events with an average velocity of deformation at the source of up to a few metres per second. Slow type events have a long time duration at the source and thus radiate predominantly lower frequency waves, as opposed to dynamic sources of the same size. Since excitation of seismic energy can be represented in terms of temporal derivatives of the source function one may infer that a slower source process implies less seismic radiation. In terms of fracture mechanics, the slower the rupture velocity, the less energy is radiated; the quasi-static rupture would radiate practically no energy.

To assess the physical sources of radiated energy let us consider the formula for seismic energy for the single fracture-type source, expressed in terms of source parameters (Kostrov, 1974; Rudnicki and Freund, 1981):

$$E = -2\gamma_{eff} A + \tfrac{1}{2}\int_A \Delta\sigma_{ij}\, u_i\, n_j\, dA + \int_0^{t_{cs}} dt \int_{A(t)} \dot{\sigma}_{ij}\, u_i\, n_j\, dA \tag{10.9}$$

where:

- $\gamma_{eff}$ effective surface energy, which includes the total loss of mechanical energy, in particular inelastic work, and heat flow from the fracture edge;
- $A$ fracture area with the displacement $u_i$;
- $\Delta\sigma_{ij}$ difference between the final (at the end of the event) and the initial stress;
- $n_j$ unit vector normal to the fracture plane;

$t_{cs}$ source duration;
$\dot{\sigma}_{ij}$ traction rate.

From dimensional analysis it follows that the last two terms in equation (10.9) vary with the fracture area as $A^{3/2}$, whereas the fracture work, $2\gamma_{eff}A$, is proportional to $A$ (Kostrov and Das, 1988). Thus, the relative contribution of the fracture work to seismic energy increases with a decrease in the size of the fracture. Consequently for sufficiently small fractures the first term may almost cancel the second term, suppressing the acoustic emission so that 'silent' fracture would occur.

The second term in equation (10.9) is determined by the static quantities – stress drop and final slip. The last term, containing the traction rate, strongly depends on how the fracture propagates and correlates with slip. The faster the oscillations of stress, the greater their contribution to the radiated energy due to the presence of the time derivative of stress. The energy due to radiation of high frequency waves during accelerating and decelerating rupture is called radiational friction. It usually occurs when the moving zone of slip pulse reaches regions of differing resistance to deformation. If traction rate and slip are uncorrelated the third term will vanish.

There is ample evidence that real rupture propagates in a discontinuous fashion, controlled by the strength and distribution of barriers and asperities. It is unclear, however, what such incoherent rupture contributes to the total radiated energy.

Equation (10.9) shows that it is impossible to obtain an unambiguous expression for radiated energy only in terms of changes from the initial equilibrium state of the medium to the final equilibrium.

In the far field the P and S wave contributions to the total radiated energy are proportional to the integral of the square of the P and S velocity spectrum. For a reasonable signal to noise ratio in the bandwidth of frequencies available on both sides of the corner frequency, the determination of that integral from waveforms recorded by seismic networks is fairly objective.

The high frequency component of seismic radiation needs to be recorded by the seismic system if a meaningful insight into the stress regime at the source region is to be gained.

### 10.1.3 Apparent stress, energy index and apparent volume

*Apparent stress* Apparent stress, $\sigma_A$, is the product of apparent strain and the ratio of seismic energy $E$ over seismic moment $M$, rigidity $\mu$,

$$\sigma_A = \xi\bar{\sigma} = \mu \frac{E}{M} = \frac{E}{\Delta\epsilon V} = \frac{E}{P} \tag{10.10}$$

where $\xi$ is so-called seismic efficiency and $\bar{\sigma}$ is average or effective stress, which is equal to half the sum of the initial and final stress, before and after the event (Wyss and Brune, 1968; 1971) is recognized as a model independent measure of the dynamic stress release in the source region.

Apparent stress is proportional to the integral of the square of the velocity spectrum divided by the amplitude of the low frequency asymptote to the displacement spectrum or, when the acceleration spectrum is considered, apparent stress depends linearly on the product of the corner frequency and the amplitude of the high frequency asymptote to the acceleration spectrum. It is then a more reliable and less model dependent parameter describing an average stress release at the source than the corner frequency cubed dependent static stress drop (see equation (10.8)).

Although seismic waveforms do not have direct information about absolute stress (for a bit of hope in inferring absolute stress from seismic radiation, see Spudich, 1992), but merely about the dynamic stress drop at the source, a number of seismological studies and numerous underground observations suggest that a reliable estimate of stress drop, or preferably apparent stress, can be used as an indicator of the local level of stress. Gibowicz *et al.* (1990) considered apparent stress an independent parameter of stress release in the case when P and S wave contribution to the seismic energy is included. Mendecki (1993) showed an example where apparent stress associated with seismic events of magnitudes between 1.3 and 1.5 varies from 0.2 to 40 bar, being generally higher at greater depth and within less faulted rock. McGarr (1994) states that if seismic efficiency $\xi \leq 0.06$ – which is an average value deduced from laboratory stick-slip experiments – then measurements of apparent stress can be used to infer the absolute level of deviatoric stress at the hypocentre. In general, the apparent stress expresses the amount of radiated seismic energy per unit volume of inelastic coseismic deformation.

Let us imagine the source of a seismic event associated with a relatively weak geological feature or with a soft patch in the rockmass. Such a source will yield slowly under lower differential stress producing larger seismic moment and radiating less seismic energy, resulting in a low apparent stress event. The opposite applies to a source associated with a strong geological feature or hard patch in the rockmass.

In the case of a so-called complex event or multiple events, the rapid deformation process at the initial source can push the stresses in the adjacent volume to a level much higher than could normally be maintained by the rock, producing higher apparent stress sub event(s) that need not be an indication of a generally high ambient stress prior to the event. Although the estimate of apparent stress does not depend on the rupture complexity (Hanks and Thatcher, 1972), the complexity of the event should be tested before meaningful interpretation in terms of stress can be given.

Urbancic *et al.* (1993) associated low apparent stress of mining related microseismic events with their low rupture velocities, of approximately 0.3 to 0.6 times shear wave velocity, implying the presence of source complexity in these events.

Since radiated seismic energy broadly increases with increasing moment, to gain insight into the stress regime one would need to compare the radiated seismic energies associated with seismic events of similar seismic moments.

Figure 10.4 shows the apparent stress associated with seismic events of similar moment magnitude, between 0.8 and 1.0, in the Welkom area of South Africa. One can observe considerably higher values of apparent stress associated with seismic events in area A, where mining of small pillars takes place approximately 300 m deeper than scattered mining of larger blocks in area B. The ratio of the largest to the smallest apparent stress in Fig. 10.4 is well over 100. Figure 10.5 shows log energy versus log moment of seismic events for the *entire* magnitude range that occurred in the areas A and B and the whole area, including A and B.

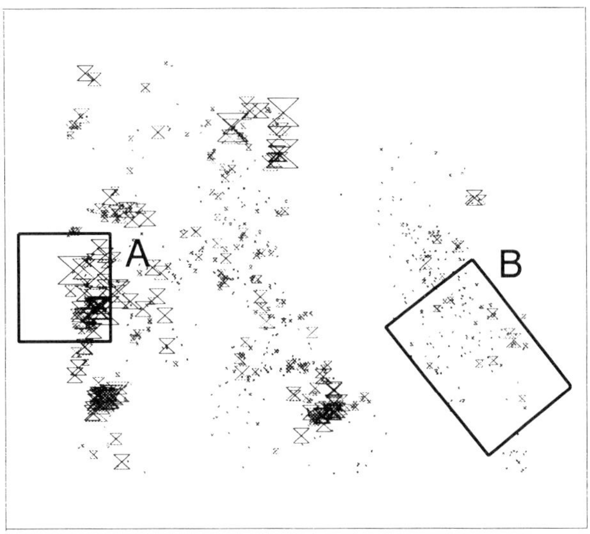


**Fig. 10.4** Apparent stress of selected seismic events within the restricted magnitude range of 0.8 to 1.0 in the Welkom area, South Africa, for the period 1 January to 31 October 1993. A (~1850 m) and B (~1550 m) indicate examples of areas (at different depths) of consistently high (box A) and consistently low (box B) apparent stress.

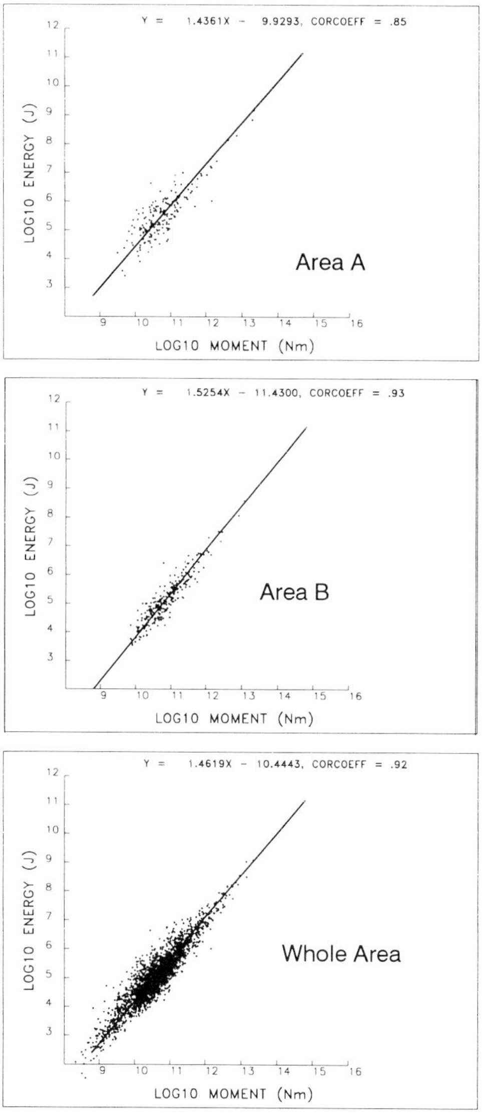


**Fig. 10.5** Seismic moments and radiated seismic energies of 2146 events of unrestricted magnitude, which occurred in areas A and B and the whole area (including areas A and B) as depicted in Fig. 10.4. Note the higher energies (3.5 times on average) for the same seismic moments associated with the deeper mining in area A. The greater scatter in the case of A could be explained by greater inhomogeneities brought about by the presence of faults.

*Energy index* The notion of comparing the radiated energies of seismic events of similar seismic moments can conveniently be translated into a practical tool called energy index, *EI* (van Aswegen and Butler, 1993). The energy index of an event is the ratio of the observed radiated seismic energy of that event to the average energy radiated by events of the same seismic moment $\bar{E}(M)$ taken from the log*E* versus log*M* relation for the area of interest (see Fig. 10.6):

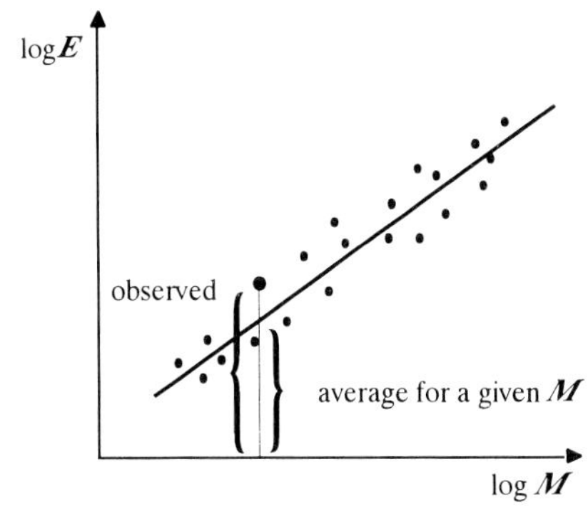


**Fig. 10.6** Energy index concept.

$$EI = \frac{E}{\bar{E}(M)}, \quad \bar{E}(M) = 10^{c_3 \log M + c_4} \tag{10.11}$$

where $c_3$ and $c_4$ are constants for a given area.

Figure 10.7 shows contours of energy indices associated with 2146 seismic events of unrestricted magnitude which occurred in the area shown in Fig. 10.4 and whose plots are shown in Fig. 10.5 (whole area).

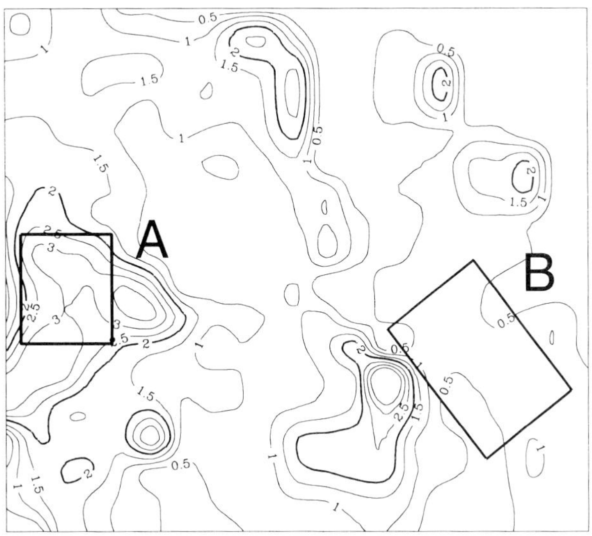


**Fig. 10.7** Contours of energy index of all 2146 events (see Fig. 10.5, whole area) which occurred between 1 January and 31 October 1993, in the area shown in Fig. 10.4.

*Apparent Volume* The strain drop produced by a seismic event is localized in a zone that is small in comparison with the length of the recorded seismic waves. However, seismic events that occur close to the underground

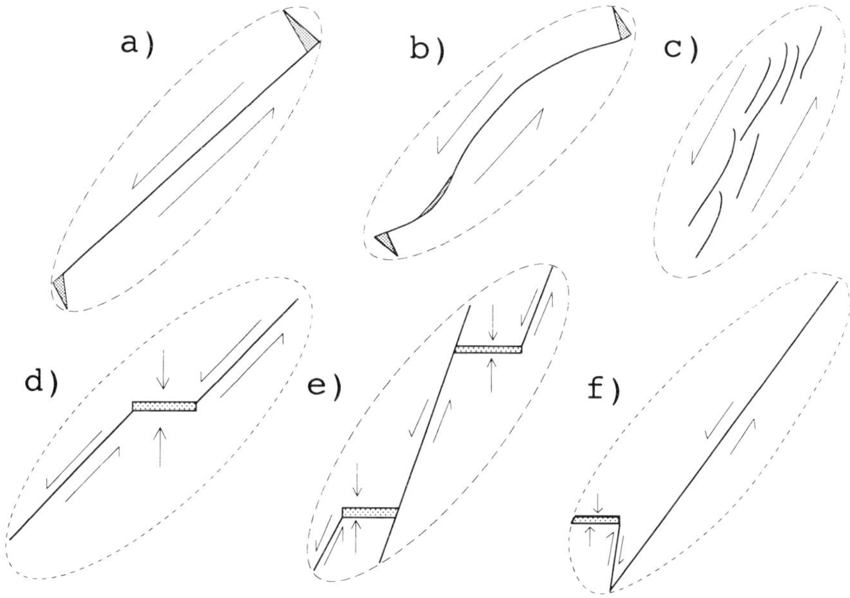


**Fig. 10.8** Example of displacement patterns at seismic sources in mines where arrows indicate movement(s) at the source:
a) simple shear and two tensiles;
b) shear over non-planar fault, dilation and two tensiles;
c) multiple shears;
d) two fault related shears and implosion, associated with stope closure;
e) multiple shears and stope related implosions;
f) fault related shear, implosion associated with stope closure and shear along newly created discontinuity.

excavations and/or in a complex geological setting frequently exhibit a volumetric character, with many zones of permanent deformation and complex geometry accompanied by a local volume change. Fig. 10.8 shows a few examples of possible displacement patterns at the sources of smaller (Fig. 10.8a,b,c) and larger (Fig. 10.8d,e,f) seismic events associated with mining (Teisseyre, 1980a,b; Ortlepp, 1994).

Source volume $V$ – the volume of coseismic inelastic deformation that radiated the recorded seismic waves – can be estimated from $V = M / \Delta\sigma$. Since apparent stress $\sigma_A$ scales with stress drop, and since there is less model dependence in determining the apparent stress than there is in determining corner frequency cubed dependent static stress drop, and because, in general, $\Delta\sigma \geq 2\sigma_A$, one can define the apparent volume $V_A$ as follows (Mendecki, 1993):

$$V_A = \frac{M}{2\,\sigma_A} = \frac{M^2}{2\,\mu E} \tag{10.12}$$

The apparent volume for a given seismic event measures the volume of rock with coseismic inelastic strain with accuracy in the order of magnitude of apparent stress divided by rigidity.

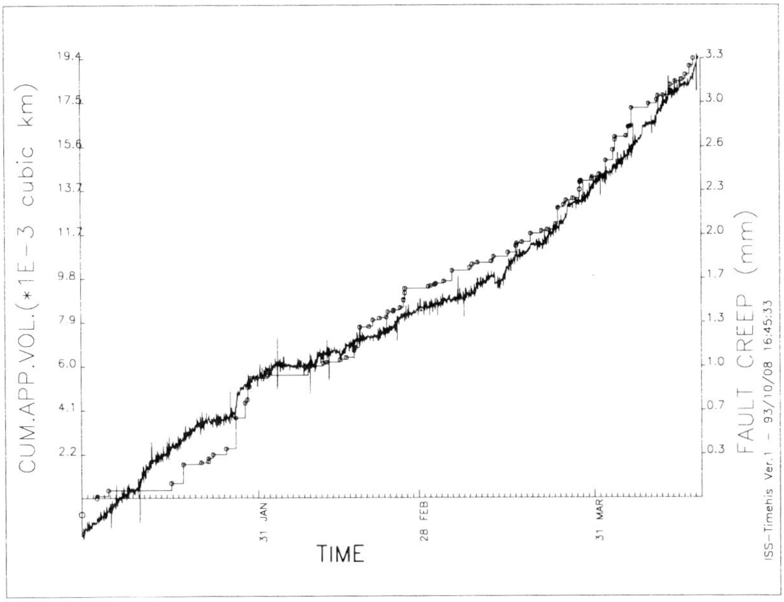


**Fig. 10.9** Plot of cumulative apparent volume of events around the fault (line with circles) with superimposed, measured fault creep (darker, noisy line).

Apparent volume, like apparent stress, depends on seismic moment and radiated seismic energy, and, because of its scalar nature, can easily be manipulated in the form of cumulative or contour plots etc., providing insight into the rate and the distribution of coseismic deformation and/or stress transfer in the rockmass.

Figure 10.9 shows the correlation of cumulative apparent volume of seismic events associated with the Ararat fault in the Welkom area and creep on this fault over a period of three and a half months (see also van Aswegen and Butler, 1993). It confirms the notion of a positive correlation between coseismic and aseismic deformation in brittle rock.

Apparent volume, or source volume, gains sensitivity to the inelastic coseismic deformation from its two dimensional description of the seismic source, namely, by seismic moment and seismic energy or by seismic moment and stress drop.

Figure 10.10 shows cumulative plots of radiated seismic energy, seismic moment and apparent volume versus time for the area surrounding the source of a large seismic event of energy $E$=2×10$^{10}$ J. There is no precursory behaviour on cumulative energy or cumulative moment plots, both being one dimensional descriptions of the source. One can, however, see a considerable increase in the rate of the cumulative apparent volume for the last three months before the event.

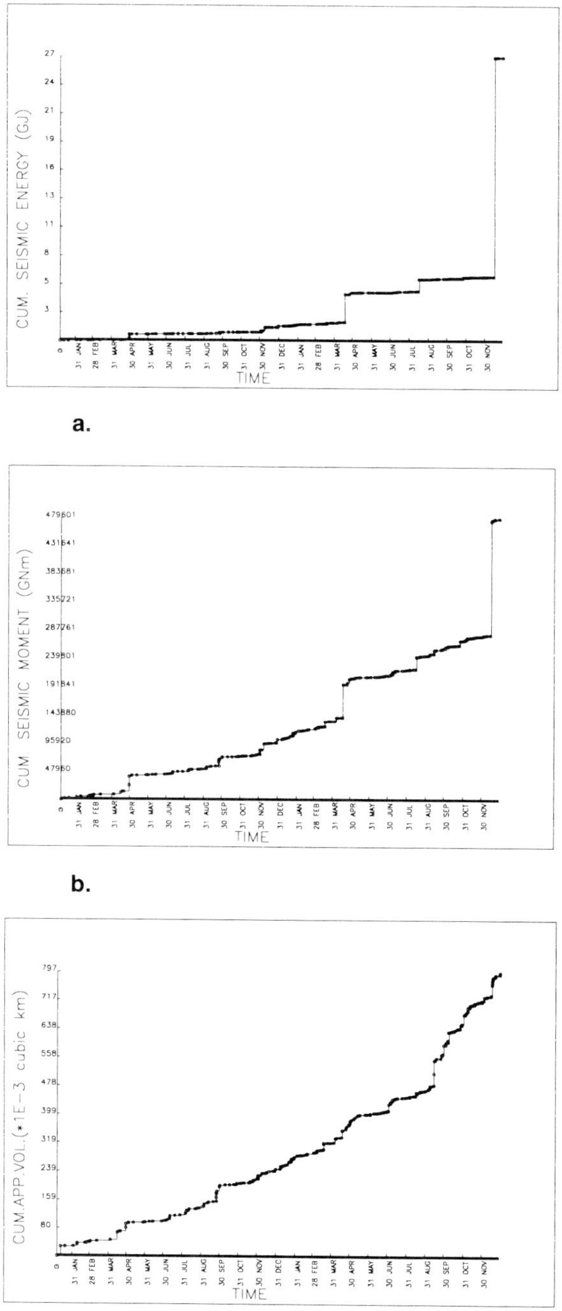


**Fig. 10.10** Cumulative plot of radiated seismic energy (a), cumulative seismic moment (b) and cumulative apparent volume (c) versus time for the area surrounding the source of a seismic event of energy $E=2\times10^{10}$ J. Note an increase in the gradient in the cumulative apparent volume plot for the last three months prior to the major event (after Mendecki, 1993).

## 10.2 Quantitative description of seismicity

Mining excavations create considerable stress gradients which the rockmass constantly reduces by means of inelastic deformation with associated transport of mass, momentum and energy – rock flows towards mining excavations or, more precisely, in its phase space it flows towards local equilibrium, reducing and transferring stresses.

The seismic component of that deformation is often called the seismic flow of rock (Riznichenko, 1965; Kostrov 1974). While aseismic rockmass deformation or flow is smooth and quiet, the seismic flow is associated with intermittent changes of, or fluctuations in, all mechanical characteristics of rock in space and in time. Such flows are called turbulent and they possess far greater ability to transfer momentum and, thus, are of greater significance to many engineering problems.

Kagan (1992) lists the following similarities between seismicity and turbulence of fluid flow:

- they both share inherent randomness;
- their major statistical ingredients are scale invariant; scale invariance is a signature of criticality, i.e. the physical state of the system or process that constrains it to evolve according to a fundamentally different type of behaviour than before;
- they both have hierarchically organized structures with maximum size comparable to the size of the region;
- the scale of self-similar structures extends over many orders of magnitude; self-similarity means reproducing itself on different time and space scales;
- both phenomena are intermittent in space and time, although the degree of seismic intermittency, as reflected in its fractal exponent, is higher than that of fluid turbulence (see Chapter 9 of this book);
- both are complex systems with a large number of degrees of freedom.

There are many differences between seismicity and turbulence in fluid flow. One of the most interesting here is the significance of vorticity (propensity to rotate) in deformation processes and its influence on local instabilities in rock, as they are the primary vehicles of deformation in the turbulent motion of fluids.

One can observe the following general features of seismicity, all of them driven by or resulting from the inelastic nature of seismic deformation and associated time dependent transfer of stresses:

1. Size distribution – the number of seismic events increases exponentially with their decrease in size, so-called 'power-law' growth (Ishimoto and Iida, 1939; Gutenberg and Richter, 1944). In nature, power-law growth or 'scaling' never appears by accident; it reveals a very important property of the phenomenon under investigation, namely its self-similarity. Failure can then be thought of as a scaling-up process in which failure at one scale is part of the damage accumulation of the larger instability. In seismology, the size distribution relation is usually referred to as the Gutenberg–Richter law: $\log N(m) = a - bm$, where the power-law exponent is commonly known as the b-value, $N(m)$ is the number of events of magnitude greater than or equal to $m$ and $a$ quantifies the level of seismicity that, as shown by McGarr (1976a), appears to be determined by the rate of ground deformation $\dot{\epsilon}$ as $a = \log\dot{\epsilon} +$ constant.

For b=1 and $a$ constant, the number of events $N(m)$ increases by a factor of 10 with each unit decrease in their magnitude.

In laboratory experiments on rocks, the b-value decreases with increasing stress (e.g. Scholz, 1968). Meredith *et al.* (1990), Main *et al.* (1992) and Liakopoulou-Morris *et al.* (1994) argue that the b-value decreases with increasing stress in all phases of rock deformation except strain softening, where they observed the opposite behaviour. For some earthquake sequences the b-value is reported to decrease before the main shocks (e.g. Suyehiro, 1966; Gibowicz, 1973, 1979; Smith, 1981; Jones, 1994), meaning an increase in the number of intermediate size events before larger instabilities.

Unfortunately, this type of precursory behaviour of b-value has not been observed consistently, specifically in mining where the dynamics of the fracturing processes is enhanced through blasting or by other forms of rock excavation. In general, the size distribution is the least informative of all the power laws that govern the occurrence of seismicity since there are a number of other physical phenomena that yield a power law size distribution with similar exponent value (Kagan, 1994). There is a large body of literature concerning the size distribution (or the b-value) and its possible correlation with geotectonics, stress, rock properties etc. This author concurs with Kagan (1994), that the proliferation of these investigations has not contributed significantly to a better understanding of the earthquake or fracturing processes.

2. Time distribution – every seismic event has its foreshocks and aftershocks. The number of foreshocks increases as the main shock approaches and the frequency of aftershocks decays with time as $(time)^{-p}$, where the exponent $p$ is slightly larger then 1.0 (Omori, 1895; Kagan and Jackson, 1991). Both of the sequences obey a power law. In general, there are fewer foreshocks than aftershocks.

3. Clustering in space and time – this is a consequence of the size and the time distributions of seismic events. Although clustering may result from an increase in differential stress in certain areas, it eventually leads to the softening of that part of the rockmass.

4. Migration – the interaction between the applied stress, a large enough population of clustered seismic events which create the critical volume of damaged rock, and the presence of fluids, creates a condition conducive to a spatially propagating wave of seismicity, i.e. inelastic deformation might ensue (Newman, 1983). Guberman (1975) argues that there is a systematic migration of seismic activity, characterized by travelling wave properties, that has a global character. In general, seismic events tend to migrate in space, pursuing and diffusing the stress gradients.

To monitor rockmass response to mining, one must be able to quantify continuously in time and space the parameters describing the flow changes in the stress and strain regime. From seismological observations, one can measure only that portion of stress, strain or rheology of the rockmass associated with the radiation of seismic waves recorded by the monitoring system.

Seismicity can be described quantitatively by at least the following four independent parameters:

- average time between seismic events $\bar{t}$;
- average distance between consecutive events $\bar{X}$, including source size;
- sum of seismic moments $\Sigma M_{ij}$;
- sum of seismic energies $\Sigma E$.

### 10.2.1 Seismic strain and seismic stress

*Seismic strain* An average seismic strain tensor $\epsilon_{s_{ij}}$ is proportional to the sum of the seismic moments of all events that have occurred within the volume $\Delta V$ over the period of time $\Delta t = t_2 - t_1$ (Kostrov, 1974; Molnar, 1983; Jackson and McKenzie, 1988):

$$\epsilon_{s_{ij}}(\Delta V, \Delta t) = \frac{\sum_{t_1}^{t_2} M_{ij}}{2\mu \, \Delta V} \tag{10.13}$$

The average seismic strain rate tensor then is

$$\dot{\epsilon}_{s_{ij}}\ (\Delta V, \Delta t) = \frac{\sum_{t_1}^{t_2} M_{ij}}{2\mu \Delta V \Delta t} \tag{10.14}$$

Seismic strain rate geometry can be described by the normalized strain orientation tensor $\bar{\epsilon}_{ij}$

$$\bar{\epsilon}_{ij} = \frac{\Sigma\,(\bar{M}_{ij})_k}{\|\Sigma\,(\bar{M}_{ij})_k\|} \tag{10.15}$$

where $\bar{M}_{ij} = M_{ij} / \|M_{ij}\|$ is the normalized moment tensor, called the source mechanism (Silver and Jordan, 1982), and describes the geometry of the strain release for a specific seismic event. The normalized strain orientation tensor $\bar{\epsilon}_{ij}$ describes the geometry of the volumetrically averaged strain rate for a given set of data (Fischer and Jordan, 1991). Seismic events with a considerable volume change should be excluded from this type of strain rate geometry statistics.

The degree to which moment tensors of seismic events are mutually similar within a given area can be measured by seismic consistency, $C_s$, defined by Fröhlich and Apperson (1992) as

$$C_s = \frac{\|\Sigma\,(\bar{M}_{ij})_k\|}{\Sigma\,\|(\bar{M}_{ij})_k\|} \tag{10.16}$$

$C_s$ varies from 0, where the individual moment tensors cancel each other out, to unity, where the moment tensors are identical.

The 'distance' or the difference between orientations of the normalized moment tensors of two seismic events, say, $k_1$ and $k_2$, can be measured by

$$\delta_{k_1,k_2} = \cos^{-1}\left[\left(\bar{M}_{ij}\right)_{k_1} \cdot \left(\bar{M}_{ij}\right)_{k_2}\right] \tag{10.17}$$

where the inner product $\left[\left(\bar{M}_{ij}\right)_{k_1} \cdot \left(\bar{M}_{ij}\right)_{k_2}\right]$ is called coherence (Kagan and Knopoff, 1985a,b) and varies from −1, for any pair of opposite moment tensors, to +1, for any pair of identical moment tensors. $\delta$ is an angle and varies from 0, for a pair of identical moment tensors, to $\pi$, for an opposite pair. One can then analyse the space or space-time clustering of seismic moment tensor orientations (Willemann, 1993).

*Seismic stress* An average seismic stress tensor $\sigma_{s_{ij}}$ can be defined as

$$\sigma_{s_{ij}}(\Delta V, \Delta t) = \frac{\sum_{t_1}^{t_2} E}{\dot{\epsilon}_s \Delta V \Delta t} = \frac{2\mu \sum_{t_1}^{t_2} E}{\sum_{t_1}^{t_2} M_{ij}} \tag{10.18}$$

where $\Sigma E$ is the seismic energy radiated by all events recorded within the volume $\Delta V$ during the period $\Delta t$ (Kostrov and Das, 1988). As tensors, seismic strain, strain rate and seismic stress have the same principal axes.

If a volume of rock, $V_m$, is mined out at time $t_0$ and if the altered stress and strain field can readjust to an equilibrium state through seismic movements only, the sum of seismic moments released within a given period of time would be proportional to the excavation closure and in the long term at $t = t_\infty$

$$\sum_{t_0}^{t_\infty} M \simeq \mu V_m \tag{10.19}$$

where $M$ is the scalar seismic moment (McGarr, 1976a). The relative stress level in a given volume of rock $\Delta$V surrounding the excavation, shown in Fig. 10.11, at the current time $t$ can be calculated from the difference between $\mu V_m$ and the cumulative moments released to date (McGarr 1976b):

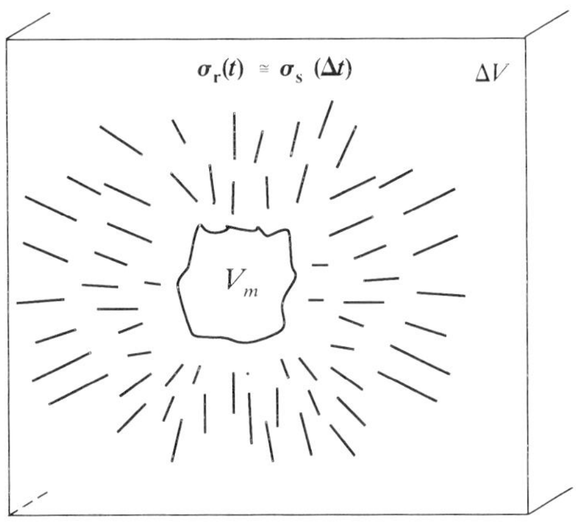


**Fig. 10.11** Seismic flow of rock into the mined out void $V_m$.

$$\sigma_r(t) = \frac{\mu V_m - \sum_{t_0}^{t} M}{\Delta V} \tag{10.20}$$

If, after removing $V_m$, the rockmass is left undisturbed the relative stress $\sigma_r$ would decay linearly with the cumulative seismic moments while, with time, it would exhibit diffusion type behaviour (see Fig. 10.12). The average seismic stress during $\Delta t = t_2 - t_1$ within a seismically flowing volume of rock $\Delta V$ for $t_2 > t > t_1$ and small $\Delta t$ is therefore approximately equal to the relative stresses at time $t$:

$$\sigma_r(t) = \frac{\mu V_m - \sum_{t_0}^{t} M}{\Delta V} \cong \frac{2\mu \sum_{t_1}^{t_2} E}{\sum_{t_1}^{t_2} M} = \sigma_s(\Delta t) \tag{10.21}$$

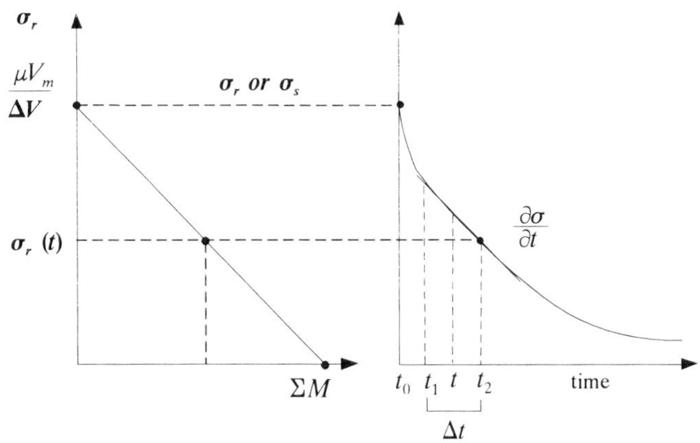


**Fig. 10.12** Relative stress versus cumulative seismic moments (left) and relative or seismic stress versus time (right).

### 10.2.2 Unstable system and seismic softening

*Unstable system* The surface which bounds all stress states that correspond to elastic deformation in six dimensional stress space is called the yield or damage surface. The evolution of the yield surface with continued deformation specifies the manner in which the material hardens or softens during the deformation. The direction of the inelastic strain increment beyond the yield surface is determined by the inelastic potential. The inelastic potential surface is always normal to the inelastic strain rate vector $\dot{\epsilon}_{in}$.

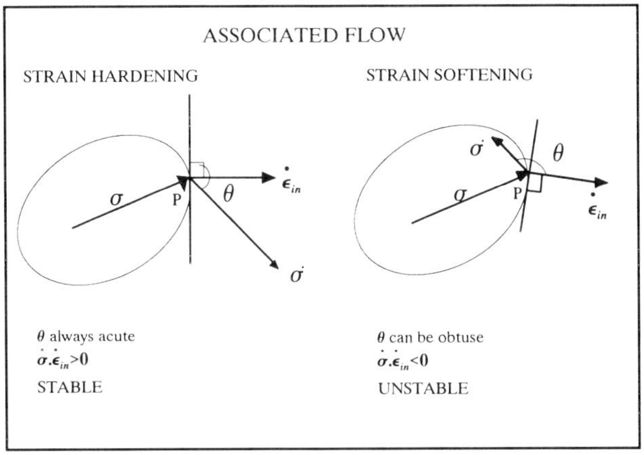


**Fig. 10.13** Yield surfaces for associated flow (the plastic potential surface coincides with the yield surface). $\sigma$ is the state of stress at P where the stress rate vector is $\dot{\sigma}$ and the plastic strain rate vector is $\dot{\epsilon}_{in}$. In the strain hardening situation the angle $\theta$ between $\dot{\sigma}$ and $\dot{\epsilon}_{in}$ is acute so that $\dot{\sigma}\cdot\dot{\epsilon}_{in}$ is positive resulting in stable deformation. In the strain softening situation, $\theta$ can be obtuse, so that $\dot{\sigma}\cdot\dot{\epsilon}_{in}$ is negative. If then $\dot{\sigma}\cdot\dot{\epsilon}_{in} > \dot{\sigma}\cdot\dot{\epsilon}_e$, where $\dot{\epsilon}_e$ is the elastic strain rate, then $\dot{\sigma}\cdot\dot{\epsilon}$, in this case, is negative, leading to unstable deformation (after Hobbs *et al.*, 1990).

The associated flow or normality rule applies to the material where yield and potential surfaces coincide, i.e. the inelastic strain rate vector is always normal to the yield surface. Under normal conditions, i.e. for associated flow with a smooth yield surface (Fig. 10.13), one would expect a strain softening behaviour before the instability.

This could be described by the classical instability criterion formulated by Hadamard in 1903 (Truesdell and Noll, 1992) and rediscovered by Pearson (1956) and Hill (1957, 1958):

$$\int_{\Delta V} \dot{\sigma}\,\dot{\epsilon}_e\, dV + \int_{\Delta V} \dot{\sigma}\,\dot{\epsilon}_{in}\, dV < 0 \qquad (10.22)$$

where:
$\dot{\epsilon}_e$; $\dot{\epsilon}_{in}$ elastic and inelastic strain rates;
$\Delta V$ volume of interest;
$\dot{\sigma}$ rate of change of stress.

Inequality (10.22) states that, for unstable deformation of $\Delta V$, the inner product of the next increment of stress with the next increment of strain should be negative:

$$d\sigma\, d\epsilon = d\sigma\, d\epsilon_e + d\sigma\, d\epsilon_{in} < 0 \qquad (10.23)$$

For the elastic strain increments, $d\sigma d\epsilon_e > 0$, the deformation therefore will be unstable only when the inelastic term balances the elastic term, so that there is a net strain softening. Near the end of the hardening regime, however, almost all further strain increments are inelastic. As stress and strain are tensors, there are many different components that may be influential in causing the instability.

One should distinguish between the stability of deformation defined by the stress–strain response of the material and the stability of the system being deformed. In some cases, or at some stage, the system ceases to deform in a homogeneous mode and undergoes bifurcation. In general, bifurcation refers to a qualitative change of the object under study due to a change of parameters on which the object depends – in this case, it is non-uniqueness of the deformation path. A bifurcation mode may or may not be amplified by continued deformation past the bifurcation point. After the displacement corresponding to the bifurcation point more than one incremental displacement field exists, satisfying the equilibrium and boundary conditions. If $\dot{\sigma}_1$ and $\dot{\sigma}_2$ are the stress rate fields corresponding to two such solutions, and $\dot{\epsilon}_1$ and $\dot{\epsilon}_2$ are the respective strain rates, then the necessary condition for bifurcation to occur is that there exists an incremental displacement field such that

$$\int_V \Delta\dot{\sigma}\, \Delta\dot{\epsilon}\, dV = 0 \tag{10.24}$$

where $\Delta\dot{\sigma} = \dot{\sigma}_1 - \dot{\sigma}_2$, $\Delta\dot{\epsilon} = \dot{\epsilon}_1 - \dot{\epsilon}_2$.

If the bifurcation grows it may be symmetrical or asymmetrical. A symmetrical bifurcation (e.g. barrelling of the specimen) is stable, so $d\sigma d\epsilon > 0$, i.e. the deformation is stable when it is easier to start deformation elsewhere rather than to pursue it where it has begun. Asymmetrical bifurcation, characteristic of materials that show frictional and dilatational responses to stress, results in strain localization in single, conjugate or quasi-periodic multiple shear bands (Rudnicki and Rice, 1975; Hobbs *et al.* 1990; Mühlhaus *et al.* 1992). The possibility of localization arises when one or more stress components within a volume are able to decrease with increasing strain (Cundall, 1990). Here the deformation may be unstable. Alternatively, strain localization can be caused by relative strain softening, where material within a localized strain zone hardens at a lower rate than the adjacent body of rock (Hobbs *et al.* 1990; Jessell and Lister, 1991). The localization is driven by the release of strain energy from an unloading region outside the localization zone.

Strain localization may occur under both strain softening and strain hardening material response and is promoted by non-associated flow (Fig. 10.14) and/or by the presence of vertices on the yield surface (Hobbs *et al.* 1990; Olsson, 1992).

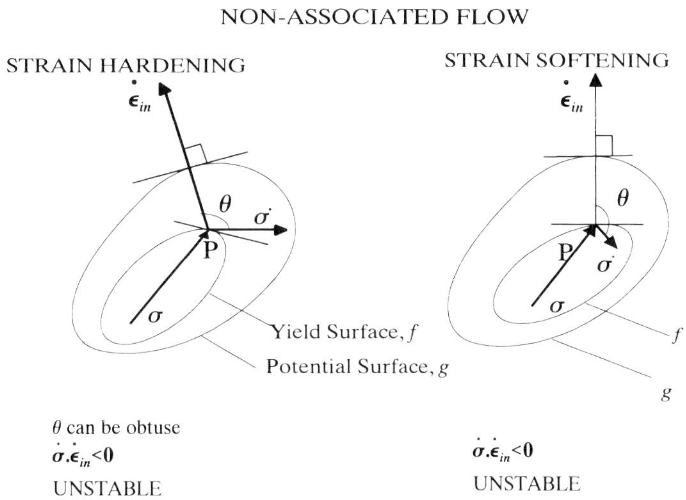


**Fig. 10.14** Yield surfaces for non-associated flow (the plastic potential surface, $g$, does not coincide with the yield surface, $f$). At the point P on the yield surface, $\sigma$ and $\dot{\sigma}$ are the stress state and stress rate respectively. At P the plastic strain rate is $\dot{\epsilon}_{in}$ and this is normal to the plastic potential surface. In both strain hardening and strain softening situations $\dot{\sigma}\cdot\dot{\epsilon}_{in}$ can be negative (after Hobbs, *et al.*, 1990).

At a vertex, the inelastic strain increment vector will tend to follow the stress increment vector (see Fig. 10.15). Vertices allow localization to occur at smaller overall deformations and at a significantly more positive tangent modulus (stiffer system).

Fredrich *et al.* (1989) conducted the first laboratory study to quantify all the constitutive parameters for pressure sensitive dilatant materials in both the brittle and semibrittle regime. They measured an internal friction coefficient, a dilatancy factor, being the ratio between the increments of inelastic volume strain and the inelastic shear strain, and a hardening modulus for Carrera marble deformed at room temperature and at different confining pressures. At a confining pressure of 5 MPa they showed a brittle failure mode and at 40–190 MPa semibrittle failure was observed. In the brittle region, shear localization was not evident until the sample was deformed well into the postfailure region. In the semibrittle regime, with increasing confining pressure, the hardening modulus increased whereas the friction coefficient, dilatancy factor and Poisson ratio decreased. Thus, the bifurcation model predicts that shear localization is inhibited in the semibrittle regime as long as the rock continues to strain harden.

Wawersik *et al.* (1990) investigated localization of deformation theoretically and in the laboratory. They found that deviation from normality (associated flow) promotes localization before peak stress for plane–strain compression. Triaxial deformation was characterized by a broad peak followed by gradual strain softening, with breakdown instability occurring only well beyond the peak stress.

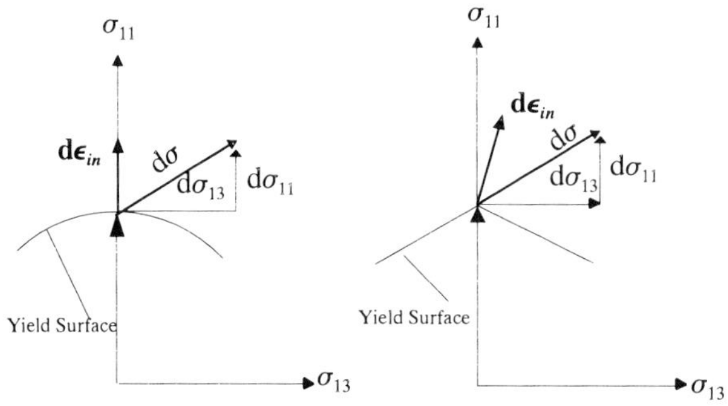


**Fig. 10.15** Relation between inelastic strain increment and stress increment for the smooth yield surface and for a vertex. (From Olsson, 1992.)

In the case where the nucleation zone is localized and is developing far from the free surface, the shear band(s) are constrained by both elastic and, to a limited extent, inelastic deformation in the less deformed rock surrounding the shear zone. In such a case, hardening of the system undergoing deformation is to be expected even if the material softens within the nucleation zone. The critical states of stability of a three dimensional body can generally occur only if the compression stress is of the same order of magnitude as the shear modulus of the material (Bazant and Cedolin, 1991).

Whether or when the process of strain softening and/or strain localization becomes unstable depends on the energy release and the dissipation inside the softening zone and on the exchange of mechanical energy between the softening zone and the ambient rock. Stress decrease associated with softening within the nucleation zone induces some elastic unstraining and redistribution of stresses in the ambient rock. Energy released by elastic rebound of the surrounding rock is fed into the softening region and accelerates its deformation. The amount of energy fed into the nucleation zone correlates positively with its size and with the strain rate or, rather, rate of unstraining in the ambient rock. As soon as an input of elastic energy exceeds the energy dissipation due to the inelastic processes in the growing nucleation zone the system becomes unstable.

One can consider an interpretation of instability in terms of the relative stiffness of the components of the system – an instability will occur when the stiffness of the nucleation volume is equal to or exceeds the stiffness of the external body. Thus instability results from the interaction of the strain weakening constitutive relation of the nucleation zone with the variable stiffness of the unloading surrounding body (see Fig. 10.16).

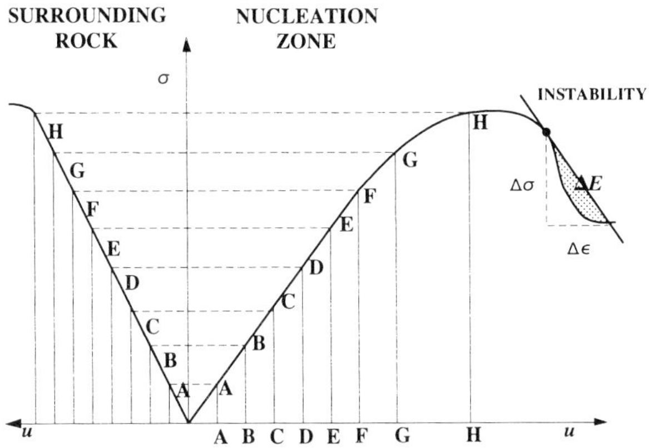


**Fig. 10.16** Interaction of the inelastic behaviour of the nucleation zone with the stiffness of the surrounding rockmass: $\Delta\epsilon, \Delta\sigma$ are the strain and stress drops during the instability, $\Delta E$ is the energy released during the instability.

It is assumed in Fig. 10.16 that the surrounding rockmass is more dilatant (thus having stiffer response) than the nucleation zone and undergoes strain hardening at the time when the nucleation zone softens. If the surrounding rock maintained its stiffness the system would be stable. Figure 10.16 also illustrates the acceleration of deformation preceding instability - see increasing increments of displacement within the nucleation volume, points FG to GH, corresponding to equal increments of stress within the surrounding rockmass (see also Rudnicki, 1988).

In general, the development of instability within rock type material is a complicated process involving changes in its stiffness and strain rate, and the influence of temperature and water. Hobbs *et al.* (1986), analysed the following constitutive law

$$\sigma = \sigma\,(\epsilon, \dot{\epsilon}, T, f_{H_2O}), \tag{10.25}$$

where $T$ is the temperature and $f_{H_2O}$ is the fugacity of water. Fugacity measures the chemical potential of a substance and correlates positively with changes in temperature, pressure and concentration.

If, for any small fluctuation in any or all the parameters in equation (10.25), the stress decreases with continued deformation, i.e. $d\sigma d\epsilon<0$, then the potential exists for the system to become unstable. From equation (10.25) we then get

$$\frac{d\sigma}{d\epsilon} = \left(\frac{\partial\sigma}{\partial\epsilon}\right) + \left(\frac{\partial\sigma}{\partial\dot{\epsilon}}\right)\left(\frac{d\dot{\epsilon}}{d\epsilon}\right) + \left(\frac{\partial\sigma}{\partial T}\right)\left(\frac{dT}{d\epsilon}\right) + \left(\frac{\partial\sigma}{\partial f_{H_2O}}\right)\left(\frac{df_{H_2O}}{d\epsilon}\right) \tag{10.26}$$

One can now consider each term in equation (10.26) and its influence on stability: positive terms have a stabilizing effect on the deformation, negative terms are destabilizing:

$d\sigma/d\epsilon$ conventional strain hardening rate θ, which is positive during the strain hardening and negative during the strain softening regime.

$\partial\sigma/\partial\epsilon$ related to the so-called stress sensitivity coefficient of the strain rate $n = \partial(\ln\dot{\epsilon})/\partial(\ln\sigma)$. Sufficiently high strain rate sensitivity (low $n$) can counterbalance the tendency to instability due to strain softening (negative θ). An instability can then be delayed because the short time response of the material surrounding the nucleation zone is stiffer than its long time response and/or because the volume of the nucleation zone increases faster than the rate fluid can flow in. This in turn increases the friction and decreases the pore pressure, making rockmasses much harder, and inhibits further dilatancy (Frank, 1965; Rudnicki, 1977; 1988); see Fig. 10.17. The amount of dilatant strengthening which can be affected is limited by diffusive instability. If the rockmass is fluid infiltrated, dilatant hardening (and elastic stiffening) in an initially very narrow weakened zone will cause adjacent regions to sustain greater stress and consequently to undergo inelastic dilatant deformation – dilatancy diffusion (Rice 1975).

$\partial\sigma/\partial\dot{\epsilon}$ the change in stress with strain rate, positive for power law materials.

$d\dot{\epsilon}/d\epsilon$ the change in strain rate with continued straining; for homogeneous steady state deformation this is zero, but for deformations in which strain is localized it becomes non-zero.

$\partial\sigma/\partial T$ stress response to change in the temperature; this has a destabilizing effect since stress tends to relax with an increase in temperature.

$dT/d\epsilon$ the temperature increase due to straining; in general this is positive.

$\partial\sigma/\partial f_{H_2O}$ the stress response to changes in the fugacity of water. An increase in fugacity decreases the stress so this parameter is destabilizing. Fluctuations in $f_{H_2O}$ would arise from local fluctuations in pressure or temperature or both.

$df_{H_2O}/d\epsilon$ the water fugacity response to change in strain. This could arise from local regions of dilatation developing during deformation where a transient pressure decrease could give rise to the vapour phase. This leads to a decrease in $f_{H_2O}$ and an associated increase in $\sigma$ and stabilizes the deformation.

The accelerating inelastic deformations within the nucleation zone would lead to different modes of failure depending on the presence and the magnitude of the confining pressure (Horii and Nemat-Nasser, 1986):

- axial splitting or unstable crack growth coupled with surface buckling, in the absence of, or at low, confining pressure and in the proximity of a free surface;

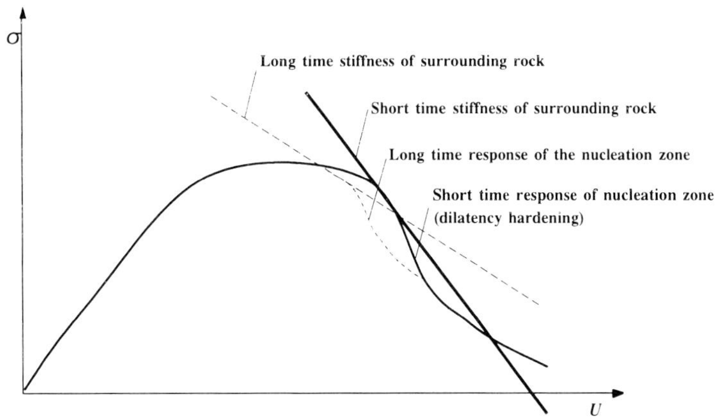


**Fig. 10.17** The delay of instability and enhancement of precursory deformation due to the stiffer short time response of the surrounding rock and dilatancy hardening within the nucleation zone. (Compiled after Rudnicki, 1977.)

- macroscopic shear failure, when axial compression is accompanied by moderate confining pressure;
- ductile flow, in the presence of large confining pressure.

The duration and magnitude of mechanical precursory phenomena, specifically the deformation and the deformation rate, increase with an increase in mechanical heterogeneity of the rockmass (Mogi, 1985; Tang *et al.* 1993).

The deformation jump at the point of instability relates to the properties of the rock within and, to a certain extent, outside the nucleation zone. Thus strain drop, stress drop and energy release (shaded area of Fig. 10.16) during the instability depend on the strength and degree of inhomogeneity of the rock at the hypocentre and in the surrounding rockmass. The stronger and more homogeneous the rockmass the higher the stress drop and the higher the ratio of the stress drop over strain drop associated with instability.

The amount of postpeak deformation in the weakened nucleation zone prior to instability strongly depends on the deviatoric state of stress induced within the nucleation zone and on the degree of inhomogeneity. In particular, instability is predicted much nearer the peak load for very narrow weakened zones and for states of deviatoric simple shear than for states of axisymmetric compression (Rudnicki, 1977).

*Seismic Softening* Figure 10.18 presents schematically the typical stress–strain diagram of a rock sample under compression. After passing through the elastic and strain hardening regimes, the material softens and further increments of strain are associated with the loss of stress.

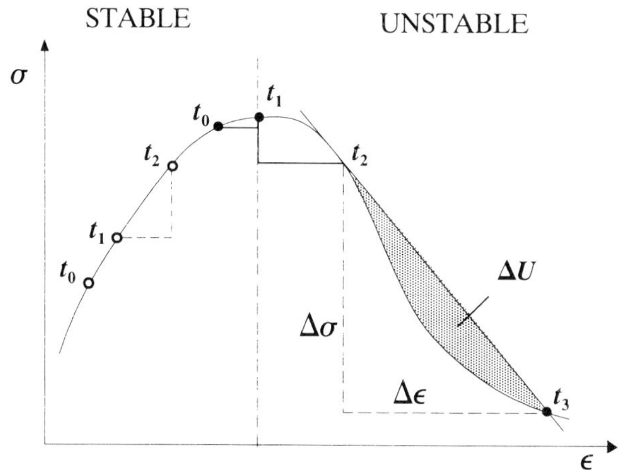


**Fig. 10.18** Typical stress–strain diagram of a rock sample. A stable sequence ($t_0$ – $t_2$) is shown on the left hand side and an unstable sequence ($t_0$ – $t_3$) is shown on the right hand side.

For finite stress and strain departures from the current state at $t_1[\sigma(t_1), \epsilon(t_1)]$ to the next state at $t_2[\sigma(t_2), \epsilon(t_2)]$, the definition of instability of the volume element $\Delta V$ (see Fig. 10.18 and equation (10.22)), is

$$\int_{\Delta V} \int_{\epsilon(t_1)}^{\epsilon(t_2)} [\sigma(t_2) - \sigma(t_1)] \, d\epsilon(t) \, dV < 0 \qquad (10.27)$$

One can consider the seismic version of the instability criterion (10.27) with the following assumptions:

$$\sigma_{ij}(t_2) \approx \sigma_{s_{ij}}(\Delta V, t_2 - t_1) = 2\mu \sum_{t_1}^{t_2} E / \sum_{t_1}^{t_2} M_{ij}, \text{ for } \sum_{t_1}^{t_2} M_{ij} > 0$$

$$\sigma_{ij}(t_1) \approx \sigma_{s_{ij}}(\Delta V, t_1 - t_0) = 2\mu \sum_{t_0}^{t_1} E / \sum_{t_0}^{t_1} M_{ij}, \text{ for } \sum_{t_0}^{t_1} M_{ij} > 0$$

$$\int_{\epsilon(t_1)}^{\epsilon(t_2)} d\epsilon(t) \approx \epsilon_{s_{ij}}(\Delta V, t_2 - t_0) = \sum_{t_0}^{t_2} M_{ij} / (2\mu \Delta V)$$

and for $t_2 > t_1 > t_0$,

$$S_s(\Delta V, \Delta t_1, \Delta t_2) = \frac{\left(\sum_{t_1}^{t_2} E \sum_{t_0}^{t_1} M_{ij} - \sum_{t_0}^{t_1} E \sum_{t_1}^{t_2} M_{ij}\right) \sum_{t_0}^{t_2} M_{ij}}{\sum_{t_1}^{t_2} M_{ij} \sum_{t_0}^{t_1} M_{ij}} \tag{10.28}$$

where $S_s$ is the measure of an average seismic softening or hardening within the volume $\Delta V$ between $t_0$ and $t_2$, where $\Delta t_1 = t_1 - t_0$ and $\Delta t_2 = t_2 - t_1$.

The energy released or work done, $\Delta U$, during the instability $(t_2, t_3)$, see Fig. 10.18, is proportional to stress drop $\Delta\sigma$ and strain drop $\Delta\epsilon$

$$\Delta U_{23} = \int_{\Delta V} \int_{\epsilon(t_2)}^{\epsilon(t_3)} [\sigma(t_3) - \sigma(t_2)] \; d\epsilon(t) \, dV \tag{10.29}$$

Figure 10.19 shows the results of a triaxial experiment at a confining pressure of 94 MPa and strain rate of $1.4 \times 10^{-7}$/s performed on a sample of fine grained granite (Yukutake, 1992). One can see the onset of potential instability as defined by equation (10.27) from 1000 s before the failure. The amount of stress drop from peak stress to fracture is more than 20 MPa. Note the dramatic stress decrease and displacement jump just prior to the breakdown instability.

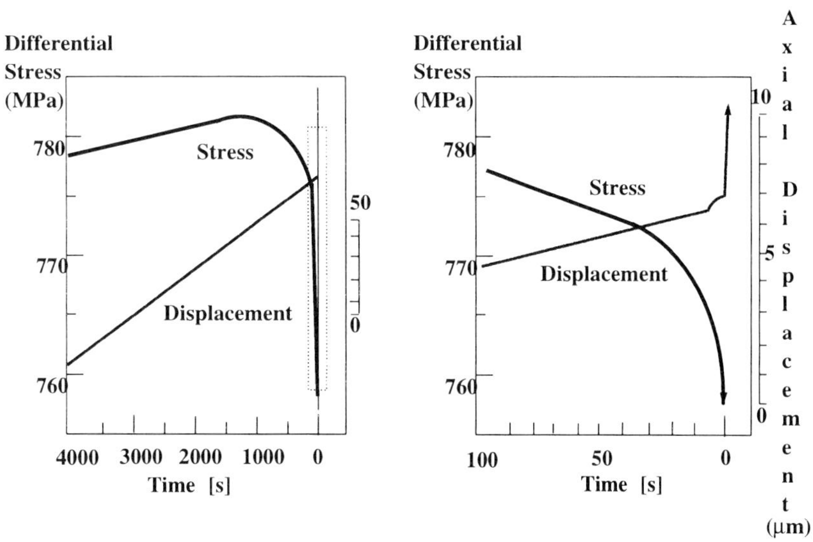


**Fig. 10.19** The change in the differential stress and in axial displacement versus time for the sample of fine grained granite during triaxial compression with a confining pressure of 94 MPa (after Yukutake, 1992).

### 10.2.3 Seismic viscosity, relaxation time and seismic Deborah number

Rockmass resistance to the flow of coseismic inelastic deformation within the volume $\Delta V$ over time $\Delta t$ can be measured by the ratio of seismic stress over seismic strain rate; this is called seismic viscosity (Kostrov and Das, 1988)

$$\eta_s(\Delta V, \Delta t) = \frac{\sigma_s}{\dot{\epsilon}_s} = \frac{4\mu^2 \Delta V \Delta t \sum_{t_1}^{t_2} E}{\left(\sum_{t_1}^{t_2} M_{ij}\right)^2} \tag{10.30}$$

If one considers seismicity as the turbulent part of the flow of rock towards equilibrium, then the concept of seismic viscosity is similar to the fluid mechanics concept of turbulent or eddy viscosity. Unlike ordinary or molecular viscosity, eddy viscosity does not describe the physical properties of the medium but characterizes the statistical properties of the flow. Consequently, it does not have to be constant but varies in time and space. Seismic viscosity would increase during quiescence, due to the increase in $\Delta t$ while all other components in equation (10.30) are constant, or during a sequence of higher than average apparent stress events; it would decrease during a sequence of softer events. Seismic viscosities obtained to date for the volumes surrounding mine excavations vary from $10^{14}$ to $10^{19}$ Pa.s. The inverse of viscosity is called fluidity.

The coseismic viscosity $\eta_{cs}$, i.e. the viscosity of the flow of rock during an individual seismic event within its source, could be estimated from $\eta_{cs} = \Delta\sigma / \Delta\dot{\epsilon}_{cs}$ and since $\Delta\dot{\epsilon}_{cs} = \Delta\sigma / (\mu t_{cs})$, then $\eta_{cs} = \mu t_{cs}$, where $\Delta\sigma$ and $\Delta\dot{\epsilon}$ are stress drop and coseismic strain rate respectively and $t_{cs}$ here is the duration of the source. Since the duration of mine seismic sources $t_{cs} \cong 10^{-3}$ to $10^0$ s and $\mu \cong 10^{10}$ Pa then $\eta_{cs} = 10^7$ to $10^{10}$, Pa.s, which is a few orders of magnitude less than the calculated average seismic viscosity $\eta_s$ for a given volume $\Delta V$ and over a given time $\Delta t$. The reason is in the intermittent nature of seismic flow of rock, where sometimes large portions of space $\Delta V$ and time $\Delta t$ are not affected by seismic events but are accounted for in equation (10.30).

Lower seismic viscosity (or high fluidity) implies the easier flow of seismic inelastic deformation, or greater stress transfer due to seismicity, and promotes softer rockmass response to mining.

Having determined seismic viscosity, one can estimate seismic relaxation time $\tau_s$, which governs the rate of change of stress during the seismic flow of rock,

$$\tau_s(\Delta V, \Delta t) = \frac{\eta_s}{\mu} \tag{10.31}$$

Seismic relaxation time defines the usefulness of past data in the predictability of the seismic flow of rock. The lower the relaxation time the shorter the time-span of useful past data and the less predictable the seismic flow of rock. The ratio of the dynamic viscosity $\eta$ to density $\rho$ is called the kinematic viscosity. The seismic kinematic viscosity is then

$$\nu_s(\Delta V, \Delta t) = \frac{\eta_s}{\rho} \tag{10.32}$$

A given rockmass may appear to flow seismically on a time-scale, $\Delta t$, close to or larger than the relaxation time, $\tau_s$, whereas it will behave as a rigid or elastic solid on shorter time-scales. The ratio of the relaxation time $\tau_s$ to the time of observation or flow time is called the Deborah number, $De$, (Reiner, 1969). The seismic Deborah number $De_s$ is then

$$De_s(\Delta V, \Delta t) = \frac{\tau_s}{\Delta t} = \frac{4\mu \Delta V \sum_{t_1}^{t_2} E}{\left( \sum_{t_1}^{t_2} M_{ij} \right)^2} \tag{10.33}$$

The Deborah number, like the relaxation time, may be interpreted as the ratio of elastic to viscous forces, with $De$ going to infinity for a perfectly elastic medium. For $De<1$ viscous effects dominate, whereas for a large Deborah number the system will essentially behave as an elastic solid. Seismic Deborah number can be used as a criterion to delineate clusters of seismicity and is sensitive to clusters associated with soft, or low apparent stress, events, e.g. soft clusters would include volume(s) of rock with a seismic Deborah number below a certain threshold, say, $\Delta V$ ( $De< 10$).

At some stage, the process of localization of inelastic deformation that leads to the breakdown instability may be largely controlled by the most critical cluster of events, in the sense that events outside this critical cluster would make little or no contribution to the nucleation processes (Huang and Hutchinson, 1989; Huang, 1993). The potentially critical cluster would be characterized by the lower Deborah number and by non-uniform shape. This idea needs to be vigorously tested, since clustering of microseismic activity has been observed before larger failures in mines.

In design problems, say, of mine stabilizing pillars, one may replace the flow time $\Delta t$ in equation (10.33) with the time of expected service of that structure and then, for a given relaxation time determined from modelling or backanalysis, one can evaluate the potential stability of that pillar over time $\Delta t$. The higher the seismic Deborah number the more stable it is. Crush pillars, on the other hand, should be designed for low relaxation time so that they would soften and drop off the load within the designed $\Delta t$.

### 10.2.4 Seismic dissipation and seismic diffusion

From the point of view of energy balance, the viscous term can be divided into two parts, of which one is diffusive in character, and vanishes when integrated, and the other is dissipative. The dissipative term is negative definite since kinetic energy cannot be gained from friction. The seismic energy dissipation rate per unit mass within a given volume $\Delta V$ can be estimated by

$$e_s(\Delta V, \Delta t) = \nu_s \cdot (\dot{\epsilon}_s)^2 = \frac{\sum_{t_1}^{t_2} E}{\rho \Delta V \Delta t} \tag{10.34}$$

A non-equilibrium situation which is moving towards equilibrium at the rate governed by its distance from equilibrium (see Fig. 10.12), is described by the diffusion equation

$$\frac{\partial \phi}{\partial t} = D \nabla^2 \phi \tag{10.35}$$

where $\partial \phi / \partial t$ is the change of $\phi$ with time, $D$ is the diffusivity and

$$\nabla^2 \phi \equiv \frac{\partial^2 \phi}{\partial x^2} + \frac{\partial^2 \phi}{\partial y^2} + \frac{\partial^2 \phi}{\partial z^2}$$

is called the divergence of gradient. The value of the scalar $\phi$ depends upon the point in space at which it is measured. If a small cube of $L \times L \times L$ is constructed around a point where $\phi = \phi_0$ then the average value of $\phi$ over the cube volume is

$$\bar{\phi} = \frac{1}{L^3} \iiint_{-L/2}^{L/2} \phi \, dx \, dy \, dz$$

Expanding $\phi$ about the point where $\phi = \phi_0$ by a Taylor series and integrating gives

$$\bar{\phi} - \phi_0 = \frac{L^2}{24}\left(\frac{\partial^2 \phi}{\partial x^2} + \frac{\partial^2 \phi}{\partial y^2} + \frac{\partial^2 \phi}{\partial z^2}\right)_{\phi=\phi_0} \tag{10.36}$$

The diffusion equation (10.35) then states that the change of $\phi$ with time at a point in the volume is proportional to the difference $\bar{\phi} - \phi_0$.

The diffusivity $D$ has dimension m$^2$/s and is interpreted in terms of a characteristic distance of the process which varies only with the square root of time. The average seismic diffusivity, $D_s$, of the seismic flow of rock within the volume $\Delta V = L \times L \times L$ over time $\Delta t = t_2 - t_1$ can then be defined as:

$$D_s(\Delta V, \Delta t) = \frac{L^2}{\tau_s} = \frac{L^2\mu}{\eta_s} = \frac{L^2\left(\sum_{t_1}^{t_2} M_{ij}\right)^2}{4\mu\Delta V\Delta t\sum_{t_1}^{t_2} E} = \frac{\left(\sum_{t_1}^{t_2} M_{ij}\right)^2}{4\mu L\Delta t\sum_{t_1}^{t_2} E} \tag{10.37}$$

where $\tau_s$ and $\eta_s$ are seismic relaxation time and seismic viscosity respectively.

In terms of seismic source parameters, diffusivity increases with the square of seismic moments and decreases with an increase in seismic energies for the same seismic moment events, i.e. diffusivity is higher when the same size events are softer (i.e. with low apparent stress) and decreases with an increase in viscosity of the seismic flow of rock. Seismic diffusivity, like seismic viscosity, is to be considered as a function of space and time.

The average velocity of seismic diffusion is

$$v_{D_s}(\Delta V, \Delta t) = \sqrt{D_s/\tau_s} = \frac{\left(\sum_{t_1}^{t_2} M_{ij}\right)^2}{4\mu\Delta t L^2\sum_{t_1}^{t_2} E} \tag{10.38}$$

The average acceleration of that diffusion is

$$a_{D_s}(\Delta V, \Delta t) = \sqrt{D_s/\tau_s^3} = \frac{\left(\sum_{t_1}^{t_2} M_{ij}\right)^4}{16\mu^2\Delta t^2 L^5\left(\sum_{t_1}^{t_2} E\right)^2} \tag{10.39}$$

One can consider the statistical definition of seismic diffusion $d_s$ for a cluster of seismic events:

$$d_s = \frac{\bar{X}^2}{\bar{t}} \tag{10.40}$$

where $\bar{X}$ is an average distance between consecutive sources of interacting seismic events, $\bar{t}$ is an average time between events. Here, $\bar{X}$ measures the characteristic scale of seismic flow of rock within a given cluster of seismic events. Figure 10.20 shows an influence of the source size on seismic diffusion given by equation (10.40).

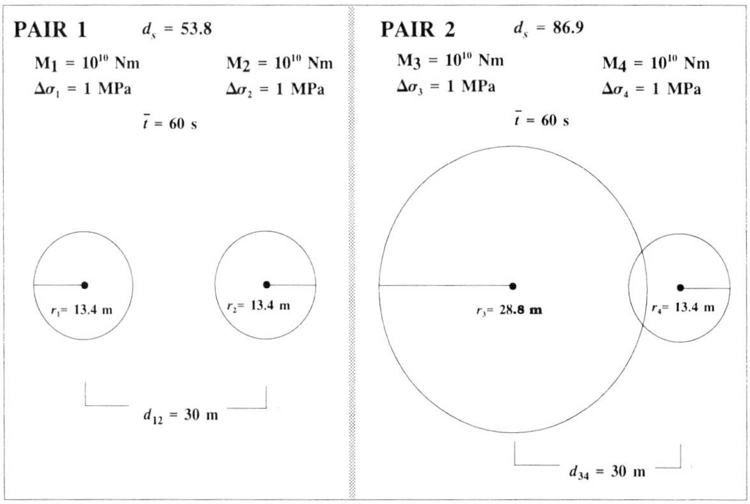


**Fig. 10.20** An illustration of the difference in seismic diffusion $\bar{X}^2/\bar{t}$ between two pairs of events due to the inclusion of the event sizes. The size of the event has been calculated as a diameter of the sphere of source volume $V = M/\Delta\sigma$.

The average velocity and average acceleration of seismic diffusion, given by equation (10.40), are

$$v_{ds} = \frac{\bar{X}}{\bar{t}} \tag{10.41}$$

$$a_{ds} = \frac{\bar{X}}{\bar{t}^2} \tag{10.42}$$

The diffusion parameters, given by equations (10.40) – (10.42), can be decomposed into *x*, *y* and *z* components, which could be very useful in quantifying the magnitude, directions, velocities and accelerations of migration of microseismic activity and associated fracturing processes. Such a travelling deformation could be responsible for triggering of larger seismic events or other types of rockmass instabilities (Scholz, 1977; Brune, 1979; Lehner *et al.* 1981). On the other hand, equations (10.37) – (10.39) are particularly useful if one wants to contour diffusivity over the area or volume of interest.

### 10.2.5 Seismic Schmidt number

The ratio of kinematic viscosity to diffusion is called the Schmidt number. When the flow parameters, as opposed to the molecular parameters, are divided the so-called turbulent Schmidt number (McComb, 1990) measures the degree of turbulence in a given flow – the lower the Schmidt number the more turbulent the flow. The seismic Schmidt number will be given by

$$Sc_{sd}(\Delta V, \Delta t) = \frac{\nu_s}{d_s} = \frac{4\mu^2 \Delta V \Delta t(\bar{t}) \sum_{t_1}^{t_2} E}{\rho\, (\bar{X})^2 \left( \sum_{t_1}^{t_2} M_{ij} \right)^2} \tag{10.43}$$

The seismic Schmidt number measures the spatiotemporal complexity of the seismic flow of rock. The lower the seismic Schmidt number the less stable the flow. Note that equation (10.43) encompasses four independent parameters which describe seismicity, namely: $\bar{t}$, $\bar{X}$, $\Sigma E$ and $\Sigma M_{ij}$.

## 10.3 Nucleation of instability and time to failure

*Softening* The growth of the deformation processes up to the point of instability is called nucleation. The breakdown instability will only take place once a quasi-static and/or quasi-dynamic inelastic deformation has occurred within the critical volume of rock.

The entire nucleation process prior to overall instability includes aseismic creep (e.g. Dieterich, 1992; Ohnaka, 1992), subcritical crack growth and dynamic instabilities of local to small scale. Experimental data on subcritical crack growth in synthetic quartz crystals indicate that this process starts only above certain stress levels, which could be of the order of 50% of the rupture stress (Darot and Gueguen, 1986). During subcritical crack growth, crack advance occurs by discrete, individually dynamic events, but with slow average rupture velocity. Though individual microseismic events may be dynamic, their evolution can be modelled as a continuous, quasi-static process because they release only small amounts of stress or seismic moment compared to the breakdown instability. Alternatively, the larger the event the stronger its influence on the stress and strain environment in the area and the more likely it will affect both the time and the size of the next seismic event.

Since the development of the nucleation zone is associated with overall strain softening, sources of seismic events located within the nucleation zone would be, on average, of a slower or softer nature. This effect would be magnified in cases where the nucleation zone interfaces through the fractured zone with the opening. In the laboratory experiments, crack branching and distributed damage associated with the fracture are observed only if the strength variation in the source region is greater than the stress concentration (Labuz *et al.* 1985; Cox and Patterson, 1990).

Outside, or at the interface of, the nucleation zone one would expect seismic events of a harder or faster nature characterized by higher energy indices. The overall size of the nucleation zone increases with the size (seismic moment) of the potential instability (Scholz, 1990) and with the degree of inhomogeneity, and decreases with the increase in the strain rate (Kato *et al.* 1992).

*Accelerating deformation* It is the nature of rock fracture and friction that the breakdown instability does not occur without some preceding phase of accelerating deformation (e.g. Hirata *et al.* 1987; Rudnicki, 1988; Scholz, 1990; Mendecki, 1993; Dieterich, 1994).

The frequently observed increase in the rate of coseismic deformation before instability can be due to the increase in the rate of microseismicity, or, for the same rate and sizes of events, due to the softer nature of individual events occurring within the nucleation volume.

A small but statistically significant decrease in seismic activity (quiescence) has been observed at the beginning of the strain softening stage, followed by an increase just before the instability (Brady, 1977a,b; Main *et al.* 1992).

The following seismic behaviour has frequently been observed at some stage before unstable rockmass behaviour in South African gold mines (Mendecki, 1993; Chapter 11 of this book).

- An overall softening, i.e. a decrease in the average value of the energy index (*EI*), seismic softening ($S_s$) or seismic viscosity ($\eta_s$), with time, spatially associated with seismic events close to the hypocentre of impending instability. The softened (fractured) zone is characterized by low seismic Deborah number. However, in a number of cases, short periods of increase and/or oscillations in these parameters just prior to instability have been reported.

- An increase in the rate of coseismic deformation, as measured by the cumulative apparent volume $\Sigma V_A$ or by seismic diffusion $D_s$, due to the increase in seismic activity rate (decrease in $\bar{t}$) and/or due to the softer nature of events occurring within the already fractured zone.

If softening coincides with accelerating deformation, then a decrease in seismic viscosity and an increase in seismic diffusion $d_s$ would result in a sharp decrease in seismic Schmidt number before instability.

In general, the success rate for detecting the nucleation of instabilities will depend on the sensitivity of the seismic system in the area of interest. The stronger and the more homogeneous the rockmass, the smaller the nucleation zone and thus the more sensitive the seismic system needs to be to detect it.

Figure 10.21 shows cumulative apparent volume $V_A$ (line with dots indicating seismic events) and moving average energy index *EI* (thin line) versus time for an area surrounding a dyke in the Welkom goldfield. Each event above local magnitude 1.7 is marked with a vertical line capped with a square. In August 1992 a significant increase in the slope of $\Sigma V_A$ (indicating accelerating coseismic deformation) and an associated decrease in average energy index (indicating a drop in stress) led to significant instability in a form of a number of larger seismic events. Note an increase in and the oscillations of the energy index during unstable rockmass behaviour at the end of August. The recovery in *EI* and flattening-off in $\Sigma V_A$ during September were interpreted at the time, and latterly positively confirmed, to be the return of that part of the rockmass to more stable behaviour.

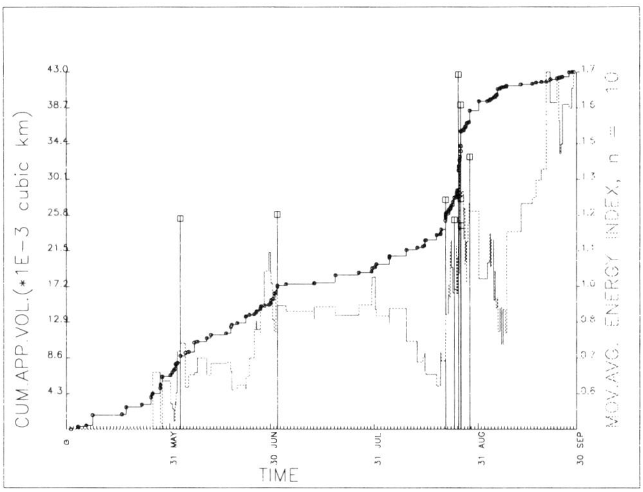


**Fig. 10.21** Cumulative apparent volume (thick dotted line) and moving average energy index (thin dashed line) versus time for mining area surrounding a dyke in Welkom, South Africa. The largest seismic events are marked by square capped vertical lines.

*Time to failure* The behaviour of materials in the terminal stages of failure under conditions of approximately constant stress and temperature can be described by a simple relation:

$$\dot{\Omega}^{-\gamma}\,\ddot{\Omega} = c \tag{10.44}$$

where $\Omega$ is strain or other appropriate measurable quantity and $\gamma$ and $c$ are constants (Voight, 1989). Equation (10.44) is derived from experimental observations of a linear relation between the logarithm of velocity and the logarithm of acceleration of deformation just before failure. A similar relation can be devised from models of nucleation and critical crack propagation (Das and Scholz, 1981; Varnes, 1983; Main, 1988) and from the constitutive equations describing damage under uniaxial tests (Kachanov, 1961; Rabotnov, 1969). For $1<\gamma<2$, $\Omega$ becomes very large at a finite time and failure may be predicted. Recently, Sykes and Jaumé (1990) observed that the larger earthquakes tended to occur within a few years of the time that $\Sigma M$ reached $6\times10^{18}$ Nm, but they were not successful in using equation (10.44) to estimate the time of the main shocks.

One can consider nucleation of instability as being similar to critical phenomena where progressive correlations develop, leading to a cascade of events at increasingly large scales, culminating in failure throughout the entire structure (Sornette and Sammis, 1995). If one can view instability as a critical point, then precursors should follow characteristic power laws in which the rate of $\Omega$ is proportional to the inverse power of remaining time to failure

$$\frac{d\Omega}{dt} = \frac{k}{|t_f - t|^{\alpha}} \tag{10.45}$$

where $k$ and $\alpha$ are constants and $t_f$ is time at failure. Integration of equation (10.45) gives

$$\Omega(t) = A + B\,|t_f - t|^{m} \tag{10.46}$$

where $A, B$ and $m = 1 - \alpha$ are constants.

By recognizing equations (10.45) and (10.46) as a universal scaling law near a critical point, Sornette and Sammis (1995), using the normalization group framework, derived the leading corrections to scaling in the form of a periodic function of $\log|t_f - t|$, which is superimposed on the power law in equation (10.46)

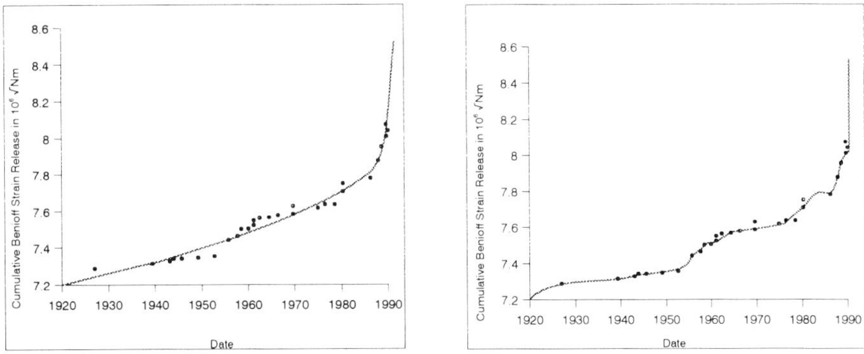


a b

**Fig. 10.22a-b** Cumulative Benioff strain released by magnitude 5 and greater earthquakes in the San Francisco Bay area prior to the 1989 Loma Prieta earthquake (from Bufe and Varnes, 1993). In (a), the data have been fitted to the powerlaw equation (10.46), as in Bufe and Varnes, (1993). In (b), the data have been fitted to equation (10.47) which includes the first order correction to scaling.

$$\Omega(t) = A + B\,|t_f - t|^m \left[1 + C\cos\left(2\pi\,\frac{\log(t_f - t)}{\log D} + E\right)\right] \tag{10.47}$$

where $C$, $D$ and $E$ are constants and, if $C = 0$, equation (10.47) reduces to equation (10.46). Since $\Omega(t_f) = A$, from equations (10.46) and (10.47) one can obtain an estimated upper bound for $\Delta\Omega$ associated with the impending instability:

$$\Delta\Omega(t_f) = A - \Omega(t) \tag{10.48}$$

Bufe and Varnes (1993) fitted equation (10.46) with $\Omega = \Sigma E^{1/2}$, called cumulative Benioff strain release, to the seismic catalogue of Northern California for the period 1927–1988. They predicted $t_f = 1990$ for the Loma Prieta earthquake, which had a magnitude between 6.7 and 7.1 and occurred on 18 October 1989 (1989.8); see Fig. 10.22a. Sornette and Sammis (1995) fitted equation (10.47) to the same data, getting $t_f = 1989.9 \pm 0.8$ years (see Fig. 10.22b) since from early 1980 enough of the log-periodic structure has developed to constrain the prediction.

By way of illustration, a good example of precursive behaviour, as described by the phenomenological model on the left hand side of the following equation and quantified by time to failure $(t_f - t)$, is presented in Figs. 10.23 and 10.24,

$$\frac{\Sigma V_A}{\bar{t}\;\overline{EI}\;\Delta V} = \frac{k}{(t_f - t)^{\alpha}} \tag{10.49}$$

where $\bar{t}$ and $\overline{EI}$ are running means.

Figure 10.23 shows a section through contours of average seismic Deborah number in the hypocentral area constructed from 450 events which occurred over 150 days, prior to the main event. The location of the log$E$=9.43 event and its size (diameter of its apparent volume sphere) are also shown.

Figure 10.24a shows the behaviour of time to failure ($t_f - t$), derived from the equation (10.49), before a log$E$=9.43 event.

Figure 10.24 b–g shows changes in the averaged values of energy index $EI$, seismic softening $S_s$, activity rate $1/\bar{t}$, log seismic viscosity, log$\eta_s$, seismic diffusion $d_s$ and log seismic Schmidt number log$Sc_s$ against the background of cumulative apparent volume $\Sigma V_A$, of the seismicity that occurred over the final 18 days in the volume surrounding the hypocentre of the log$E$=9.43 event.

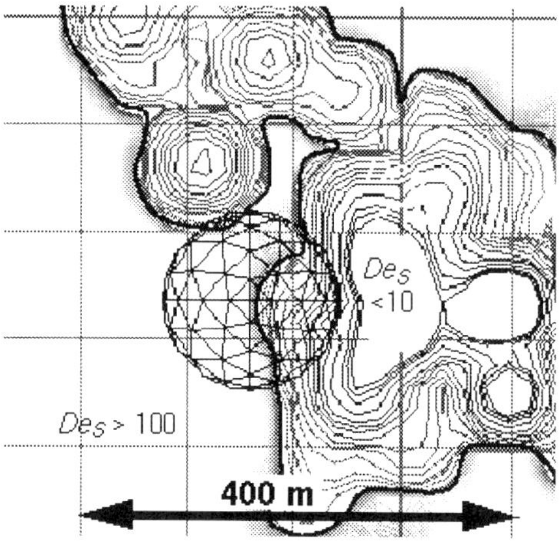


**Fig. 10.23** A section through contours of seismic Deborah number in the hypocentral area of a log$E$=9.43 event, whose apparent volume is represented by the mesh sphere. Note the steep gradient around the event location.

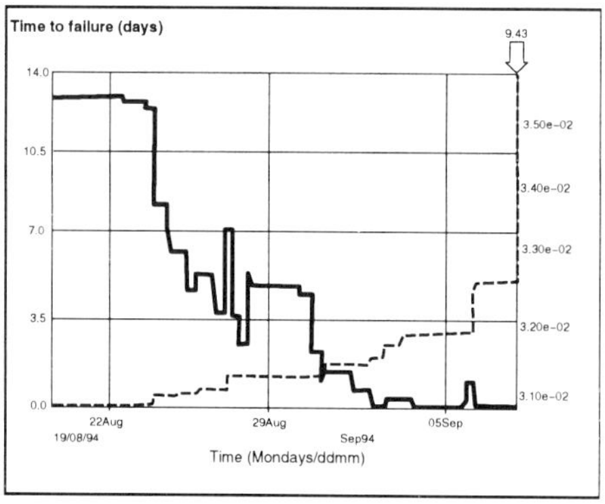


**Fig. 10.24a** Time to failure $(t_f - t)$ quantifying a phenomenological model of equation (10.49) for a log$E$=9.43 event. Cumulative apparent volume $\Sigma V_A$, is plotted with the broken line with values indicated on right hand side axis.

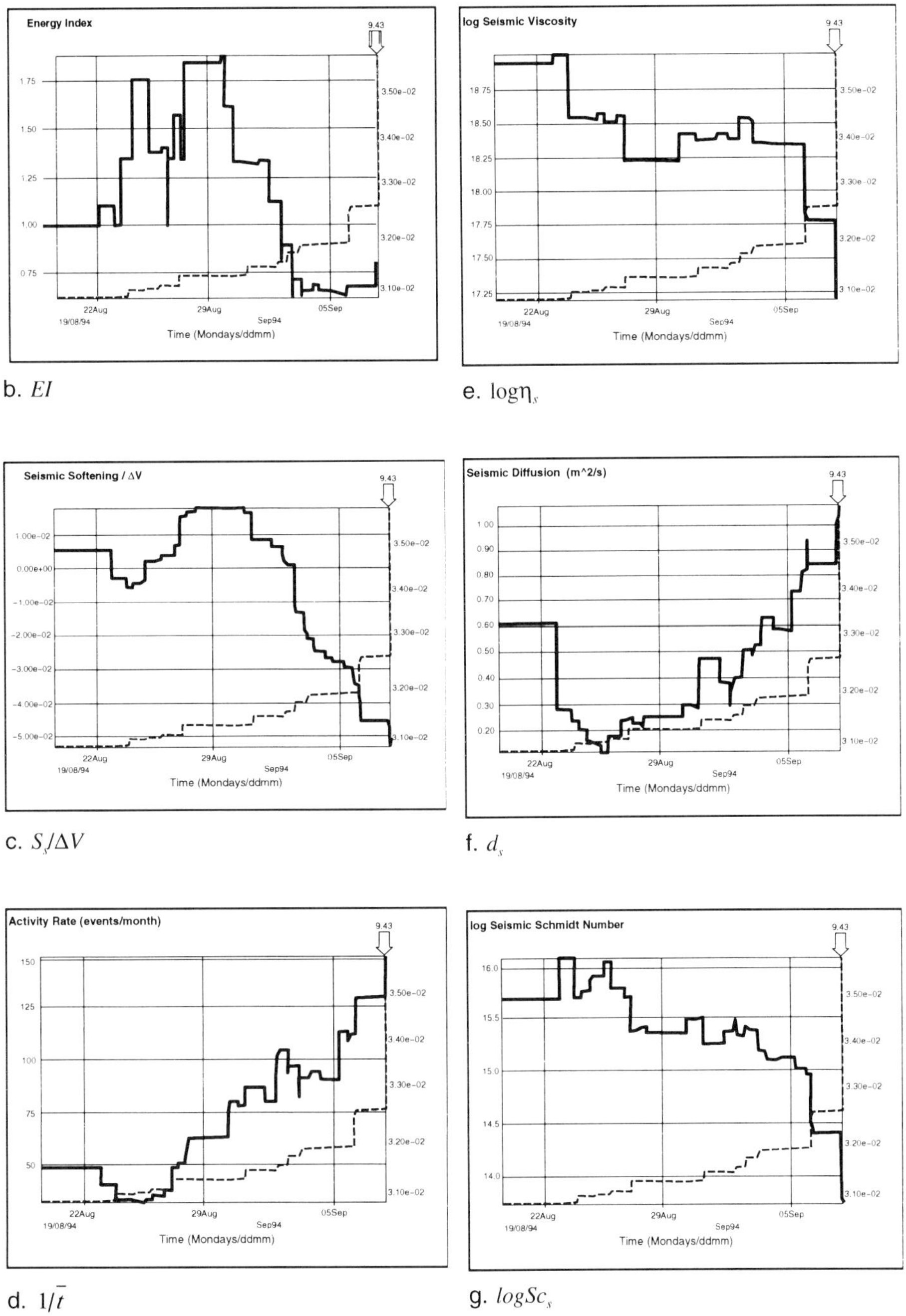


**Fig. 10.24b–g** Changes in the average values of energy index $EI$, seismic softening $S_s$, activity rate $1/\bar{t}$, log seismic viscosity $\log\eta_s$, seismic diffusion $d_s$ and log seismic Schmidt number $\log Sc_s$ against the background of cumulative apparent volume $\Sigma V_A$, (broken line, values indicated on right hand side axis) that have occurred over the final 18 days in the volume surrounding the hypocentre of the $\log E$=9.43 event.

# 11 Application of Quantitative Seismology in Mines

## 11.1 Introduction

The development of the technology and the appropriate theory of quantitative seismology allows applications far beyond the aims of the first seismic monitoring in mines when hypocentre location and some kind of magnitude determination was considered satisfactory.

The aim here is to show, through example cases, how quantitative seismology can be applied to fulfil the main objectives of seismic monitoring in mines. The first author of this chapter has spent more than 10 years analysing seismicity in one of the major gold mining regions of South Africa, colloquially referred to as the Welkom goldfield, and most example cases are based on this experience.

Since the physical laws governing centimetre scale to kilometre scale discontinuous deformation processes are the same, at least some scale invariance must apply in seismology. This allows, to a certain extent, the projection of crustal seismological theory to microtremors and laboratory results to earthquakes. Mines provide a convenient intermediate step between the laboratory and crustal seismology and elements of both environments can be found. Some fault slip events observed in mines are true earthquakes, with Richter magnitudes > 5.0 being recorded in the global seismological catalogues. Some of the smallest seismic events recorded are of the same order as laboratory failures. It may be informative to briefly consider two cases where some information about the actual physical extent of recorded seismic events available.

*The 1989 Brand earthquake* The Brand fault in the Welkom goldfield yielded under mining induced shear stress in January 1989, causing a local magnitude 4.7 earthquake. The fault is intersected by several tunnels and fresh slip could be measured over a strike length of at least 1600 metres, ranging from a few millimetres at the extremities to 370 mm near the hypocentre (van Aswegen, 1990); see Fig.11.1.

Scattered falls of ground occurred over an area of some 4 km$^2$. Damage to stopes was explained by Tinuci and Spearing (1993) who calibrated a numerical model to replicate the ground motions at the nearest seismometer. This instrument, which was approximately 750 m from the source, recorded a maximum velocity of ground motion of 170 mm/s.

A surprising aspect was the minimal damage to tunnels in the source region. Conventional mesh and lacing contained the rock sufficiently to keep the tunnels operational, even where a maximum of 370 mm displacement was measured. This is explained by the 'slow' or 'soft' nature of the tremor. The fault zone is characterized by thick (up to 200 mm wide in places) soft fault rock consisting mainly of pyrophyllite, reducing the shear strength of the structure and thus reducing the shear stress required to cause slip. The event was caused by a regional pillar left against the fault and clamping it along a strike distance of 200 metres, creating an artificial asperity.

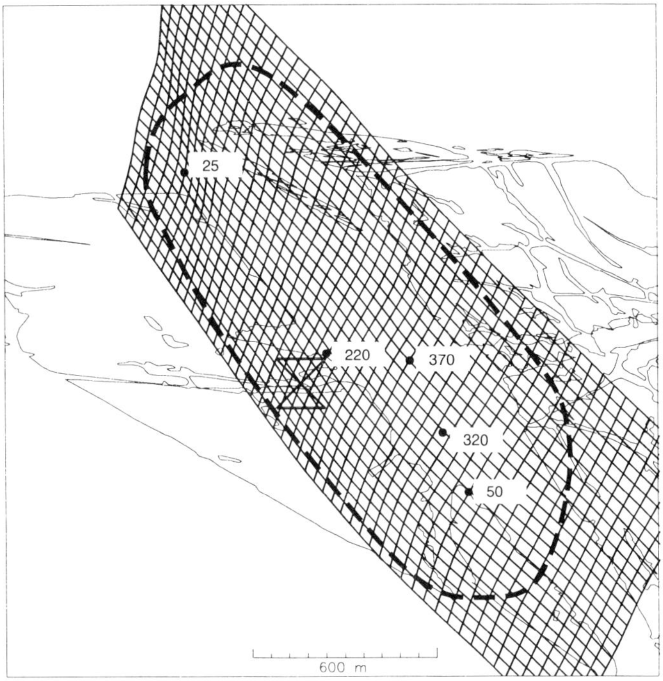


**Fig. 11.1** Mesh representing the Brand fault in the Welkom gold field with the location of the January 1989 magnitude 4.6 event and the fresh slip displacements in mm associated with the tremor.

*The URL experiment* A unique opportunity to very accurately monitor microseismicity associated with tunnelling presented itself to researchers at the so-called mine-by experiment at the Underground Research Laboratory of Atomic Energy of Canada Limited in Manitoba province (e.g. Talebi and Young, 1992). Results of rock mechanics research at this facility have been described by several papers, one of the latest being that of Martin *et al.*(1995).

A 3.5 m diameter circular tunnel was driven horizontally at a depth of 420 m in massive granite (virtually no joints or other discontinuities at this level). The tunnel was developed mechanically with no blasting so that all microseismicity could be attributed to stress induced fracturing around the tunnel. With minimum principal stress $\sigma_3$ near vertical, the tunnel was developed parallel to $\sigma_2$. Dog-earing developed with notches both in the hangingwall and the footwall of the tunnel, consistent with the expected stress concentration. This fracturing was recorded by an extremely sensitive and accurate microseismic monitoring system. The moment magnitudes of the events associated with the dog-earing typically ranged between −4.0 and −3.0, but moment magnitudes down to −6.0 have been recorded. The maximum dimensions of the dog-ear notches were around 0.5 m, i.e. the cracks associated with these small events (moment between 400 and 40 000 Nm) would be of the order of 5 to 25 cm in radius.

Most seismic events in mines recorded today are somewhere in between the above extremes. The range of event sizes recorded depends mainly on the characteristics of the seismic system. The 'rock talk' commonly observed under high stress conditions is the sound made by the creation of fractures ranging in size from 0.1 to 10 m in diameter. In practice the lower limit for the event size recorded simply depends on the sensitivity of the seismic system. An upper limit is a function of the rock strength, geological structure, virgin stress conditions and mine layout.

## 11.2 Benchmark case studies

### 11.2.1 Brunswick Mining and Smelting

When sufficient numbers of moment tensor analyses are available, information about the general mechanism of rockmass deformation can be deduced. One consistent problem, of course, is that the lower order moment tensor analysis always leaves two equally 'correct' shear planes and it is not possible to make the appropriate choice without *a priori* information. If a cluster of events all have the same mechanism, the same will apply. If, however, the cluster represents slip along structures of the same orientation (e.g. bedding planes, joint sets) or the same structure (e.g. fault), but some variation in slip direction occurs due to spatial variation in mining induced stress orientation, then a hemispherical plot of poles to both slip planes for all events in the cluster would map out the orientation of the active plane(s) quite clearly. Such a plot would show:

i) a point concentration being the normal to the active plane(s); and
ii) a great circle concentration representing the orientation of the active plane(s).

In general, the most stable parameters from moment tensor analyses are the orientations of the *P*- and *T*-axes, i.e. the directions of compression and extension at the model point source. Individually, if the seismic event is caused by slip along a pre-existing plane, there could be a significant difference between, say, the *P*-axis and $\sigma_1$. Statistically these directions should, however, coincide with the orientations of the maximum and minimum principle stress axes. A classic example is described by Peterson and Simser (in prep.). Figure 11.2 (figure kindly provided by the authors) shows lower hemisphere projections of the *P*- and *T*-axis orientations for a cluster of small seismic events at the No. 12 Mine of the Brunswick Mining and Smelting Corporation in Canada. In this case the *P*-axis and *T*-axis orientations correlate well with modelled and measured orientations of the principle stress axes in the area of the cluster.

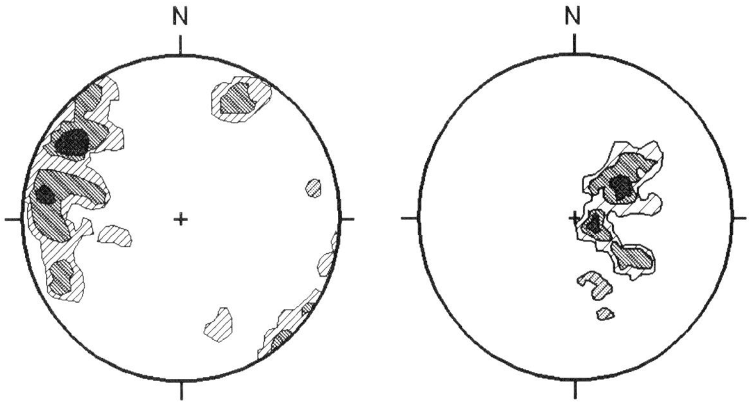


**Fig. 11.2** Lower hemispherical projections of (left) the *P*-axes and (right) the *T*-axes from a cluster of 126 small seismic events recorded at a New Brunswick mine in Canada. The concentrations depicted by the contour highs coincide with the respective principle stress directions in the area of the cluster.

### 11.2.2 Tanton fault

Fault stability remains one of the most important rock engineering problems in mines. Numerical modelling for predicting the response of particular structures to planned mining is common practice. Albeit very useful, the modelling results remain qualitative due to the lack of detailed information about the spatial and temporal variations in rockmass properties. Seismic parameters pertaining to stress and strain rate could go a long way towards overcoming such problems because each event spatially associated with the structure and properly recorded yields information about changes in rheological properties of that structure associated with seismic radiation.

An example of a fault around which mining could have a destabilizing

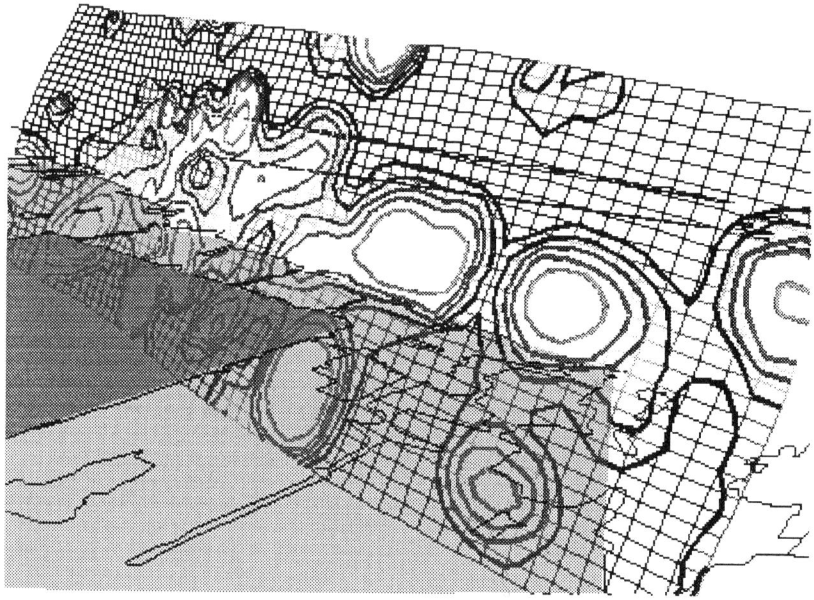

**Fig. 11.3a** Perspective view looking down on the hangingwall side of the Tanton fault (mesh) with energy index contours based on the source parameters of seismicity within 20 m of the fault surface. Grey shading depicts a major Basal reef block on the hangingwall side of the fault, with the shaft pillar in a darker shade. The reef blocks on the footwall side, dipping parallel to the line of sight, show up as straight lines behind the fault surface. The grid spacing on the fault mesh is about 20 m.

effect is the Tanton fault at President Steyn Mine No. 4 Shaft in Welkom. Figure 11.3a shows part of the fault with energy index, *EI*, contours based on seismic events located on the fault surface (±20 m) over a period of 19 months, ending 31 July 1995. It indicates several areas of relatively high stress along the structure, particularly noteworthy being the concentrations at the lower part of the fault loss area and the area beneath that corner of the shaft pillar closest to the fault.

An earlier research project (Dennison and van Aswegen, 1993) resulted in a particular mining strategy being applied on an experimental basis, by which a pillar block of reef against the Tanton fault (referred to as the 7874 block) was mined against the fault first, with a final remnant planned to be away from the structure. The concept was to allow the fault to gradually yield from the beginning rather than allowing the build up of high elastic energy through the decreasing size of the pillar.

The progress of seismicity along the fault is well illustrated by seismic Deborah number contours, shown in Fig. 11.3b. The fault shows two main areas of progressive 'softening', one in the fault loss area in the south (due to mining the upthrown reef block 'behind' the fault) and the second, more significant one, in the footwall of the ore body in the north. The latter happens to be exactly below the 7874 block.

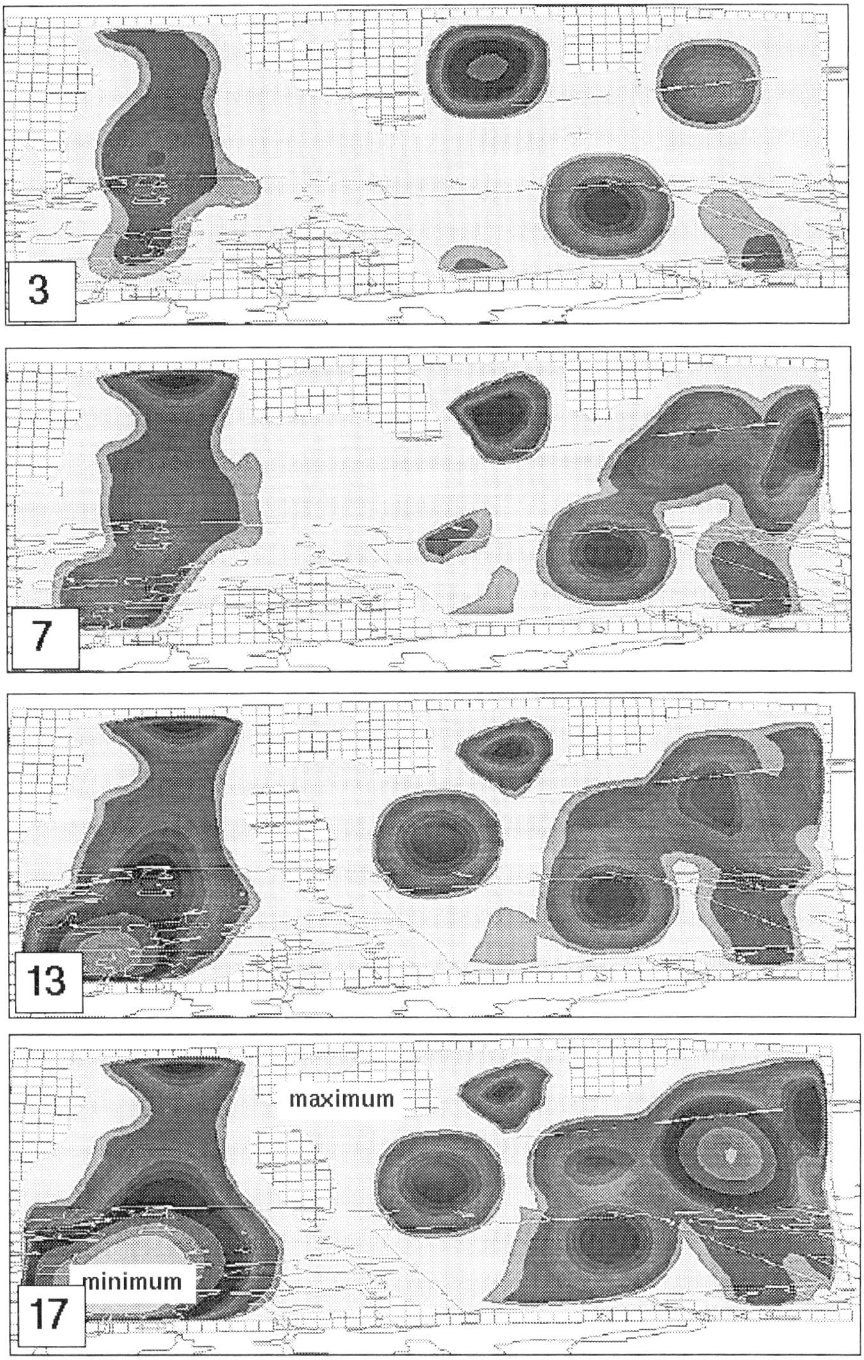


**Fig. 11.3b** Contours of seismic Deborah number along the Tanton fault showing the situations at the end of the 3rd, 7th, 13th and 17th months after 1/1/94. Note that the decrease in seismic Deborah number is not 'cumulative'. In at least two areas the value of this parameter has increased.

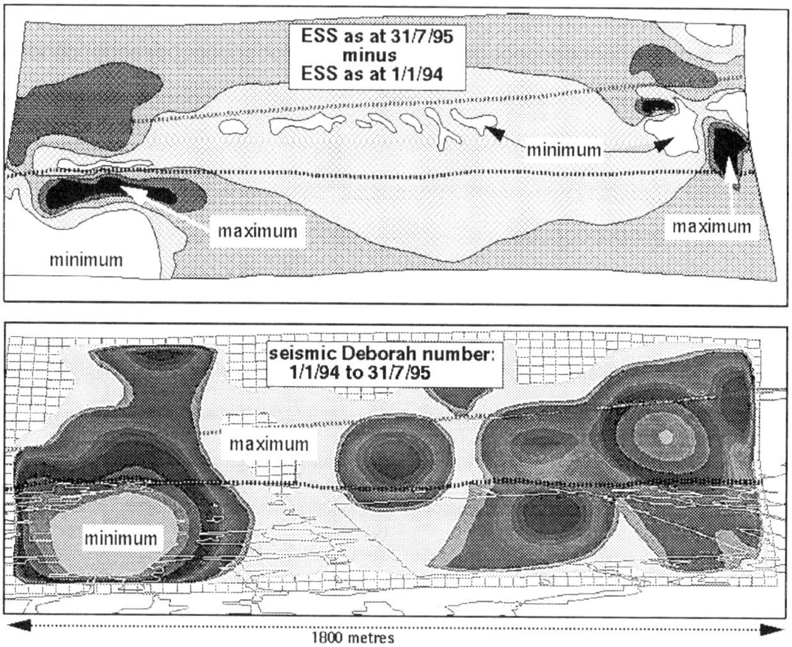


**Fig. 11.3c** Contours of the difference in modelled 'excess shear stress' on the Tanton fault for the mine layouts as at 1/1/94 and at 31/7/95, compared with seismic Deborah number contours. The fault dips westerly and the view is down towards the east. The broken lines indicate the intersections of the Basal reef with the fault. Note that the minima in seismic Deborah number (i.e. maximum seismic flow) coincide with minima in ESS (sites where the numerical modelling suggests minimum seismicity).

The experiment in this block was successful, as can be seen by the fact that, as mining against the fault progressed, the fault yielded gracefully, with the largest seismic flow in the reef footwall – no rockbursts occurred in the 7874 block. This in spite of the fact that numerical modelling showed maximum ESS ('excess shear stress' – see, for example, Ryder, 1988) for the mining period of interest in the immediate vicinity of this block (Fig. 11.3c).

### 11.2.3 Western Holdings No. 6 shaft pillar

The Western Holdings No. 6 Shaft (WH6#) pillar extraction provided a remarkable mimic of a rock failure test and serves as a good example case to show the physical meaning of temporal changes in certain seismic parameters, particularly in terms of their value in quantifying the degree of rockmass stability. The mining layout and event distribution over the period 1992–1994 are shown in Fig. 11.4a.

Seismically, the shaft pillar displayed behaviour similar to that generally experienced in triaxial rock strength tests, namely an initial period of hardening followed by softening and accelerated deformation. Figure 11.4b shows this history in terms of cumulative apparent volume, $V_A$, and moving

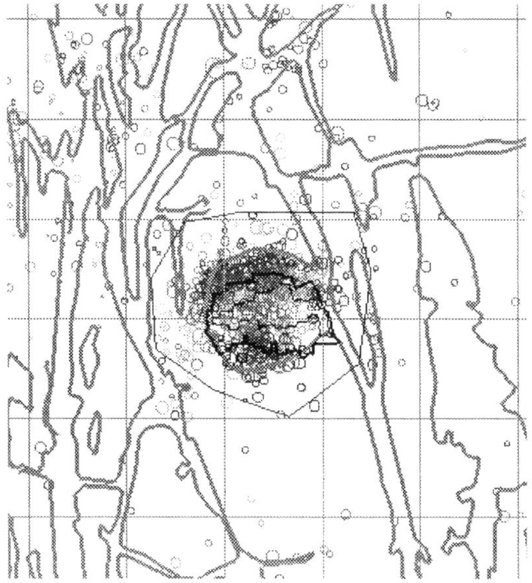

**Fig. 11.4a** Distribution of seismic events at and around the Western Holdings No. 6 Shaft (WH6#). The grey lines outline old mine faces; note the circular shape of the shaft pillar. The black lines at the centre show the positions of the pillar mining faces during two stages. The polygon around the shaft was used for time history analysis event selection. The grey scales depict the age of the events starting from May 1992 (light) to September 1994 (dark). The grid spacing is 250 m.

median energy index, $\overline{EI}$. Due to the limited sensitivity of the seismic network in this area, the moving time windows of $EI$ are relatively large – 50 days. An overall increase in $EI$ and relatively low seismic deformation rate were observed until May 1993, when a drop in $\overline{EI}$, an incease in the rate of $\Sigma V_A$ signified the softening (crushing) of the remnants of the shaft pillar. Note that all larger seismic events, log$E$>7.5, occurred after the drop in $\overline{EI}$ and during the increased rate in coseismic inelastic deformation.

To imitate a stress/strain curve, Fig. 11.4c is a plot of moving median $EI$ versus cumulative $V_A$ showing a 'peak strength' followed by weakening and unstable behaviour.

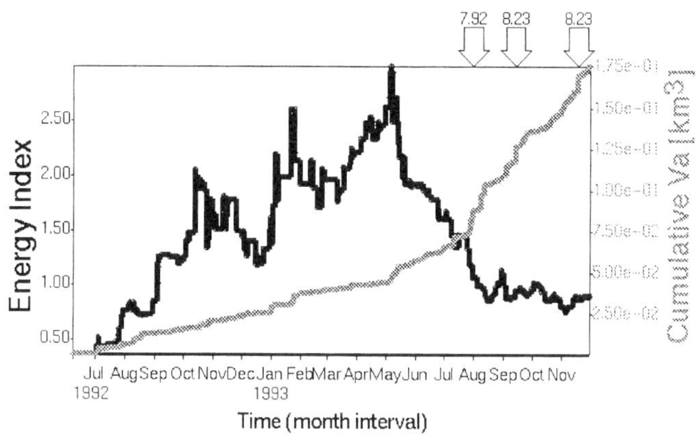


**Fig. 11.4b** Time history of $\Sigma V_A$ and moving median $EI$ for the seismicity associated with the WH6# pillar mining, June 1992 to December 1994. Note that all seismic events larger than log$E$>7.5 occurred after the drop in $EI$ and during the increased rate in coseismic inelastic deformation.

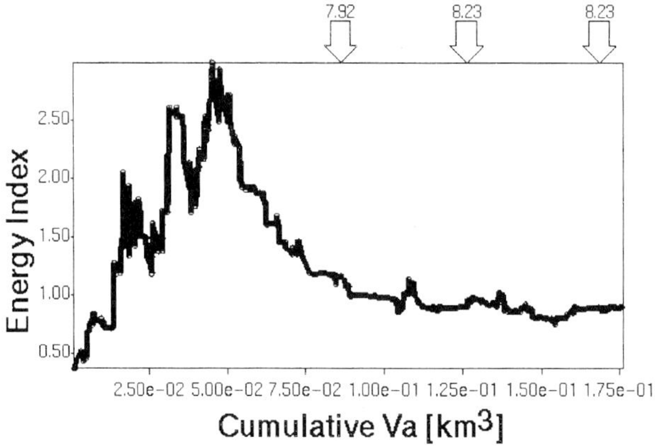


**Fig. 11.4c** The same data as shown in Fig. 11.4a, plotting moving median *EI* against $\Sigma V_A$, yielding a 'stress–strain' curve. Note that the larger seismic events occurred during softening of the system.

To establish the spatial distribution of the stress increase, as indicated by *EI* and which culminated during May–June 1993, three dimensional *EI* contours based on events during this time period are shown in Fig. 11.4d. This indicates that stress had increased throughout the volume of interest, as can be seen by the scattered nature of small volumes of high energy index.

The two strongest seismic events which occurred during the period 1 June 1993 to 1 October 1993 are well predicted by accelerated seismic deformation as shown by $\Sigma V_A$, drop in stress, as shown by moving median *EI* (Fig. 11.4e), increase in seismic diffusion, drop in seismic viscosity (Fig. 11.4f) and, as a natural consequence of the latter two, a drop or low values of seismic Schmidt number (Fig. 11.4g). The time history analysis of the velocity of seismic diffusion is presented in Fig. 11.4h. Note the added information given by

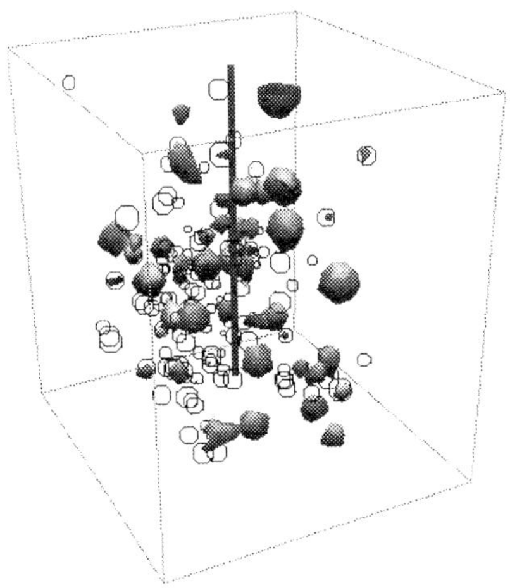

**Fig. 11.4d** Iso-surfaces of above average *EI* with the seismic events around WH6# for the period January to July 1993. Scale varies in this perspective view.

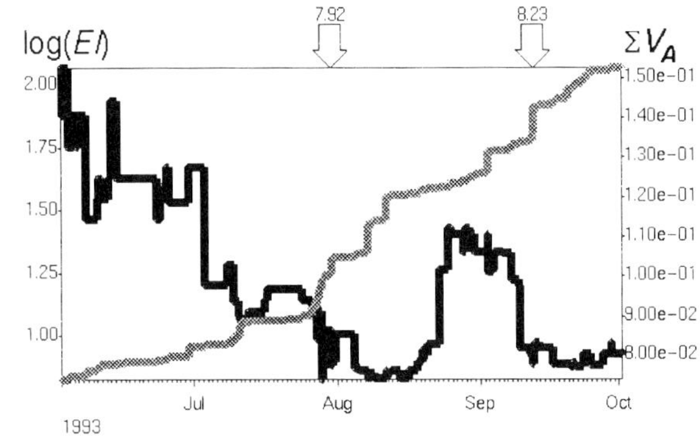


**Fig. 11.4e** The time variation of $EI$ and $\Sigma V_A$ in the WH6# shaft pillar during the period 1/6/93 to 1/10/93.

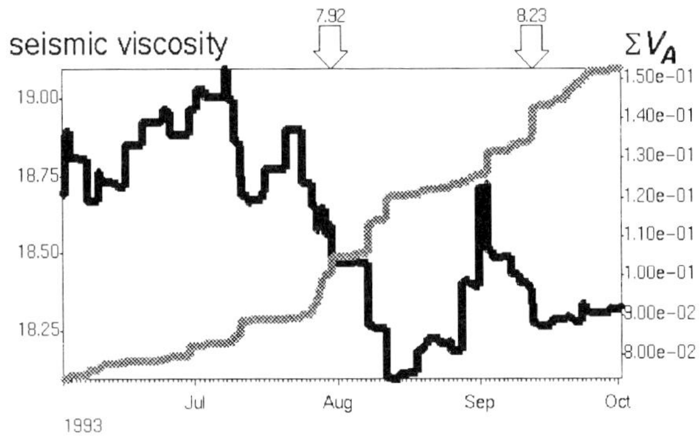


**Fig. 11.4f** The time variation of seismic viscosity for the same data set as in Fig. 11.4e.

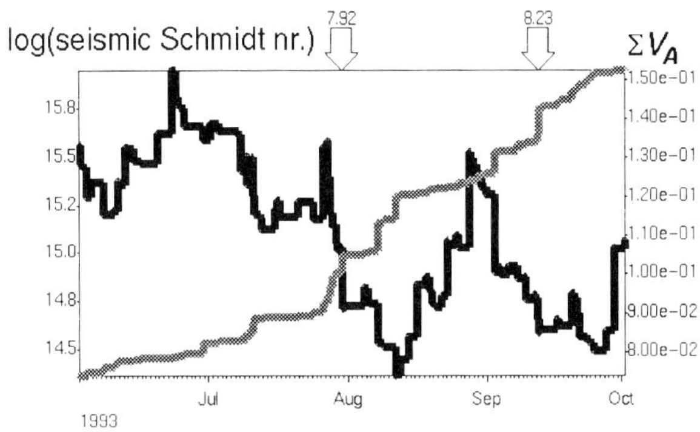


**Fig. 11.4g** The time variation of seismic Schmidt number for the same data set as shown in Fig. 11.4e.

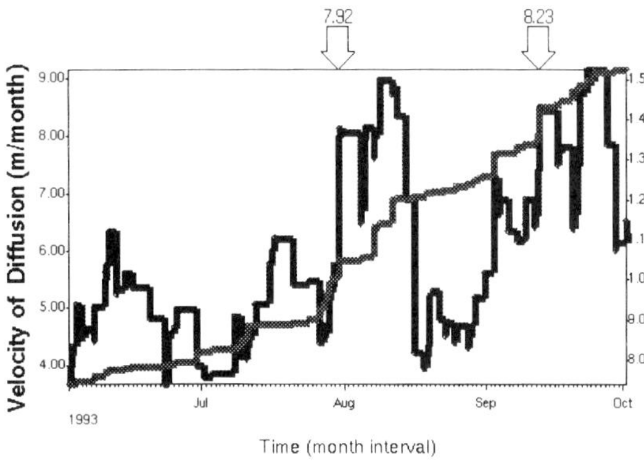


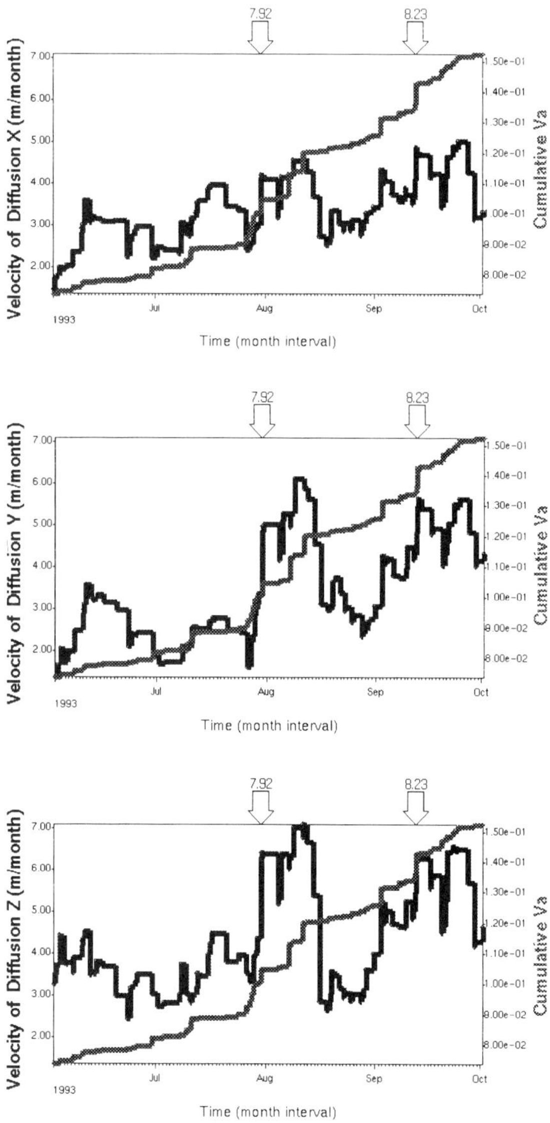


**Fig. 11.4h** The main plot (above) shows the time variation of the velocity of seismic diffusion for the same data set as in Fig. 11.4e.
The subsequent plots (right) reflect the decomposition of this parameter into the three coordinate directions *X* (N–S), *Y* (E–W) and *Z* (up–down). Note that the seismic diffusion is limited in the north–south direction and is maximum in the vertical direction. (The scale on each of the three smaller plots is the same.)

decomposing seismic diffusion between the three coordinate directions. In the case of WH6#, the relative values reflect the E–W geometry of the mining faces and probably the activation of a minor, near vertical fault zone, also E–W trending, which intersects the reef in the shaft pillar, limiting the seismic diffusivity to the *YZ* (west-down) plane.

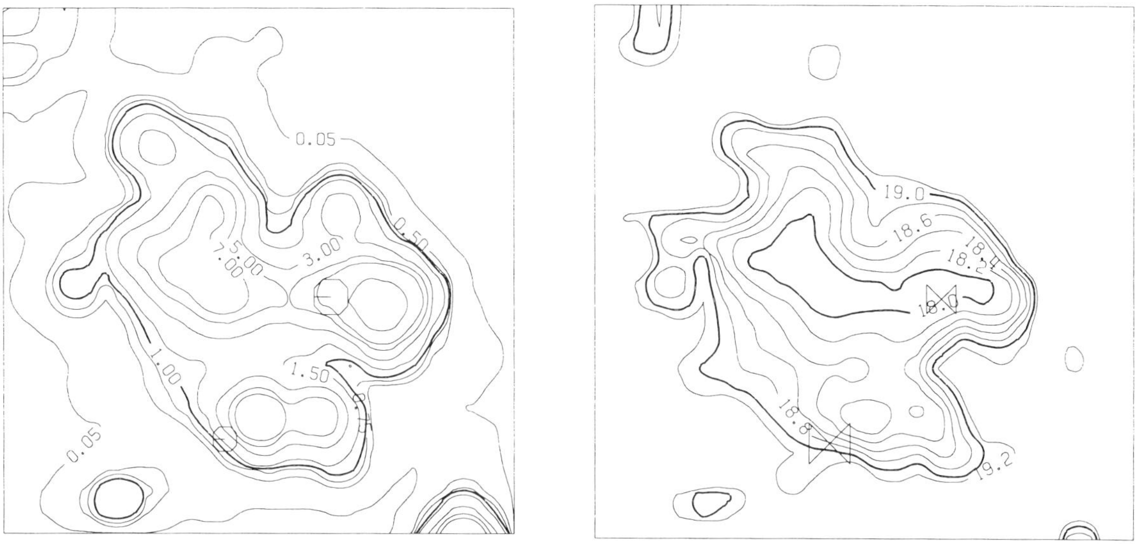


**Fig. 11.4i** Contours for the same data set based on an 18 month monitoring period ending a month before the two events shown. The box dimension is 600x600 metres.
The left hand plot shows contours of seismic strain rate [as seismic strain rate×1E$^{13}$]. The symbols depict magnitude.
The right hand plot shows contours of log (seismic viscosity). The symbols depict apparent stress.

The occurrence of two particular events in the shaft pillar furthermore serves to point out how the nature of events correlates with the precursory behaviour of the pillar for 18 months prior to the two damaging events. The events were separated in time by 54 hours and in space by 200 metres. Some of the source parameters are shown in Table 11.1 and their relative locations are shown in Fig. 11.4i.

Table 11.1 Source parameters examples

| Local Magnitude | Moment log(Nm) | Energy log(J) | Apparent Stress bar | Apparent volume radius m |
|---|---|---|---|---|
| 2.5 | 12.6 | 7.8 | 4.90 | 95 |
| 2.0 | 11.8 | 7.1 | 6.07 | 50 |

Despite the fact that the seismic moment of the first event is nearly one order of magnitude greater than that of the second, the apparent stress of the second is higher than that of the first. Damage caused by the first and larger event was mainly the collapse of the hangingwalls and side walls of a tunnel. The damage associated with the second event was more concentrated and violent, as indicated by the fragment size of broken rock and the presence of fresh brittle shears.

Figure 11.4i shows the positions of these events in relation to contours of seismic strain rate and seismic viscosity based on seismic data for the 18 month period ending a week before the events. The following observations can be made from these plots:

a. Both events occurred adjacent to local maxima in seismic strain rate.

b. The larger event with the lower apparent stress occurred in the area of lowest seismic viscosity. The smaller, but more violent, event occurred in an area with a high gradient in seismic viscosity.

### 11.2.4 Postma dyke

Seismicity, mine layout and major geological structures in the area of interest are shown in Fig. 11.5. The Postma dyke (dolerite) is about 10 m wide. It has a strike of 260°, dipping 80° towards the north. The Basal reef dips easterly at an average of 30° and is intersected by the dyke at around 1500 m below the surface. Mining operations are limited in the east by the Arrarat fault and in the west by the Basson fault, both structures dipping to the west at about 65°. To the north and south most of the reef has been mined out and *modelled*

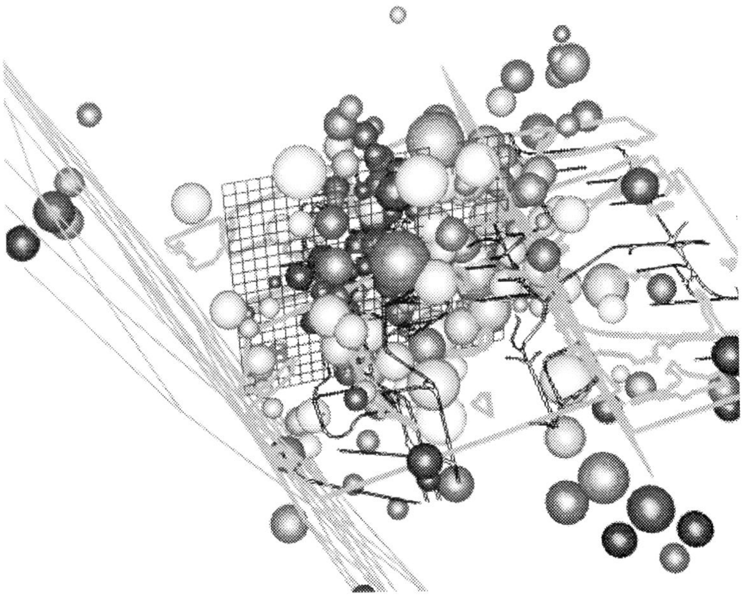

**Fig. 11.5** The setting of the Postma dyke, looking down towards the south. The dyke is represented by the vertical mesh. Also shown is the Arrarat fault to the east (left), the Basson fault to the west (right), haulages and cross-cuts (black lines) and some associated seismic events (spheres).

stress on the remaining pillar containing the Postma dyke is around 60 MPa, reaching 120 MPa where the dyke and its 10 m pillars either side form the final remnant. (The stress modelling was done with the Minsim-D code of MiningTek, South Africa.) From the latter part of 1993 to the end of 1995, the sensitivity of the seismic network in the area is quantified as $\log E_{min} \geq 2.4$.

*Event* WSN-9404241126 The local magnitude 2.7 event which occurred on the Postma dyke on 24 April 1994 is one of the successfully predicted cases. On 22 April 1994 an official warning was issued by mine seismologists based on an increased slope in cumulative $V_A$ and an associated drop in moving median *EI*. For completeness, time histories of all the most useful seismicity parameters are given in Figs. 11.6a - 11.6e. Note that the event of 24 April as well as one on 11 March 1994 are well predicted by the different parameters (as explained in Chapter 10 of this book).

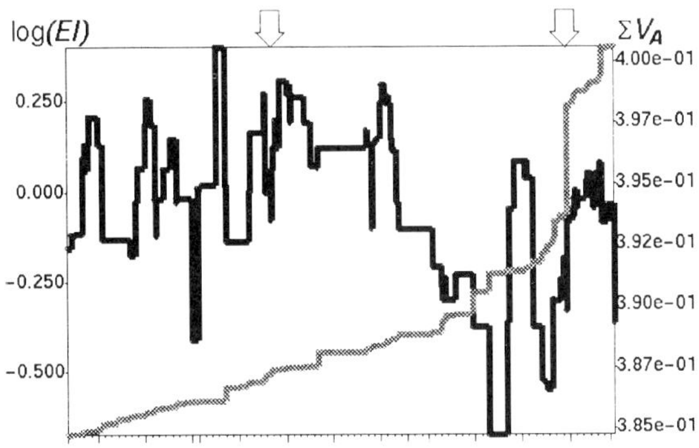


**Fig. 11.6a** Time history of $\log(EI)$ and $\Sigma V_A$ for seismicity along the Postma dyke for the period 14 February to the end of April 1994.

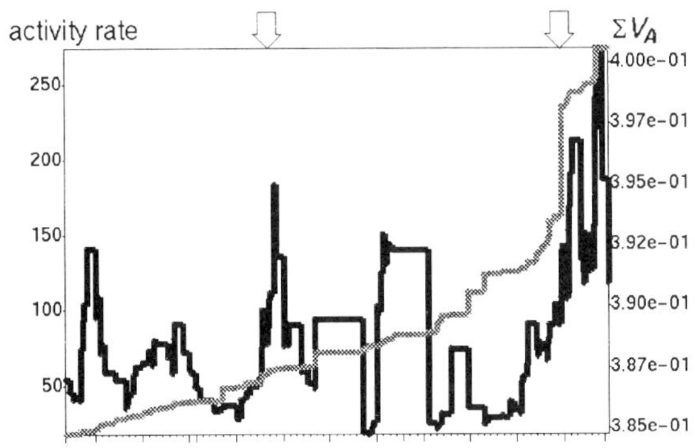


**Fig. 11.6b** Time history of activity rate for seismicity along the Postma dyke for the period 14 February to the end of April 1994.

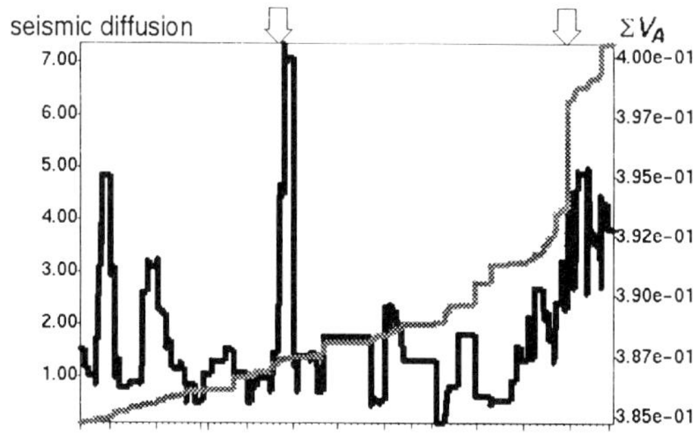


**Fig. 11.6c** Time history of seismic diffusion for seismicity along the Postma dyke for the period 14 February to the end of April 1994.

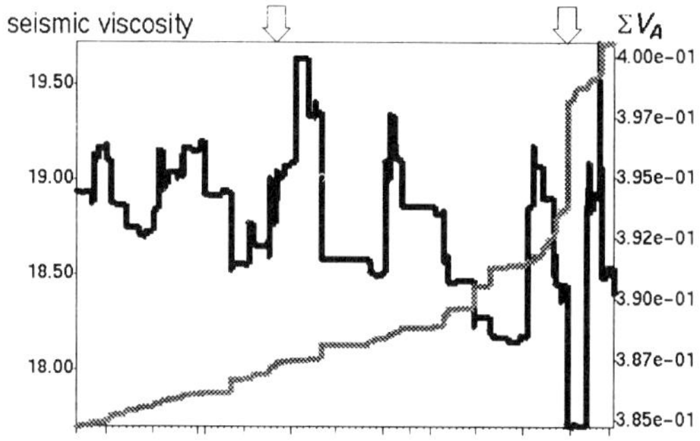


**Fig. 11.6d** Time history of seismic viscosity for seismicity along the Postma dyke for the period 14 February to the end of April 1994.

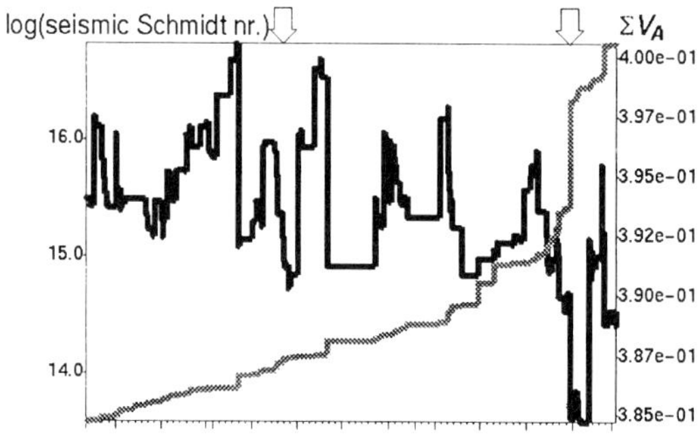


**Fig. 11.6e** Time history of log(seismic Schmidt number) for seismicity along the Postma dyke for the period 14 February to the end of April 1994.

*Event* WSN-9406270610  Two months after WSN-9404241126, on 27 June 1994, a local magnitude 3.7 event rocked the mining area around the Postma dyke. The time history analysis prior to this event suffers from a lack in sensitivity of the seismic system during the weeks prior – damage to underground cables could not be repaired in time and two of the nearest seismometers were not operational. This lack of sensitivity artificially inhibits seismic activity rate and seismic diffusion, which limits the sensitivity of seismic Schmidt number. Nevertheless, the classic parameters $\Sigma V_A$ and *EI* plus

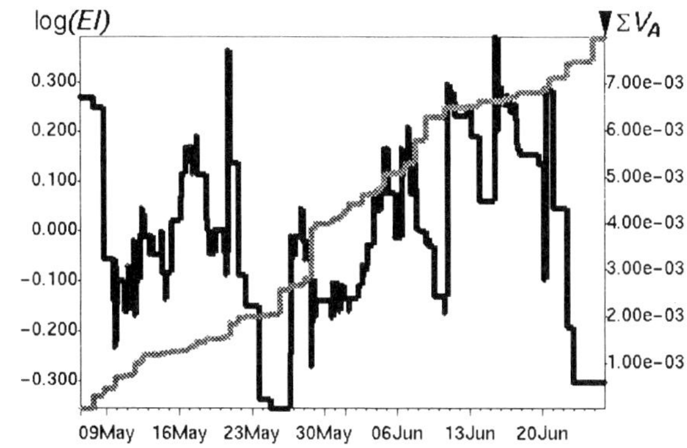


**Fig. 11.7a** Time history plot for $\log(EI)$ and $\Sigma V_A$ for seismicity along the Postma dyke from 9 May to 27 June 1994. The arrow indicates the time of event WSN-9406270610.

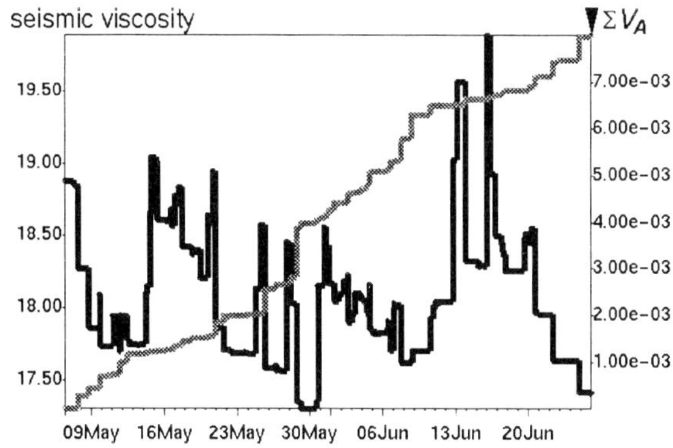


**Fig. 11.7b** Time history plot for seismic viscosity for seismicity along the Postma dyke from 9 May to 27 June 1994.

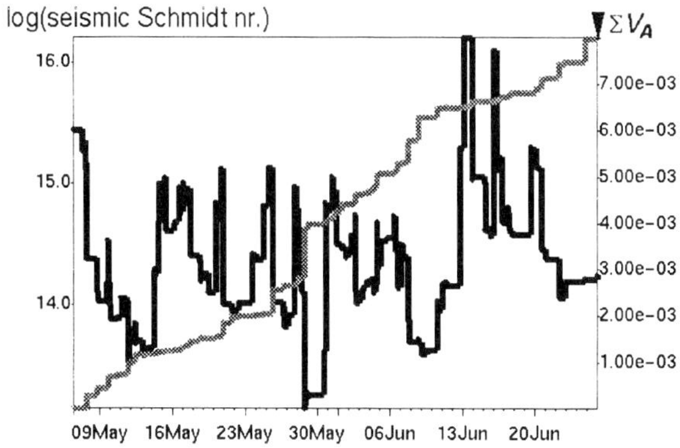


**Fig. 11.7c** Time history plot for log(seismic Schmidt number) for seismicity along the Postma dyke from 9 May to 27 June 1994.

seismic viscosity do predict the event quite well (Figs. 11.7a–11.7c). Figure 11.7d shows an iso-surface of seismic viscosity reflecting the situation prior to the large tremor (based on data for the period 1/1/94 to 26/6/94). Figure 11.7e shows the energy index contours along the Postma dyke, based on the last six weeks before the strong tremor. High energy index characterizes the

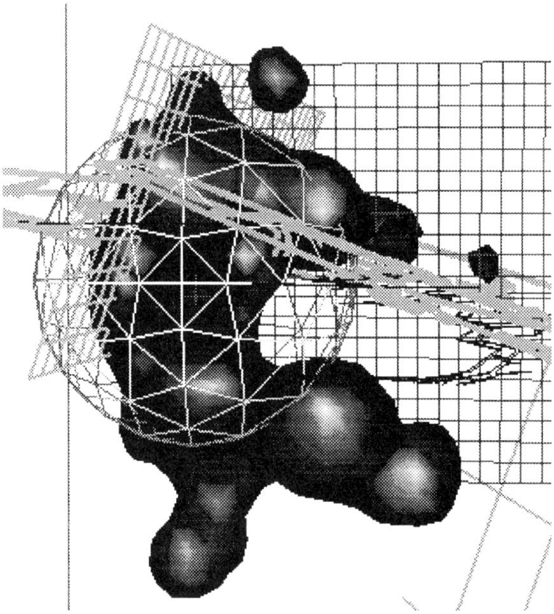

**Fig. 11.7d** Iso-surface of seismic Deborah number =10 for the Postma dyke area, based on six months' data prior to event WSN-9406270610. The mesh sphere represents the event $V_A$.

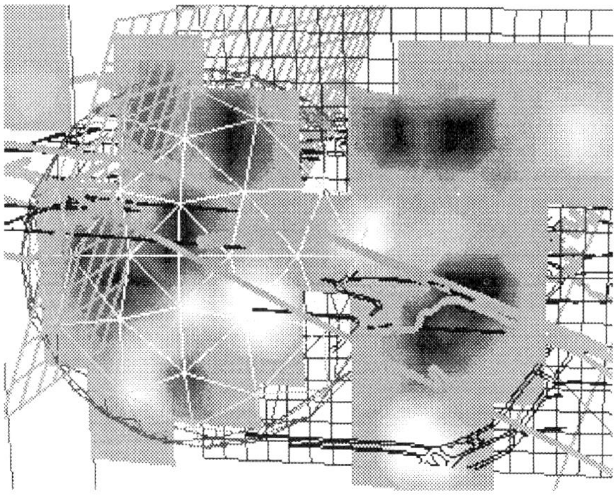

**Fig. 11.7e** Contours of log($EI$) (dark is high) along the Postma dyke for the last six weeks prior to the major tremor.

intersection of the Basson fault and the Postma dyke as well as a seismic gap on the dyke east of the intersection between the dyke and the fault.

Underground damage distribution and fresh shearing observed after event WSN-9406270610 indicated that slip had taken place both along the Basson fault and the northern contact zone of the Postma dyke. The observations on the dyke were made where the 42 level haulage from WH3# transects the dyke, approximately 95 m from the gap seen in Fig. 11.7d (west of and above the gap). The observations on the fault were made in the 42/14A cross-cut further west. The event location is close to the intersection of the two structures. Moment tensor analysis suggested reverse slip along the dyke, but the local 'flat' network limits its reliability.

$V_A$ gives an indication of the physical size of the rock volume affected by coseismic deformation, but not of its shape. Nevertheless, the size of the sphere depicting event WSN-9406270610 in Fig. 11.7d plus the underground observations mentioned above do suggest that both the seismic gap and the fault/dyke intersections were involved during the event.

Event WSN-9406270610 appeared to have broken whatever asperities may have existed along the Postma dyke near the Basson fault intersection. Immediate stress redistribution to the east resulted in a severe decline in seismic energy release in the west and a remarkable increase in activity to the east. Figure 11.7f demonstrates this phenomenon quite clearly.

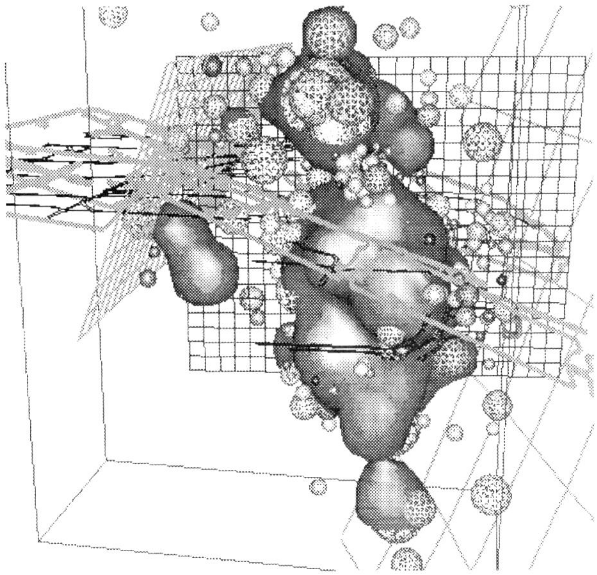

**Fig. 11.7f** Iso-surface of seismic Deborah number and the events contributing to it for six months after the WSN-9406270610 event.

*Time history analyses* The successful prediction of event WSN-9404241126 is not an isolated case. Of the 12 events above log$E$>6.8 associated with the dyke during the period October 1993 to December 1994, 11 display precursory behaviour according to the stability concept, one did not, and, in one case a lack of data does not allow any conclusions. These results are summarized in Table 11.2.

**Table 11.2** Summary of trends in log($EI$); log(seismic Schmidt number) and $\Sigma V_A$ precursory to significant events in the Postma dyke study area.

| Area = 1.70e$^5$ $N_{events}$ = 1923 $M_{min}$ = −0.6($m_M$) | | | | | | | |
|---|---|---|---|---|---|---|---|
| log$E$ | log$EI$ | | log$Sc_s$ | | $\Sigma V_A$ | Comments | Success |
| 8.11 | ↓↓ | ↑ | — | ↓↓ | ↑ | | 3.5 |
| 7.26 | ↓↓ | ↑ | — | ↓↓ | ↑ | | 3.5 |
| 7.82 | — | — | ↑↑ | ↓ | ↑ | | 2 |
| 7.34 | — | ↑ | — | ↑↑ | — | Insuff. data | −3 |
| 6.99 | — | ↓ | — | ↓↓ | ↑ | | 3.5 |
| 6.85 | ↓ | ↓ | ↓↓↓ | ↑ | ↑ | | 3 |
| 7.38 | ↓↑ | — | ↓↑ | ↓ | — | | 0 |
| 8.20 | ↓↓ | ↑ | ↑ | ↓↓ | ↑ | | 3.5 |
| 7.15 | ↓ | ↑ | ↑↑ | ↓↓ | ↑ | | 2.5 |
| 9.90 | — | ↓↓ | ↑ | ↓ | ↑ | | 3.5 |
| 6.96 | ↑ | ↓ | ↑ | ↓↓ | ↑ | | 3.5 |
| 7.26 | ↑ | ↓ | — | ↓↓ | ↑ | | 3.5 |

The arrows indicate *clear* trends compared to the average.
More than one arrow reflects a very strong trend.
Two columns are used to tabulate the precursory behaviour of $EI$ and seismic Schmidt number. The first reflects the behaviour up to five days prior to the stronger tremor and the second reflects the behaviour during the last hours prior to the stronger tremor. In this environment, an increase in $EI$ during the last hours is explained as relative strain hardening in the volume surrounding the failing nucleation zone.

Scoring of precursory behaviour ('success'):

| | |
|---|---|
| +1 | *clear steepening of* $\Sigma V_A$ *slope* |
| −1 | *flat* $\Sigma V_A$ *slope* |
| +1 | *clear drop in EI* |
| −1 | *clear rise in EI* |
| +1 | *strong drop in EI and recovery during final stage* |
| +1 | *clear drop in seismic Schmidt number* |
| −1 | *clear rise in seismic Schmidt number* |
| +½ | *very strong positive trend* |

### 11.2.5 The Trough event

The morning of 3 May 1994 the mine seismologist at Western Deep Levels South Mine (near Carltonville in the far west Rand goldfield) observed that the routine contour plot of weighted energy index (averaged over four days) indicated an anomalous stress concentration around a large bracket pillar on the Trough structure. An updated contour plot was produced at midday. The contours indicated a further increase in stress and a decision was made to evacuate that part of the mine and to keep workers out until the conditions changed. At 18h53 a local magnitude 4 tremor occurred along the Trough structure, followed by several aftershocks in the immediate surroundings. The details of the procedures followed are given in an internal report (Naude, 1994). The description given below is mainly based on backanalysis.

The Trough structure consists of a west dipping mafic dyke which varies in thickness from 5 to 15 m. The dyke is flanked by faults of small (<20 m) throw. Available data allowed a simplified approximation of the Trough structure geometry as shown in Fig. 11.8a.

**Fig. 11.8a** Seismic events spatially associated with the Trough structure for the period 1/1/94 – 3/5/94. The cutaway iso-surface is of a particular seismic Deborah number.

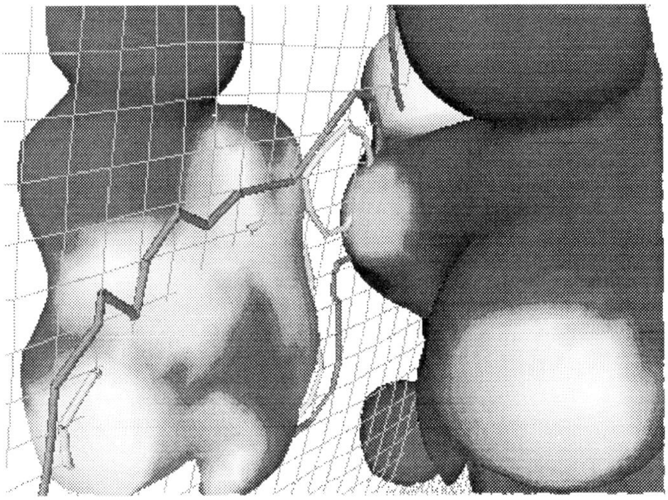

**Fig. 11.8b** Iso-surface of seismic viscosity delineating volumes of accelerated seismic flow of rock. A gap is defined where seismic deformation is retarded and it coincides with a particular part of the Trough structure. The iso-surfaces are based on seismic events from 1/1/94 to before the Trough event of 3/5/94. Greyscale shading and contour lines on the fault surface depict variation in *EI* (lighter shades reflecting high values in this case) based on seismicity from 18h00 on 25/5/94 to 10h00 on 3/5/94. The mesh lines representing the Trough structure are spaced at 40 m.

The precursory behaviour of the Trough structure and environs was studied through space-time analysis of seismic viscosity and energy index using three dimensional iso-surfaces. The most striking result from the 3D analysis was the recognition of a seismic gap between two lobes of seismic viscosity iso-surfaces, coinciding with the Trough surface. Here, the seismic gap is defined as a volume of relatively high seismic viscosity (volume resisting the seismic flow of rock). Such a volume will be characterized by high spatial gradients in seismic stress and seismic strain rates, the gap being slow to deform and accumulating elastic strain. If the gap suddenly fails, a major seismic event may result. This is exactly what happened in the case of the Trough event.

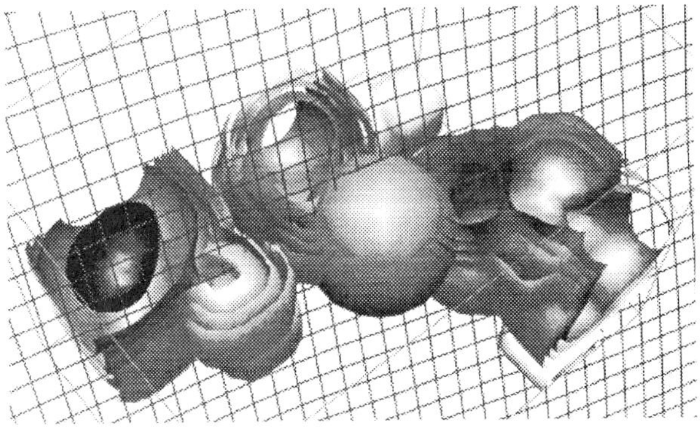

**Fig. 11.8c** Similar to Fig. 11.8b, showing iso-surfaces for different seismic Deborah number values as well as the Trough event (sphere at centre).

The stress history of the last 24 hours comprises an initial period during which stress was generally increasing in the seismic gap and along the surrounding part of the Trough structure, as shown by differentially shaded contouring of energy index on the iso-surface of seismic viscosity and additional contour lines on the Trough structure (Fig. 11.8b).

During the second period, the stress dropped in the seismic gap reflecting the softening of this part of the rockmass (nucleation volume) where the Trough event occurred (Fig. 11.8c). A time history analysis of seismic viscosity and seismic diffusion shows that precursory softening and an increase in seismic diffusivity took place a few hours before the Trough event (Fig. 11.8d).

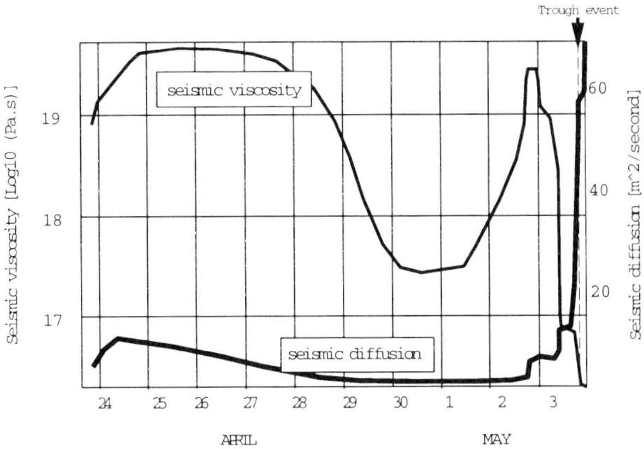


**Fig. 11.8d** A time history analysis reflecting the seismic behaviour of the fault/bracket pillar environment during the ten day period prior to the Trough event.

## 11.2.6 81/122 Longwall

The last benchmark case concerns a study area where no significant seismic events occurred during the study period. As part of a research project, the sensitivity of the mine seismic network at Western Deep Levels South Mine was increased in a part of the mine which includes the section 122 longwall at 81 level. A sensitivity of moment magnitude −0.8 was achieved, i.e. all events of moment magnitude −0.8 and larger were recorded by at least five three-component digital seismic monitoring stations and quantified in terms of seismic moment and radiated seismic energy. Mining was towards a potentially hazardous complex of geological structures and large events ahead of the mining face were anticipated. No large event occurred in the vicinity of the particular longwall of interest here, but valuable data regarding the rockmass response to production blasting were collected. The area of interest is shown in Fig. 11.9a.

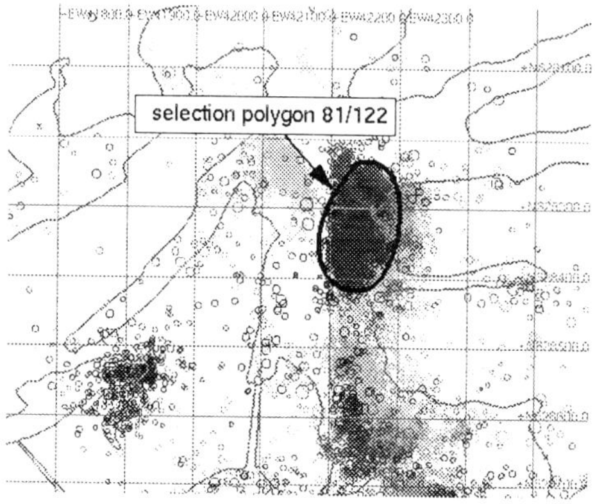


**Fig. 11.9a** The distribution of seismic events over part of Western Deep Levels Mine during 1995. The area of interest here is the dense cluster outlined by the selection polygon.

The seismic record for the area is remarkable because over a six month period nearly 4200 events were recorded of which fewer than 100 (<2.4%) occurred outside of blasting time, i.e. later than two hours after blasting. Each of the 19 events greater than moment magnitude 1.0 was triggered by blasting. The rockmass behaviour prior to two of these rather small events is considered below.

The response to daily blasting is a 'burst' of small seismic events with log(*EI*) values which vary by an order of magnitude. The general pattern involves a sudden increase in *EI*, followed by a decrease (Fig. 11.9b). This is,

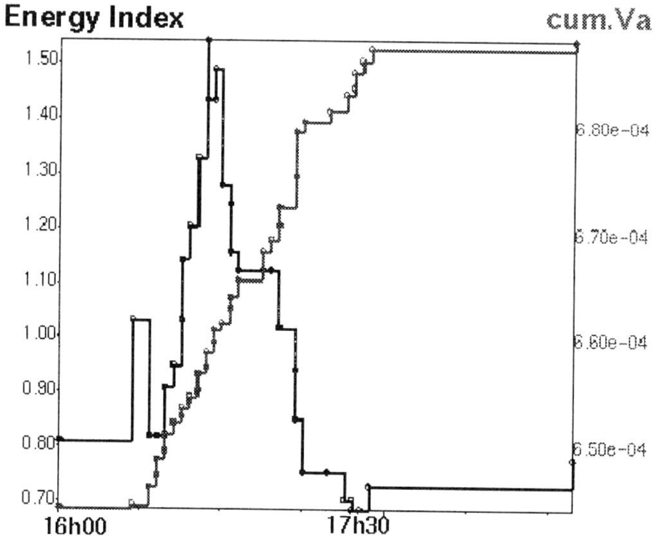


**Fig. 11.9b** A typical response to blasting of a longwall at Western Deep Levels South Mine as measured by $\Sigma V_A$ and *EI*. Note the indicated initial rise in stress level and the gradual relaxation.

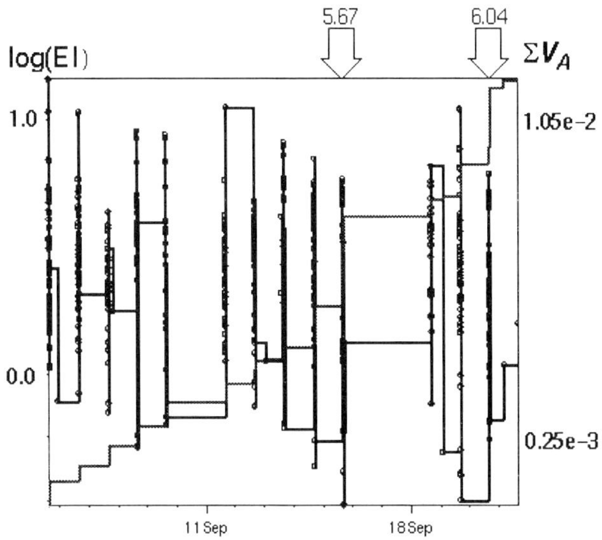


**Fig. 11.9c** Daily response to blasting of a longwall production area at Western Deep Levels South Mine as measured by log(*EI*).

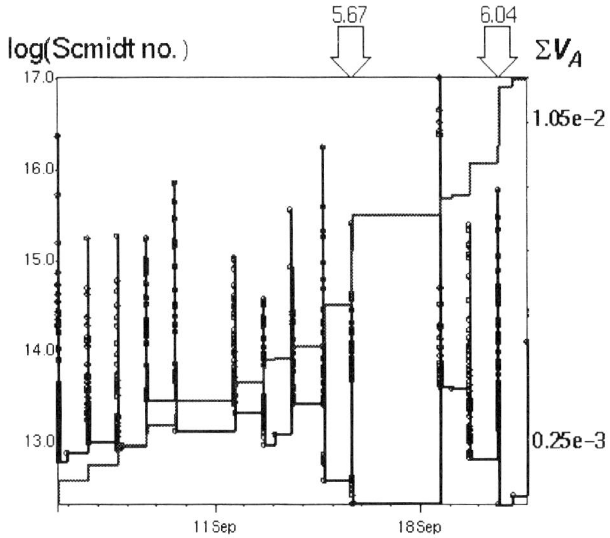


**Fig. 11.9d** Daily response to blasting in a shaft pillar mining area at Welkom as measured by log(seismic Schmidt number).

of course, consistent with what is expected: a sudden increase in the elastic strain in the rockmass ahead of the advancing mining face, followed by some relaxation. The *EI* values of the last events reflect the level of stress to which the rockmass settled after the blast. In the case of the two events larger than local magnitude 1, it was found that this daily final *EI* value decreases on a daily basis prior to the events (Fig. 11.9c). The calculation of Schmidt number is highly sensitive to the length of the moving time window. By setting this time window at less than 24 hours (20 hours in this case) the specific effect of the daily blasting is quantified. A daily decrease in the value of this parameter is also observed (Fig. 11.9d). Note that this parameter also dropped prior to the larger event.

The scale invariance of the space/time processes precursory to seismic events is illustrated by a 20 minute long seismic history in the study area on 21 October 1995. Figure 11.9e shows the distribution and size (in terms of $V_A$) of seismic events triggered by blasting on this day. The events shown in the figure are those up to and including a magnitude 0.5 event (the largest event in the figure). In space, this latter tremor occurred in a 'seismic gap' with a diameter of 50 m. The edge of the gap was characterized by high *EI*. The magnitude 0.5 event was preceded by accelerated seismic strain rate (steepened slope of the $\Sigma V_A$ curve), by a drop in *EI* and a drop in seismic Schmidt number, as shown in Figs. 11.9f and 11.9g. The seismic history prior to this event is therefore similar to that of most large events, but at a different scale in space and time.

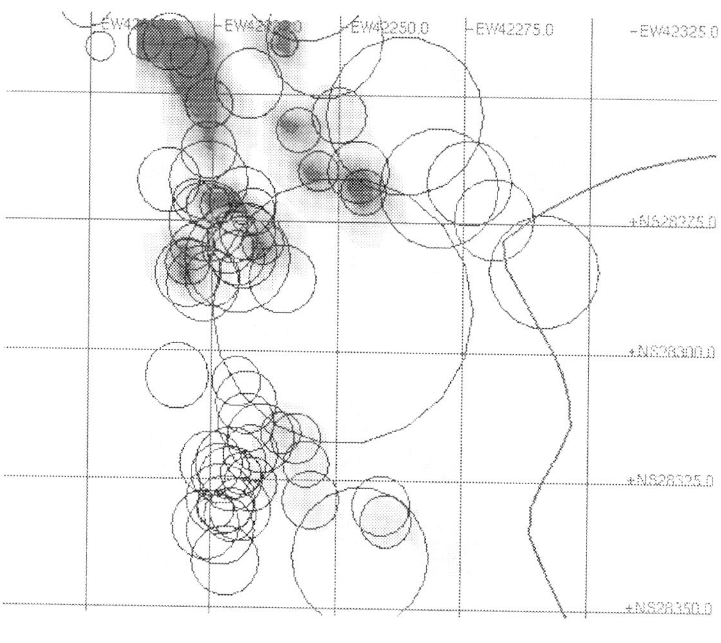


**Fig. 11.9e** The location of the magnitude 0.5 event shown in Figs. 11.9f and 11.9g, relative to the seismicity in the preceding 20 minute period. The dark shading shows high *EI* along the margin of the seismic gap which was taken out by the tremor.

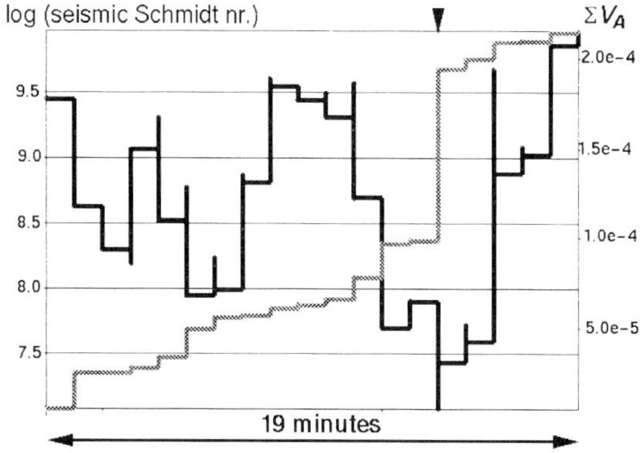


**Fig. 11.9f** Time history plot of log(seismic Schmidt number) for seismicity in the 19 minute period prior to a local magnitude 0.5 event.

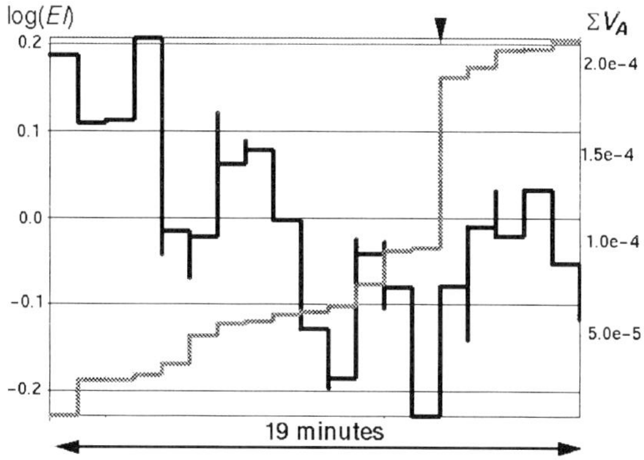


**Fig. 11.9g** Time history plot of log($EI$) for seismicity in the 19 minute period prior to a local magnitude 0.5 event.

# References


Abarbanel, H.D.I. (1990). Prediction in chaotic nonlinear systems: Methods for time series with broadband Fourier spectra. *Physical Review A*, **41**, 1782.

Abercrombie, R.E. (1995). Earthquake source scaling relationships from −1 to 5 $M_L$ using seismograms recorded at 2.5km depth. *J. Geophys. Res.* **100**, B12, 24015–24036.

Abercrombie, R.E., Agnew, D.C. and Wyatt, F.K. (1995). Testing a model of earthquake nucleation. *Bull. Seism. Soc. Am.*, **85**, 6, 1873–1878.

Aki, K., and Lee, W.H.K. (1976). Determination of three-dimensional velocity anomalies under a seismic array using first P arrival times from local earthquakes. 1. A homogeneous initial model. *J. Geophys. Res.* **81**, 4381–4399.

Aki, K., and Richards, P.G. (1980). *Quantitative Seismology. Theory and Methods.* W.H. Freeman and Company, San Francisco.

Alessandrini, B., and Perazzolo, E. (1987). An interactive technique for the deconvolution of seismograms. *Bull. Seism. Soc. Am.,* **77**, 4381–4399.

Allen, R.V. (1978). Automatic earthquake recognition and timing from single traces. *Bull. Seism. Soc. Am.,* **68**, 1521–1532.

Allgower, E.L. and Georg, K. (1990). *Numerical Continuation Methods, An Introduction.* Springer-Verlag, Berlin.

Analog Devices (1994). *Design-in Reference Manual*, Analog Devices.

Andrews, D.J. (1986). Objective determination of source parameters and similarity of earthquakes of different size. In: *Earthquake Source Mechanics* (S. Das, J. Boatwright and C.H. Scholz, eds). Am. Geophys. Union, Washington, D.C., **6**, 259–267.

Backus, G.E. and Mulcahy, M. (1976a). Moment tensor and other phenomenological descriptions of seismic sources. I. Continuous displacements. *Geophys. J.R. Astron. Soc.*, **46**, 301–329.

Backus, G.E. and Mulcahy, M. (1976b). Moment tensor and other phenomenological descriptions of seismic sources. II. Discontinuous displacements. *Geophys. J.R. Astron. Soc.*, **46**, 341–361.

Baher, H. (1990). *Analog and Digital Signal Processing*. John Wiley & Sons.

Bataille, K. and Chiu, J.M. (1991). Polarization analysis of high-frequency, three-component seismic data. *Bull. Seism. Soc. Am.,* **81**, 2, 622–642.

Bazant, Z.P. and Cedolin, L. (1991) *Stability of Structures*. Oxford University Press, New York.

Ben-Menahem, A. and Singh, J. (1981). *Seismic Waves and Sources*. Springer-Verlag, New York.

Ben-Zion, Y. and Rice, J.R. (1993). Earthquake failure sequences along a cellular fault in a three-dimensional plastic solid containing asperity and non-asperity regions, *J. Geophys. Res.*, **98**, B8, 14109–14131.

Boatwright, J. (1980). A spectral theory for circular seismic sources: Simple estimates of source dimension, dynamic stress drop and radiated energy. *Bull. Seism. Soc. Am.,* **70**, 1–27.

Brady, B.T. (1977a). Anomalous seismicity prior to rock bursts: implications for earthquake prediction. In: *Stress in the Earth* (M. Wyss, ed.), Special Issue, *Pure Appl. Geophys.*, **115**, 357–374.

Brady, B.T. (1977b). An investigation of the scale invariant properties of failure. *Int. J. Rock Mech. Min. Sci. & Geomech. Abstr.*, **14**, 121–126.

Brigham, E.O., Smith, H.W., Bostik, F.X. and Duesterhoeft, W.C. (1968). An iterative technique for determining inverse filters. *IEEE Trans. Geosci. Electronics*, **GE–2**, 86–96.

Brune, J.N. (1970). Tectonic stress and the spectra of seismic shear waves. *J. Geophys. Res.*, **75**, 4997–5009 (Correction, *J. Geophys. Res.*, **76**, 5002, 1972).

Brune, J.N. (1979) Implications of earthquake triggering and rupture propagation for earthquake prediction based on premonitory phenomena. *J. Geophys. Res.*, **84**, B5, 2195–2198.

Bufe, C.G. and Varnes, D. J. (1993). Predictive modeling of the seismic cycle of the Greater San Francisco Bay Region, *J. Geophys. Res.*, **98**, 9871–9883.

Cao, S. and Greenhalgh, S. (1994). Finite-difference solution of the eikonal equation using an efficient, first-arrival, wavefront tracking scheme. *Geophysics*, **59**, 632–643.

Červený, V. and Firbas, P. (1984), Numerical modelling and inversion of travel times of seismic body waves in inhomogeneous anisotropic media. *Geophys. J.R. Astron. Soc.*, **76**, 41–51.

Červený, V., Molotkov, I.A. and Psencik, I. (1977). *Ray methods in seismology*. University of Karlova Press, Prague.

Chander, R. (1975). On tracing seismic rays with specified end points. *J. Geophys.*, 41. 173–177.

Cichocki, A. and Unbehauen, R. (1993). *Neural Networks for Optimization and Signal Processing*. John Wiley and Sons, New York.

Cichowicz, A., Green, R.W.E. and Brink, A. van Z. (1988). Coda polarization properties of high frequency microseismic events. *Bull. Seism. Soc. Am.*, **78**, 1297–1318.

Connor, F.R. (1972) *Signals*. Edward Arnold, London.

Cox, S.J.D. and Paterson, L. (1990). Damage development during rupture of heterogeneous brittle materials: a numerical study. In: R.J. Knipe and E.H. Rutter (eds), *Deformation Mechanisms, Rheology and Tectonics*. Geological Society Special Publication No.**54**, 143–165.

Cundall, P.A. (1990). Numerical modelling of jointed and faulted rock. In: *Mechanics of Jointed and Faulted Rock*, (H.P. Rossmanith, ed.). Balkema, Rotterdam.

Cundall, P.A. (1996). *Particle Flow Code*. Itasca Consulting Group, Minneapolis, Minnesota.

Darot, M. and Gueguen, Y. (1986). Slow crack growth in minerals and rocks: theory and experiments. *PAGEOPH*, **124**, 4/5, 677–602.

Das, S. and Scholz, C. (1981). Theory of time-dependent rupture in the Earth. *J. Geophys. Res.*, **86**, 6039–6051.

Daubechies I., (1992). *Ten Lectures on Wavelets*. Society of Industrial and Applied Mathematics, Philadelphia, Pennsylvania.

Dennison, P. and van Aswegen, G. (1993). Stress modelling and seismicity on the Tanton fault: A case study in a South African gold mine. In: *Proceedings of the 3rd International Symposium on Rockburst and Seismicity in Mines* (R.P. Young, ed.), 16-18 August, 1993, Kingston, Canada. Balkema, Rotterdam, 287–296.

Di Bona M. and Rovelli, A. (1988). Effects of the bandwidth limitations on stress drops estimated from integrals of the ground motion. *Bull. Seism. Soc. Am.*, **78**, 1818–1825.

Dieterich, J.H. (1992). Earthquake nucleation on faults with rate- and state-dependent strength. *Tectonophysics*, **211**, 115–134.

Dieterich, J.H. (1994). A constitutive law for rate of earthquake production and its application to earthquake clustering. *J. Geophys. Res.*, **99**, B2, 2601–2618.

Dorize, C. and Gram-Hansen, K. (1992). Related Positive Time-Frequency Energy Distributions. In: Y. Meyer (ed.), *Wavelets and Applications*, Springer-Verlag, Berlin.

Drzezla, B. (1991). Guidelines for design of seismic network configuration. In: *Geologic Structures and Materials: Proceedings of the Fifth Conference on Acoustic Emission/Microseismic Activity, Pennsylvania State University, USA*.

Drzezla, B., and Mendecki, A. (1982a). The velocity anisotropy of rock mass and location accuracy of mining tremors. *Acta Geophys. Pol.*, **30**, 231–241.

Drzezla, B., and Mendecki, A. (1982b). Joint hypocentre location of mining tremors and determination of anisotropy parameters of P-wave velocity. *Acta Geophys. Pol.*, **30**, 321–332.

Dysart, P.S., Snoke, J.A. and Sacks. I.S. (1988). Source parameters and scaling relations for small earthquakes in the Motushiro region, southwest Honshu, Japan. *Bull. Seism. Soc. Am.*, **78**, 2, 571–589.

Ekblom, H. and Henriksson, S. (1969). $L_p$-criteria for the estimation of location parameters. *SIAM J. Appl. Math.*, **17**, 1130–1141.

Eneva, M. (1996). *Application of some concepts of nonlinear dynamics to rockburst prediction*. Report to ISS International, Welkom, South Africa, January.

Etkin, B. (1972) *Dynamics of Atmospheric Flight*, John Wiley & Sons, New York.

Essrich, F. (1996). Internal Report (unpublished), ISSI, Welkom, South Africa.

Ferber R.G. (1989). Recursive deconvolution filters for seismograph Systems. *Bull. Seism. Soc. Am.,* **79**, 5, 1629–1641.

Fischer, K.M. and Jordan, T.H. (1991). Seismic strain rate and deep slab deformation in Tonga. *J. Geophys. Res.*, **10**, 14423–14444.

Flynn, E.A. (1965). Signal analysis using rectilinearity and direction of particle motion. *Proc. IEEE*, **53**, 1874–1876.

Frank, F.C. (1965). On dilatancy in relation to seismic sources. *Rev. Geophys.*, **3**, 485–503.

Fredrich, J., Evans, B. and Wong, T.F. (1989). Micromechanics of the brittle transition in Carrara marble. *J. Geophys. Res.*, **94**, 4129–4145.

Fröhlich, C. (1993). Where do 'sweet' earthquakes occur? Characteristics of well-determined deviatoric non-double-couple shallow earthquake sources. *Supplement EOS*, 20 April.

Fröhlich, C., and Apperson, K.D. (1992). Earthquakes focal mechanisms, moment tensors and the consistency of seismic activity near plate boundaries. *Tectonic*, **11**, 279–296.

Ge, M. and Hardy, H.R. (1995). The role of transducer array geometry in AE/MS source location. In: *Proceedings of the Fifth Conference on Acoustic Emission/Microseismic Activity in Geological Structures and Materials*. (Hardy, H.R. Jr., ed.), Trans Tech Publications, Germany.

Gibowicz, S.J. (1973). Variation of the frequency-magnitude relation during earthquake sequences in New Zealand. *Bull. Seism. Soc. Am.*, **63**, 517–528.

Gibowicz, S.J. (1979). Space and time variations of the frequency-magnitude relation for mining tremors in the Szombierki coal mine in Upper Silesia, Poland. *Acta Geophys. Pol.*, **37**, 39–49.

Gibowicz, S.J. and Kijko, A. (1994). *An Introduction to Mining Seismology*. Academic Press, New York.

Gibowicz, S.J., Harjes, H.P. and Schafer, M. (1990). Source parameters of seismic events at Heinrich Robert Mine, Ruhr Basin, Federal Republic of Germany: Evidence for non-double-couple events. *Bull. Seism. Soc. Am.* **80**, 88–109.

Gonin, R. and Money, A.H. (1985a). Nonlinear $L_p$-norm estimation: Part I – On the choice of the exponent, *p*, where the errors are adaptive. *Commun. Statist. – Theor. Meth.* **14**, 827–840.

Gonin, R. and Money, A.H. (1985b). Nonlinear $L_p$-norm estimation: Part II – Asymptotic distribution of the exponent *p*, as a function of the sample kurtosis. *Commun. Statist. – Theor. Meth.*, **14**, 841–849

Gonin, R. and Money, A.H. (1987). Outliers in physical processes: $L_1$- or adaptive $L_p$-norm estimation? In: *Statistical Data Analysis Based on the $L_1$-Norm and Related Methods* (Y. Dodge, ed.), 454–464.

Guberman, S.H.A. (1975). On some regularities in earthquake occurence. *Dokl. Akad. Nauk. SSSR*, **224**, 573–576. (In Russian.)

Gutenberg, B. and Richter, C.F. (1944). Frequency of earthquakes in California. *Bull. Seism. Soc. Am.*, **34**, 185–188.

Gutzwiller, M.C. (1991). *Chaos in Classical and Quantum Mechanics*. Springer, Berlin.

Hanks, T.C. and Kanamori, H. (1979). A moment magnitude scale. *EOS Trans. Am. Geophys Union*, **59**, 1128.

Hanks, T.C. and Thatcher, M. (1972). A graphical representation of seismic source parameters. *J. Geophys. Res.*, **23**, 4393–4405.

Hanyga, A. (1988). Numerical tracing of rays and wavefronts. In: *Seismological Algorithms* (D. Doornbos, ed.). Academic Press, New York.

Hanyga, A. (1991). Implicit curve tracing and point-to-curve ray tracing. *Seismol. Obsv., Univ. Bergen, Seismo-ser.*, **56.**

Hanyga, A. (1995). Point-to-curve ray tracing *PAGEOPH*, **148**, 290–1323.

Harter, H.L. (1972). The Method of Least Squares and Some Alternatives. ARL Technical Report 72–0129, Aerospace Research Laboratories, Wright-Patterson Air Force Base, Ohio. AD 752211. (Abstract, *Inst. Math. Statist. Bull.* **1**, 257–258.)

Heiss, W.D., Nazmitdinov, R.G. and Radu, S. (1995). Periodic orbits and shell structure in Octupole deformed potentials. *Physical Review B*, **51**, 1874.

Hill, R. (1957). On uniqueness and stability in the theory of finite elastic strain. *J. Mech. Phys. Solids*, **5**, 229–241.

Hill, R. (1958). A general theory of uniqueness and stability in elastic-plastic models. *J. Mech. Phys. Solids*, **6**, 236–249.

Hirata, T., Satoh, T. and Ito, K. (1987). Fractal structure of spatial distribution of micro-fracturing in rock. *Geophys. J.R. Astron. Soc.*, **90**, 369–374.

Hobbs, B.E., Ord, A. and Teyssier, C. (1986). Earthquakes in the ductile regime? *Pure and Applied Geophysics*, **124**, 1/2, *Internal Structure of Fault Zones*. (Chi-yuen Wang, ed.) Birkhauser Verlag, Basel.

Hobbs, B.E., Mühlhaus, H.B. and Ord, A. (1990). Instability, softening and localization of deformation. In: Geological Society Special Publication (R.J. Knipe and E.H. Rutter eds) *Deformation Mechanisms, Rheology and Tectonics*, No.**54**, 143–165.

Hopkins, D.L., Cook, N.G.W. and Myer, L.R.. (1990). Normal joint stiffness as a function of spatial geometry and surface roughness. In: *Rock Joints*, (Eds. N. Barton and O. Stephansson, eds). Balkema, Rotterdam, 203–210.

Horii, H. and Nemat-Nasser, S. (1986). Brittle failure in compression: Splitting, faulting and brittle-ductile transition. *Phil. Trans. R. Soc., Lond., A*, **319**, 337–374.

Huang, Y. (1993). The role of nonuniform particle distribution in plastic flow localization. In: *Mechanics of Materials*. **16**, 3, 265–279. Elsevier, New York.

Huang, Y. and Hutchinson, J.W. (1989) A model study of the role of nonuniform defect distribution on plastic shear localization. In: *Role of modeling in materials design*. (J.D. Embury, ed.), American Institute for Mining Engineers.

Ishimoto, M. and Iida, K. (1939). Observations sur les seimes arregistrés par le microseismographe construit dernièrement. *Bull. Earthquake Res. Inst. Tokyo Univ.*, **17**, 443–478. (In Japanese.)

Jackson, D.D. (1979). The use of a priori data to resolve non-uniqueness in linear inversion, *Geophys. J. R. Astron. Soc.*, **57**, 137–158.

Jackson, J. and McKenzie, D. (1988). The relationship between plate motions and seismic moment tensors, and the rates of active deformation in the Mediterranean and Middle East. *Geophys. J.*, **93**, 45–73.

Jech, J. (1989). Seismic tomography in the Ostrava-Karvina mining region. In: *Seismicity in Mines*, (S.J. Gibowicz, ed.). Special Issue, *Pure Appl. Geophys.* **129**, 597–608.

Jeffreys, H. (1932). An alternative to the rejection of observations. *Proc. R. Soc. Lond. Ser A*, **137**, 78–87.

Jessell, M.W. and Lister, G.S. (1991). Strain localisation behaviour in experimental shear zones, *PAGEOPH*, **137**, 4, 421–438

John, R.C.S. and Draper, N.R. (1975). D-optimality for regression designs: a review. *Technometrics*, **17**, 15–23.

Jones, L.M. (1994). Foreshocks, aftershocks and earthquake probabilities: Accounting for the Landers earthquake. *Bull. Seism. Soc. Am.,* **84**, 3, 892–899.

Jones R.H. and Rayne, C.M. (1994). The use of a genetic algorithm for the optimal design of microseismic monitoring networks. In: *Proceedings of Eurock '94*, Balkema, Rotterdam.

Jost, M.L. and Herrmann, R.B. (1989). A student's guide to and review of moment tensors. *Seism. Res. Lett.* **60**, 37–57.

Julian, B. and Gubbins, D. (1977). Three-dimensional seismic ray tracing. *J. Geophys. Res.*, **43**, 95–114.

Kachanov, L.M. (1961). Rupture time under creep conditions. In: *Problems of Continuum Mechanics* (J.R.M. Radok, ed.). Society for Industrial and Applied Mathematics, Philadelphia, 202–218.

Kagan, Y.Y. (1992). Seismicity: Turbulence of solids. *Nonlinear Sci. Today*, **2**, 1–13.

Kagan, Y.Y. (1994). Observational evidence for earthquakes as a nonlinear dynamic process. *Physica D.*, **43**.

Kagan, Y.Y. and Jackson, D.D. (1991). Long-term earthquake clustering. *Geophys. J. Int.*, **104**, 117–133.

Kagan, Y.Y. and Knopoff, L. (1985a). The first order statistical moment of the seismic moment tensor. *Geophys. J.R. Astron. Soc.*, **81**, 429–444.

Kagan, Y.Y. and Knopoff, L. (1985b). The two-point correlation function of the seismic moment tensor. *Geophys. J.R. Astron. Soc.*, **83**, 637–656.

Kato, N., Yamamoto, K., Yamamoto, H. and Hirasawa, T. (1992). Strain rate effect on frictional strength and the slip nucleation process. *Tectonophysics*, **211**, 269–282.

Keilis-Borok, V.I. (1959). On estimation of the displacement in an earthquake source and of source dimensions. *Ann. Geofis.*, **12**, 205–214.

Kijko, A. (1977). An algorithm for the optimum distribution of a regional seismic network. *Pure Appl. Geophys.* **115**, 999–1009.

Kijko, A. (1993). *Seismological outliers:* $L_1$ *or adaptive* $L_p$ *norm application*. Internal Report. ISS International, Welkom, South Africa. Report Id. ISS/SR/7/VER1.

Kijko, A. (1994). Seismological outliers: $L_1$ or adaptive $L_p$ norm application. *Bull. Seism. Soc. Am.*, **84**, 473–477.

Kijko, A. and Sciocatti, M. (1993). *Optimum spatial distribution of seismic stations in mines*. Internal Report. ISS International, Welkom, South Africa. Report Id. ISS/SR/11/VER1.

Kijko, A. and Sciocatti, M. (1995). Optimal spatial distribution of seismic stations in mines. *Int. J. Rock Mech. and Mining Soc.*, **32**, 607–615.

Kostrov, B.V. (1974). Seismic moment and energy of earthquakes and seismic flow of rock. *Izv. Phys. Solid Earth*, **13**, 13–21.

Kostrov, B.V. and Das, S. (1988). *Principles of Earthquake Source Mechanics*. Cambridge University Press, Cambridge.

Krajcinovic, D. and Mastilovic, S. (1995). Some fundamental issues of damage mechanics. *Mechanics of Materials*, **21**, 217–230.

Kumar P. and Foufoula-Georgiu, E. (1994). Wavelet analysis in geophysics: An introduction. In: *Wavelets in Geophysics, Wavelet Analysis and Its Applications*. (E. Foufoula-Georgiu and P. Kumar, eds.) Academic Press, New York.

Labuz, J.F., Shah, S.P. and Dowding, C.H. (1985). Experimental analysis of crack propagation in granite. *Int. J. Rock Mech. Min. Sci. Geomech. Abstr.*, **22**, 85–98.

Lawrence, D. (1984). Seismicity in the Orange Free State gold-mining district. In: *Proceedings of the 1st international Congress on Rockbursts and Seismicity in Mines* (N.C. Gay and E. H. Wainwright, eds.), Symp. Ser. No. **6**, 121–130. S. Afr. Inst. Min. Metal., Johannesburg.

Lee, W.H.K. and Stewart, S.W. (1981). *Principles and Applications of Microearthquake Networks*, Academic Press, New York.

Lees, J.M. and Crosson, R.S. (1989). Tomographic inversion for the three-dimensional velocity structure at Mount St. Helens using earthquake data. *J. Geophys. Res.* **94**, 5716–5728.

Lehner, F.K., Li, V.C. and Rice, J.R. (1981). Stress diffusion along rupturing plate boundaries. *J. Geophys. Res.*, **86**, 6155–6169.

Liakopoulou-Morris, F., Main, I.G., Crawford, B.R. and Smart, B.G.D. (1994). Microseismic properties of a homogeneous sandstone during fault nucleation and frictional sliding. *Geophys. J. Int.*, **119**, 219–230.

Lichtenberg, A.J. and Liebermann, M.A. (1983). *Regular and Stochastic Motion*. Springer, Berlin.

Lilly, J.M. and Park, J. (1995). Multiwavelet spectral and polarization analyses of seismic records, *Geophys. J. Int.*, **122**, 1001–1021.

Ljung,L. (1987). *System Identification: Theory for the User*. Prentice Hall, New Jersey.

Lucio, P.S., Lambare, G. and Hanyga, A. (1995). 3-D multivalued dynamic ray tracing for migration/inversion. *Proc. SPIE*, San Diego, California, **2569**, 442–446.

Madariaga, R. (1979). Dynamics of an expanding circular fault. *Bull. Seism. Soc. Am.*, **66**, 639–666.

Madariaga, R. (1983). Earthquake source theory: a review. In: *Earthquakes: Observation, Theory and Interpretation*. (H. Kanamori and E. Boschi, eds). North Holland.

Magotra, N.; Ahmed, N. and Chael, E. (1987). Single station seismic event detection and location. *IEEE Trans. Geoscience Remote Sensing*, **27**, 15–23.

Main, I. (1988). Prediction of failure times in the earth for a time-varying stress. *Geophys. J.*, **92**, 455–464.

Main, J.G., Meredith, P.H. and Sammonds, P.R. (1992). Temporal variations in seismic event rate and b-values from stress corrosion constitutive lows. *Tectonophysics*, **211**, 233-246.

Martin, C.D., Young, R.P. and Collins, D.S. (1995). Monitoring progressive failure around a tunnel in massive granite. In: *Proceedings of the 8th ISRM International Conference*, Tokyo, Japan. Balkema, Rotterdam.

Maxwell, S.C. and Young, R.P. (1993a). Associations between temporal velocity changes and induced seismicity, *Geophys. Res. Lett.*, **20**, 2929–2932.

Maxwell, S.C. and Young, R.P. (1993b). A comparison between active source and passive source images. *Bull. Seism. Soc. Am.*, **83**, 1814–1831.

Maxwell, S.C. and Young, R.P. (1994). Application of seismic tomography to induced seismicity investigations. In: *Proceedings of Eurock '94*. Balkema, Rotterdam.

Maxwell, S.C. and Young, R.P. (1996). Seismic imaging of rock mass responses to excavation. Int. *J. Rock Mech. Min. Sc.* (in press).

McComb, W.D. (1990). *The Physics of Fluid Turbulence*. Clarendon Press, Oxford.

McClellan, J.M., Parks, T.W. and Rabinek, L.R. (1973). A computer program for designing optimum FIR linear phase digital filters. *IEEE Trans. on Audio and Electroacoustics*, **A4–31**, 6, 506–576, December 1973.

McGarr, A. (1976a). Dependence of magnitude statistics on strain rate. *Bull. Seism. Soc. Am.*, **66**, 1, 33–44.

McGarr, A. (1976b). Seismic moments and volume change. *J. Geophys. Res.* **81**, 1487–1494.

McGarr, A. (1991). Observations constraining near-source ground motion estimated from locally recorded seismograms. *J. Geophys. Res.*, **96**, 16495–16508.

McGarr, A. (1993). Keynote address: Factors influencing the strong ground motion from mining-induced tremors. *Proceedings of the 3rd International Symposium on Rockbursts and Seismicity in Mines*,(R.P. Young, ed.), 16-18 August, 1993, Kingston, Canada. Balkema, Rotterdam.

McGarr, A. (1994). Some comparisons between mining-induced and laboratory earthquakes. *Abstracts of the 89th Annual Meeting, Seismological Society of America.*

Mendecki, A.J. (1981). Methods of joint hypocentre location for mining tremors and determination of parameters of velocity anisotropy. Ph.D. thesis, Silesian Technical Univ. Gliwice, Poland. (In Polish)

Mendecki, A.J. (1987). Rock mass anisotropy modelling by inversion of mine tremor data. In *Proceedings of 6th International Congress on Rock Mechanics*, (G. Herget and S. Vongpaisal, eds.), Balkema, Rotterdam, 1141–1144.

Mendecki, A.J. (1993). Real time quantitative seismology in mines. Keynote lecture. In: *Proceedings of the 3rd International Symposium on Rockburst and Seismicity in Mines* (R.P. Young, ed.), Kingston, Canada. Balkema, Rotterdam, 287–296.

Mendecki, A.J. (1994). Quantitative seismology and rock mass stability. In: *Guide to Seismic Monitoring in Mines*, 1994 edition. (A.J. Mendecki, ed.). ISS International, Welkom, South Africa.

Mendecki, A.J. and van Aswegen, G. (1986). A method for the optimal design of mine seismic networks in respect to location errors and its application. Research Report, AAC Research and Development Services, Rock Mechanics Department, 56.

Menke, W., Lerner-Lam, A.L., Dubendorf, B. and Pacheco, J. (1990). Polarization and coherence of 5 to 30 Hz seismic wave fields at hard rock site and their relevance to velocity heterogeneities in the crust. *Bull. Seism. Soc. Am.*, **80**, 430–449.

Meridith, P.G., Main, I.G. and Jones, C. (1990). Temporal variations in seismicity during quasi-static and dynamic rock failure. *Tectonophysics*, **175**, 249–268.

Michael, A.J. (1987). Use of focal mechanisms to determine stress: a control study. *J. Geophys. Res.*, **92**, 357–368.

Milev, A.M., Spottiswoode, S.M. and Stewart, R.D. (in prep.) Automatic location of seismic event multiplets by crosscorrelation criteria. Submitted *Bull. Seism. Soc. Am.* (1996).

Mogi, K. (1985). *Earthquake Prediction*, Academic Press, Tokyo.

Molnar, P. (1983). Average regional strain due to slip on numerous faults of different orientations. *J. Geophys. Res.*, **88**, B8, 6430–6432.

Money, A.H., Affleck-Graves, J.H., Hart, M.L. and Barr, G.D.I. (1982). The linear regression model: $L_p$ norm estimation and the choice of *p*. *Commun. Statist. Comput.*, **11**, 89–109.

Montalbetti, J.F. and Kanasewich, E.R. (1970). Enhancement of teleseismic body phases with a polarization filter. *Geophys. J. R. Astron. Soc.*, **21**, 413–416.

Mood, A.M., Graybill, F.A. and Boes, D.C. (1974). *Introduction to the Theory of Statistics* (3rd Edition). McGraw-Hill International.

Mróz, Z. (1985). Current problems and new directions in mechanics of geomaterials in *Mechanics of Geomaterials*. (Z. Bažant, ed.). John Wiley & Sons. Chichester.

Mühlaus, H.-B., Hobbs, B.E. and Ord, A. (1992). Evolution of fractal geometries in deforming material. In: *Rock Mechanics*. (J.R. Tillerson and W.R. Wawersik, eds). Balkema, Rotterdam, 681–690.

Naude, F.C. (1994). Seismic index: a method of illuminating hazardous areas on the mine. A case study for the sequence of events on 2 and 3 May 1994. Internal Report, Western Deep Levels, South Mine, May.

Newman, W.I. (1983). Nonlinear diffusion: Self similarity and traveling-waves. *PAGEOPH*. **121**, 3, 417–441.

Nolet, G. (1987). *Seismic Tomography*. Reidel, Boston.

Oh, E. (1996). Development of a computer-based calibration facility for low-frequency acoustic emission transducers. In: *Procceedings of 6th Acoustic Emission/Microseismic Activity in Geological Structures and Materials Conference*, Pennsylvania State University (H.R. Hardy ed.). Trans Tech Publications, Clausthal-Zellerfield.

Ohnaka, M. (1992). Earthquake source nucleation: a physical model for short term precursors. *Tectonophysics*, **211**, 149–178.

Olsson, W.A. (1992). The formation of a yield-surface vertex in rock. In: *Rock Mechanics*. (H.R. Tillerson and W.R. Wawersik, eds). Balkema, Rotterdam, 701–705.

Omori, F. (1895). On the after-shocks of earthquakes. *J. Coll. Sci.*, Tokyo Imp. Univ., **7**, 111–200

Ortlepp, W.D. (1994). High ground displacement velocities associated with rockburst damage. In: *Rockbursts and Seismicity in Mines,* (R.P. Young, ed.). Balkema, Rotterdam, 449.

Paige, C.C. and Saunders, M.A. (1982). LSQR: an algorithm for sparse linear equations and sparse least squares. *ACM Transactions on Mathematical Software*, **8**, 43–71, 195–209.

Park, J., Vernon III, F.L. and Lindberg, C.R. (1987a). Frequency dependent polarization analysis of high-frequency seismograms. *J. Geophys. Res.*, **92**, 12664–12674.

Park, J., Lindberg, C.R. and Vernon III, F.L. (1987b). Multitaper spectral analysis of high-frequency seismograms. *J. Geophys. Res.*, **92**, 12675–12684.

Pavlis, G.L. and Booker, J. (1980). The mixed discrete continuous inverse problem: Application to the simultaneous determination of earthquake hypocentres and velocity structure. *J. Geophys. Res.*, **85**, 4801–4810.

Pearson, G.E. (1956). General theory of elastic stability. *Quart. Appl. Math.*, **14**, 133–144.

Peterson, D.A. and Simser B.P. (in prep.) Source mechanisms of mining induced seismic events at Brunswick Mine. Submitted to: 4th International Symposium for Rockbursts and Seismicity, Krakow, Poland, 1997.

Pereyra, V., Lee, W.H.K. and Keller, H.B. (1980). Solving two-point seismic ray tracing problems in a heterogeneous medium. Part 1. A general adaptive finite difference method, *Bull. Seism. Soc. Am.*, **70**, 79–99.

Podvin, P. (1992). Finite difference computation of multivalued traveltimes, *EAEG* annual meeting, Paris, 502–503.

Podvin, P. and Lecomte, I. (1991). Finite difference computation of travel times in very contrasted velocity models: a massively parallel approach and its associated tools, *Geophys. J. Int.*, **105**, 271–284.

Poupinet, G., Ellsworth W.L. and Frechet, J. (1984). Monitoring velocity variations in the crust using earthquake doublets: an application to the Calaveras fault, California, *J. Geophys. Res.* **89**, 5719–5731.

Press, W.H., Flannery, B.P., Teukolsky, S.A. and Vetterling, W.T. (1989). *Numerical Recipes: The Art of Scientific Computing*. Cambridge University Press, New York.

Prothero, W.A., Taylor, W.J. and Eickemeyer, J.A. (1988). A fast, two point, three dimensional raytracing algorithm using a simple step search method. *Bull. Seism. Soc. Am.*, **78**, 1190–1198.

Prugger, A.F. and Gendzwill, D.J. (1988). Microearthquake location: A non-linear approach that makes use of a simplex stepping procedure. *Bull. Seism. Soc. Am.* **78**, 799–815.

Qin, F., Luo, Y., Olsen, K.B., Cai, W. and Schuster, G.T. (1992). Finite-difference solution of the eikonal equation along expanding wavefronts, *Geophysics*, **57**, 479–487.

Rabiner, L.R. and Gold, B. (1975). *Theory and Applications of Digital Signal Processing*. Prentice Hall, New Jersey.

Rabotnov, Y.N. (1969). *Creep Problems in Structured Members*. (F.A. Lackie, ed.). North-Holland, New York.

Radu, S. (1995). Order and chaos in nuclear and metal cluster deformation. PhD Thesis, University of the Witwatersrand, South Africa.

Reiner, M. (1969). *Deformation, Strain and Flow*. H.K. Lewis, London.

Rice, J.R. (1975). On the stability of dilatant hardening for saturated rock masses. *J. Geophys. Res.*, **80**, 1531–1536.

Rice, J.R. and White, T.S. (1964). Norms for smoothing and estimation. *SIAM Rev.*, **6**, 243–256.

Rikitake, T. (1976). *Earthquake Prediction*. Elsevier, New York.

Riznichenko, Y.V. (1946). Geometrical seismics of layered media, *Trudy Inst. Theor. Geophysics*, **2**, Moscow. (In Russian.)

Riznichenko, Y.V. (1965). Seismic rock flow. In: *Dynamics of the Earth's Crust*. Nuka, Moscow. (In Russian.)

Rudnicki, J.W. (1977). The inception of faulting in a rock mass with a weakened zone. *J. Geophys. Res.*, **82**, 844–854.

Rudnicki, J.W. (1988). Physical models of earthquake instability. *PAGEOPH*, **126**, 531–554.

Rudnicki, J.W. and Rice, J. (1975). Conditions for the localization of deformation in pressure sensitive dilatant materials. *J. Mech. Phys. Solids*, **23**, 371–394.

Rudnicki, J.W. and Freund, L.B. (1981). On energy radiation from seismic sources. *Bull. Seism. Soc. Am.*, **71**, 583–595.

Ryder, J.A. (1988). Excess shear stress in the assessment of geologically hazardous situations. *J. S. Afr. Inst. Min. Metal.*, **88**, 1.

Sambridge, M. and Gallagher, K. (1993). Earthquake hypocentre location using genetic algorithms, *Bull Seism. Soc. Am.*, **83**, 1467–1491.

Samson, J.C. (1977). Matrix and Stokes vector representation of detectors for polarized waveforms: theory, with some applications to teleseismic waves. *Geophys. J. R. Astron. Soc.*, **51**, 583–603.

Samson, J.C. and Olson, J.V. (1980). Some comments on the description of polarization states of waves. *Geophys. J.*, **21**, 115–129.

Scholz, C.H. (1968). The frequency-magnitude relation of microfracturing in rock and its relation to earthquakes. *Bull. Seism. Soc. Am.*, **58**, 399–415.

Scholz, C.H. (1977). Transform fault system of California and New Zealand: Similarities in their tectonic and seismic styles. *J. Geol. Soc. London.*, **133**, 215–229.

Scholz, C.H. (1990). *The Mechanics of Earthquakes and Faulting.* Cambridge University Press, Cambridge.

Shanks, J.L. (1967). Recursion filters for digital processing. *Geophysics*, **32**, 33–51.

Silver, P. and Jordan, T. (1982). Optimal estimation of scalar seismic moment. *Geophys. J.R. Astron. Soc.*, **70**, 755–787.

Slepian, D. (1978). Prolate spheroidal wavefunctions, Fourier analysis and uncertainty, V - The discrete case. *Bell Syst. Tech. J.*, **57**, 1371–1429.

Smith, W.D. (1981). The *b*-value as an earthquake precursor. *Nature*, **289**, 136–139.

Sornette, D. and Sammis, C.G. (1995). Complex critical exponents from renormalisation group theory of earthquakes: implications for earthquake predicitions. *J. Phys. I. France*, **5**, 607–619.

Spence, W. (1980). Relative epicenter determination using P wave arrival-time differences. *Bull. Seism. Soc. Am.*, **70**, 171–183.

Spencer, C. and Gubbins, D. (1980). Travel-time inversion for simultaneous earthquake location and velocity structure determination in laterally varying media, *Geophys. J. R. Astron. Soc.*, **63**, 95–116.

Spottiswoode, S.M. (1993). Seismic attenuation in deep level mines. In: *Proceedings of the 3rd International Symposium on Rockburst and Seismicity in Mines*, (R.P. Young, ed.), Kingston, Canada. Balkema, Rotterdam, 409–414.

Spudich, P.K.P. (1992). On the interference of absolute stress levels from seismic radiation. *Tectonophysics*, **211**, 99–106.

Sun, Y. (1992). Computation of 2D multiple arrival travel time fields by an interpolative shooting method. *Expanded Abstracts of the 62-nd SEG Annual Meeting* (New Orleans, 1992), 1320–1323.

Suyehiro, S. (1966). Differences between aftershocks and foreshocks in the relationship of magnitude to frequency of occurrence for the great Chilean earthquake of 1960. *Bull. Seism. Soc. Am.*, **56**, 185–200.

Sykes, L.R. and Jaumé, S.C. (1990). Seismic activity on neighbouring faults as a long-term precursor to large earthquakes in the San Francisco Bay area. *Nature*, **348**, 595–599.

Takens, F. (1984). *Proceedings of the Dynamical Systems and Bifurcation Conference, Groningen, 1984.* Springer, Berlin.

Talebi, S. and Young, R.P. (1992). Design of a microseismic system for the URL mine-by experiment. Report to Atomic Energy of Canada Limited.

Tang, C.A., Hudson, J.A. and Xu, X. (1993). *Rock failure instability and related aspects of earthquake mechanisms.* China Coal Industry Publishing House, Beijing.

Taub, H. and Schilling, D.L. (1971) *Principles of Communication Systems*, McGraw-Hill, New York.

Teisseyre, R. (1980a). Some remarks on the source mechanism of rockbursts in mines and on the possible source extension. *Acta Montana ČSAV Praha*, **58**, 7–13.

Teisseyre, R. (1980b). Earthquake premonitory sequence – dislocation processes and fracturing. *Boll. Geof. Teor. Appl.*, **XXII**, N88, 245–254.

Thiran J. (1971). Recursive digital filters with maximally flat group delay. *IEEE Transactions of Circuit Theory*, **CT–18**, 6.

Thomson, D.J. (1982). Spectral estimation and harmonic analysis. *IEEE Proc.*, **70**, 1055–1096.

Thorbjarnardottir, B.S. and Pechmann, J.C. (1987). Constraints on relative earthquake locations from crosscorrelation of waveforms, *Bull. Seism. Soc. Am.*, **77**, 1626–1634.

Thornburgh, H.R. (1930). Wavefront diagrams in seismic interpretation, *AAPG Bull.*, **14**, 185–200

Tinucci, J.P. and Spearing, A.J.S. (1993). Strategies for clamping faults and dykes in high seismicity tabular mining conditions. In: *Proceedings of the 3rd International Symposium on Rockburst and Seismicity in Mines.* (R.P. Young, ed.), Kingston, Canada. Balkema, Rotterdam, 435–440.

Truesdell, C. and Noll, W. (1992) *The Non-linear Field Theories of Mechanics.* 2nd edition. Springer-Verlag, Berlin.

Tugal, D.A. and Tugal, O. (1989). *Data Transmission*, McGraw-Hill, New York.

Um, J. and Thurber, C.H. (1987). A fast algorithm for two-point seismic ray tracing. *Bull. Seism. Soc. Am.*, **77**, 972–986.

Urbancic, T.I., Trifu, C.I. and Young, R.P. (1993). Stress release estimates, scaling behaviour, and source complexities of microseismic events. In: *Proceedings of the 3rd International Symposium on Rockburst and Seismicity in Mines.* (R.P. Young, ed.), Kingston, Canada. Balkema, Rotterdam, 255–260.

van Aswegen, G. (1990). Fault stability in SA gold mines. In: *Mechanics of jointed and faulted rock*, (H.P. Rossmanith, ed.). Balkema, Rotterdam, 261–266.

van Aswegen, G. and Butler, A. (1993). Applications of quantitative seismology in SA gold mines. *Proceedings 3[rd] International Symposium in Rockbursts and Seismicity in Mines*, Balkema, Rotterdam, 261–266.

Varnes, D.J. (1983). Time deformation relations in creep to failure of earth materials. *Proceedings of 7th Southeast Asian Conference, Hong Kong.* 107–130.

Vidale, J.E. (1986). Complex polarization analysis of particle motion. *Bull. Seism. Soc. Am.*, **76**, 1393–1405.

Vidale, J. (1988). Finite-difference calculation of traveltimes. *Bull. Seism. Soc. Am.* **78**, 2062–2076.

Vidale, J. (1990). Finite-difference calculation of traveltimes in 3D. *Geophysics*, **55**, 521–526.

Vinje, V., Iversen, E. and Gjoystadl, H. (1993). Traveltime and amplitude estimation using wavefront construction. *Geophysics*, **58**, 1157–1166.

Voight, B. (1989). A relation to describe rate-dependent material failure. *Science*, **243**, 200–203.

Walck, M.C. and Chael, E.P. (1991). Optimal backazimuth estimation for three-component recordings of seismic events. *Bull. Seism. Soc. Am.*, **81**, 643–666.

Wawersik, W.R., Rudnicki, J.W., Olson, W.A., Holcomb, D.J. and Chan, K.T. (1990). *Localization of deformation in brittle rock: Theoretical and laboratory investigations.* Proceedings of the International Conference on Micromechanics of Failure, Albuquerque, New Mexico, 6–8 June, 1990. (S.P. Shah, ed.), Elsevier, New York.

Wesson, R.L. (1971). Travel-time inversion for laterally inhomogeneous crustal velocity models. *Bull. Seism. Soc. Am.*, **61**, 729–746.

Willemann, R.J., (1993). Cluster analysis of seismic moment tensor orientations. *Geophys. J. Int.*, **115**, 617–634.

Wolf, A., Swift, J.B., Swinney, H.L. and Vastano, J. (1985). Determining Lyapunov exponents from a time series. *Physica D*, **16**, 285–317.

Woźniakowski, H. (1977). Numerical stability for solving nonlinear equations, *Numer. Math* **27**, 373–390.

Wyss, M. and Brune, J.N. (1968). Seismic moment, stress and source dynamics for earthquakes in the California–Nevada region. *J. Geophys. Res.*, **73**, 4681–4694.

Wyss, M. and Brune, J.N. (1971). Regional variations in source properties in Southern California estimated from the ratio of short-to-long period amplitudes. *Bull. Seism. Soc. Am.*, **61**, 1153–1167.

Xu, S., King, M.S. and Worthington, M.H. (1990). Elastic wave propagation and hydraulic properties of cracked rocks. In: *Mechanics of Jointed and Faulted Rock*, (H.P. Rossmanith, ed.). Balkema, Rotterdam.

Yukutake, H. (1992). Fracture nucleation process in intact rock. *Tectonophysics*, **211**, 247–257.

Zimmerman, R.W. and King, M.S. (1985). Propagation of acoustic waves through a cracked rock. *Proc. 26th U.S. Symposium on Rock Mechanics*, Rapid City. 739–745.

# Index

Page numbers appearing in **bold** refer to figures and page numbers appearing in *italic* refer to tables.